Sustaining Ecosystems in a Changing World

Sustaining Ecosystems in a Changing World

Edited by Sierra Adkins

SYRAWOOD
PUBLISHING HOUSE
New York

Published by Syrawood Publishing House,
750 Third Avenue, 9th Floor,
New York, NY 10017, USA
www.syrawoodpublishinghouse.com

Sustaining Ecosystems in a Changing World
Edited by Sierra Adkins

© 2019 Syrawood Publishing House

International Standard Book Number: 978-1-68286-764-8 (Hardback)

Cataloging-in-Publication Data

Sustaining ecosystems in a changing world / edited by Sierra Adkins.
   p. cm.
Includes bibliographical references and index.
ISBN 978-1-68286-764-8
1. Ecosystem management. 2. Sustainability. 3. Biodiversity conservation.
4. Nature conservation. 5. Applied ecology. I. Adkins, Sierra.
QH75 .S87 2019
333.95--dc23

TABLE OF CONTENTS

PREFACE

A healthy ecosystem is vital for the survival of all life on Earth. Human activities have a significant impact across all ecosystems. Due to the rise in human population and per capita consumption, the problems of environmental pollution, biodiversity loss and climate change are increasingly being observed. Terrestrial ecosystems face threats that include deforestation, air pollution and soil degradation. Overfishing, microplastics pollution, encroachment into coastal areas, etc. cause severe damage to the aquatic ecosystem. The primary objective of ecosystem management is the sustainable and ethical use of natural resources. Natural resource management, adaptive management, strategic management, and command and control management are few approaches to effective ecosystem management, both at a local and landscape level. This book includes some of the vital pieces of work being conducted across the world, on various topics related to sustainable ecosystems. Different approaches, evaluations, methodologies and advanced studies on ecosystem management have been included herein. It will serve as an elaborate reference text for environmentalists, ecologists, researchers and students engaged in this area of study.

The researches compiled throughout the book are authentic and of high quality, combining several disciplines and from very diverse regions from around the world. Drawing on the contributions of many researchers from diverse countries, the book's objective is to provide the readers with the latest achievements in the area of research. This book will surely be a source of knowledge to all interested and researching the field.

In the end, I would like to express my deep sense of gratitude to all the authors for meeting the set deadlines in completing and submitting their research chapters. I would also like to thank the publisher for the support offered to us throughout the course of the book. Finally, I extend my sincere thanks to my family for being a constant source of inspiration and encouragement.

Editor

Dynamic-landscape metapopulation models predict complex response of wildlife populations to climate and landscape change

Thomas W. Bonnot,[1,†] Frank R. Thompson III,[2] and Joshua J. Millspaugh[3]

[1]School of Natural Resources, University of Missouri, 302 Natural Resources Building, Columbia, Missouri 65211 USA
[2]United States Forest Service, Northern Research Station, University of Missouri-Columbia,
202 Natural Resources Building, Columbia, Missouri 65211 USA
[3]Wildlife Biology Program, Department of Ecosystem and Conservation Sciences, W. A. Franke College of Forestry and Conservation,
University of Montana, Missoula, Montana 59812 USA

Abstract. The increasing need to predict how climate change will impact wildlife species has exposed limitations in how well current approaches model important biological processes at scales at which those processes interact with climate. We used a comprehensive approach that combined recent advances in landscape and population modeling into dynamic-landscape metapopulation models (DLMPs) to predict responses of two declining songbird species in the central hardwoods region of the United States to changes in forest conditions from climate change. We modeled wood thrush (*Hylocichla mustelina*) and prairie warbler (*Setophaga discolor*) population dynamics and distribution throughout the central hardwoods based on estimates of habitat and demographics derived from landscapes projected through 2100 under a current climate scenario and two future climate change scenarios. Climate change, natural forest succession, and forest management interacted to change forest structure and composition over time, variably affecting the distribution and amount of habitat of the two birds. The resulting changes in habitat and metapopulation processes produced contrasting predictions for future populations. Wood thrush, a forest generalist, showed little response to climate-driven forest change but declined by >25% due to reduced productivity associated with existing forest fragmentation across much of the region. Prairie warblers initially declined due to loss of habitat resulting from current land management; however, after 2050 cumulative effects of climate change on forest structure created enough habitat in source landscapes to restore population growth. These species-specific responses were the result of interactions among climate, landscape, and population processes. We suggest relationships between climate change, succession, and land management are species specific and important determinants of future wildlife populations and that DLMPs are a comprehensive approach that can capture such processes to generate more realistic predictions of populations under climate change.

Key words: distribution; habitat; landscape; population dynamics; prairie warbler; wood thrush.

† **E-mail:** bonnott@missouri.edu

Introduction

One of the most significant questions in wildlife conservation is how climate change will affect biodiversity. Biodiversity and ecosystems are more stressed than at any comparable period of human history because they are intrinsically dependent on climate and thus impacted by its changes (Staudinger et al. 2013). Many species are at a far greater risk of extinction than in the recent geological past (Fischlin et al. 2007). Climate change has caused significant population declines and been linked to species extinctions (Monzón et al. 2011, Selwood et al. 2015). Climate change is also causing shifts in species' distributions and phenologies, which could substantially alter ecosystem structure and function (Schneider and Root 2002, Thomas et al. 2004). The proactive actions needed to prevent such outcomes have left managers and biologists across the globe looking for approaches that can predict how species and populations will respond to climate change and other aspects of global change (Staudinger et al. 2013).

A variety of methods have been employed to address such a broad question. Correlative and mechanistic models are two approaches to predicting species impacts that have been frequently contrasted in recent years (Moritz and Agudo 2013). Correlative species distribution models (SDMs; also known as ecological niche models or bioclimatic envelope models) predict changes in the ranges of species using statistical associations between climate/environmental variables and patterns of species distribution (Guisan and Thuiller 2005, Elith and Leathwick 2009, Fordham et al. 2012). These models have seen widespread use due to the availability of methods, data, and their ability to predict climate impacts over range-wide scales (Fordham et al. 2013, Moritz and Agudo 2013). Projecting potential changes in distribution is a sensible goal for guiding future conservation, given accumulating evidence of these effects (Parmesan 2006, LaSorte and Thompson 2007, Chen et al. 2011). However, SDMs have been criticized for their inability to account for the variety of processes affecting populations (Fischlin et al. 2007, Brook et al. 2008). Although SDMs often assume that climate alone drives shifts in species distribution, it is likely that responses to other environmental threats might overshadow those related to climate (Brook et al. 2008, Swab et al. 2015). Indeed, climate change is occurring against the backdrop of a wide range of land management and other environmental and anthropogenic stressors, which have caused dramatic changes to landscapes already (Staudinger et al. 2013). The lack of a direct mechanistic basis could predispose these models to suggest more extreme responses than might actually occur (Moritz and

Agudo 2013). Most importantly, a lack of mechanistic processes in SDMs prevents modeling changes in population dynamics under climate change which would provide important information about persistence (Fordham et al. 2012). Distributional change is one of the last symptoms of species decline, allowing populations to be at risk without any shifts in range or distribution (Selwood et al. 2015).

Recent mechanistic approaches represent an increased awareness of the processes that determine how species respond to climate change. Species responses to climate change are influenced by more than changes in habitat alone. Climate has numerous effects on demographic rates and population processes (Selwood et al. 2015), and species interactions and interactions between demographic and landscape dynamics all drive populations status and trends (Keith et al. 2008, Millspaugh et al. 2009). Therefore, efforts to account for these processes and how climate affects them have resulted in more robust predictions and better capture context-dependent variability in species responses (Monahan 2009, Cheung et al. 2012, Fordham et al. 2013). Recently, some have shown that still a more complete understanding of a population impacts is possible by explicitly integrating climate with demographic processes (Keith et al. 2008, Brook et al. 2008). Dynamic-landscape metapopulation models (DLMP; sensu Akçakaya 2000, Larson et al. 2004) represent this approach through an integration of landscape, habitat, and metapopulation modeling. These models have experienced renewed use in recent years because of their ability to provide a spatial representation of how landscapes change through time and how species respond to this spatially and temporally variable environment (e.g., Fordham et al. 2013, Franklin et al. 2014). Because population responses to changing landscapes can be complex and sometimes counterintuitive (Bonnot et al. 2013), DLMPs have provided an important step toward realistically predicting species impacts from climate change.

Mechanistic approaches such as DLMPs still face limitations in complexity and scope. An ideal approach would be complex enough to model important processes driving population dynamics and distribution at the scales at which those processes interact. However, if identifying climate effects on species' habitat or demographic rates is difficult, then spatially integrating those effects with other metapopulation processes across entire landscapes or regions is improbable. As a result, information on population dynamics provided by mechanistic approaches is limited to specific landscapes or study areas, not the regional or range-wide scales that can inform species distributions under climate change (Akçakaya and Brook 2009). Although there are more examples of DLMPs for plant species under climate change (e.g., Regan et al. 2012, Franklin et al. 2014), these models still lack important mechanisms by overlooking various ecosystem and landscape processes affecting plants (Wang et al. 2015). Characterizing wildlife habitat involves representing both the structure and composition of habitat. Therefore, modeling changes in habitat for animals compounds this problem by requiring data on climate-induced changes in entire vegetation communities through time. Forest landscape models such as LANDIS PRO can account for many of these processes to inform wildlife habitat models, but until recently have not integrated climate (Wang et al. 2015). Ultimately, achieving a comprehensive understanding of wildlife responses to climate change is going to require an approach that can integrate climate, landscape, habitat, and metapopulation processes across a range of scales to predict changes in dynamics and distributions overtime.

We advanced the capability of DLMPs to address climate change by incorporating two recent developments in landscape and metapopulation modeling. Recent efforts to extend landscape-based population modeling to regional scales provided the ability to link local habitat and demographics with population growth over tens of millions of hectares (Bonnot et al. 2011, 2013). The capability to model changes in forest structure and composition under climate change at similarly large, regional scales has provided an approach to predicting how wildlife habitat might change in the future (Wang et al. 2015, 2016). Our objective was to integrate Bonnot et al.'s (2011) regional population models with Wang et al.'s (2015) forest landscape projections to predict impacts of landscape and climate change on populations of two species of songbirds in the Central Hardwoods forest of the Midwestern United States under three future climate scenarios. We

picked two birds species, wood thrush (*Hylocichla mustelina*) and prairie warblers (*Setophaga discolor*), with contrasting demographics and habitat, to demonstrate how this approach can account for interactions among species demographics and landscape and climate change to predict population change. Furthermore, these species are a conservation concern in the Eastern United States because of long-term population declines.

METHODS

Study area

We studied a 39.5 million ha (395,519 km^2) portion of the Central Hardwoods forest in the center of the United States (Fig. 1). The area encompasses a variety of vegetation, terrains, soils, and climates (Cleland et al. 2007). The topography varies from relatively flat Central Till Plains to open hills and irregular plains (e.g., Interior Low Plateau), to highly dissected Ozark Highlands. The region supported a diversity of forest ecosystems, including upland oak (*Quercus* spp.)–hickory (*Carya* spp.) forests and oak-pine (*Pinus* spp.) forests, woodlands, and savannas. While a portion of the land that was historically forested in the Central Hardwoods remains so today, glades and woodlands and other communities have been lost and dramatically altered (Fitzgerald et al. 2005). Widespread logging in the early part of the 20th century and fire suppression in subsequent decades resulted in conversion of glade, barren, and pine woodland habitats to oak or oak-pine forests. Forests in this region have also been fragmented by agriculture and urban development.

The loss of these communities as habitat combined with the effects of fragmentation has likely contributed to long-term population declines of wood thrush and prairie warblers, and conservation organizations consider them as species of concern within the Central Hardwoods (Panjabi et al. 2005, U.S. Department of the Interior Fish and Wildlife Service 2008, Sauer et al. 2017). Prairie warblers are declining by an estimated 1.98% annually (Sauer et al. 2017). Prairie warblers breed in shrubby vegetation under an open or semi-open canopy such as in glades, savannas, abandoned fields, and regenerating forests. Their decline is likely the result of loss of this habitat over much of the region and reduced productivity

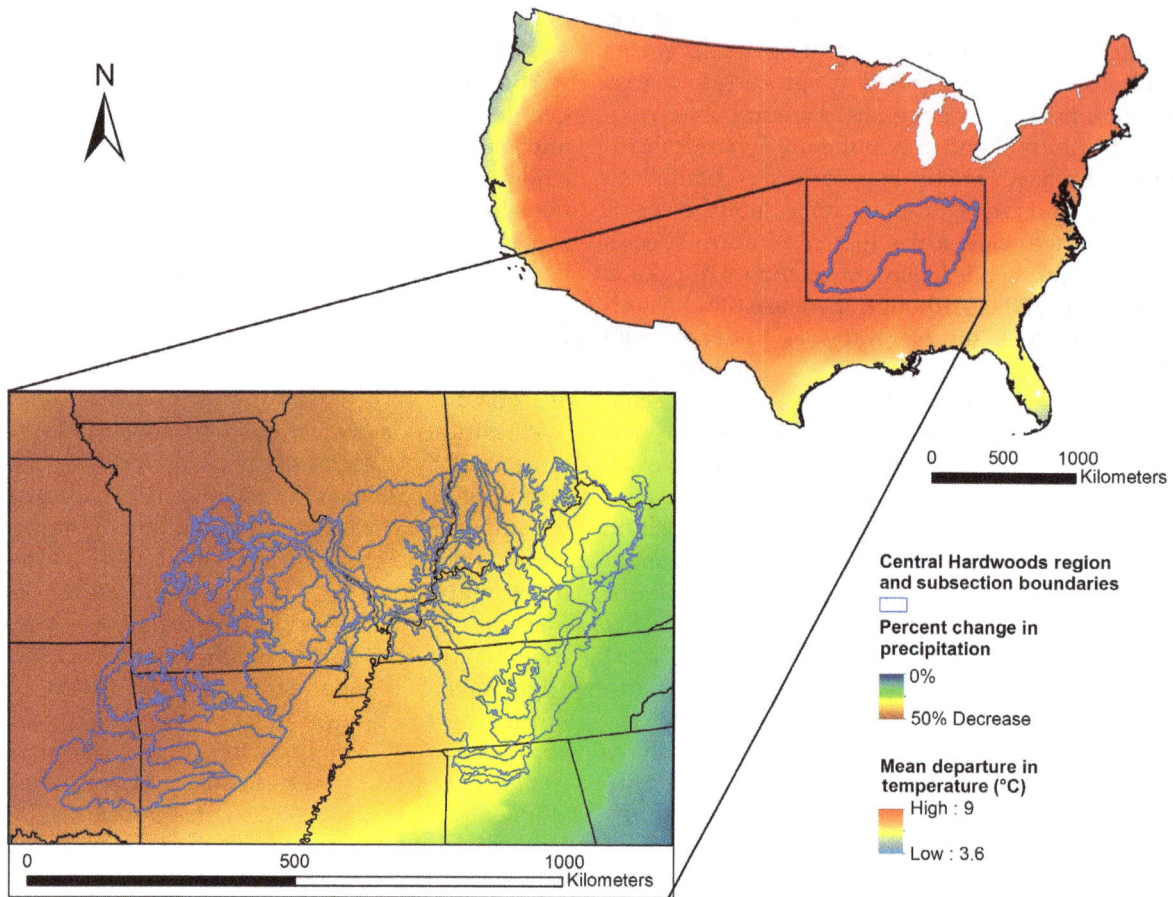

Fig. 1. Dynamic-landscape metapopulation models were developed for two species of songbirds in the Central Hardwoods forested region in central United States. The models integrate habitat with metapopulation processes to predict future changes in population dynamics and distribution throughout the region's ecological subsections. Estimates of habitat incorporated changes in the region's forest through 2100 under varying degrees of climate change, which is expected to increase temperature (main figure) and reduce summer precipitation (inset) under a GFDL-A1Fi scenario.

due to parasitism associated with fragmentation (Nolan et al. 2014). Wood thrush are much more abundant than prairie warblers because they are distributed throughout closed-canopy, mid-successional forest, which is abundant in the region (Evans et al. 2011). However, wood thrush numbers have declined 0.6% annually since 1966 (Sauer et al. 2017) and declines are at least partly due to higher predation and parasitism in fragmented forests (Robinson et al. 1995).

Modeling approach

We combined three components that are integral to DLMP approaches (Fig. 2; Bekessy et al.

2009). We begin with projections of the landscape into the future under forest management and climate change scenarios in the form of a series of spatial data grids that map the distribution, structure, and composition of forests at specified time steps. Next, we translated these landscape projections into species' habitat and demographics at each time step. We considered known relationships between habitat and population processes (e.g., abundance, reproduction, survival, or dispersal). Finally, we incorporated these spatially and temporally varying demographics in a metapopulation model that included stochasticity and uncertainty. The resulting model provided a

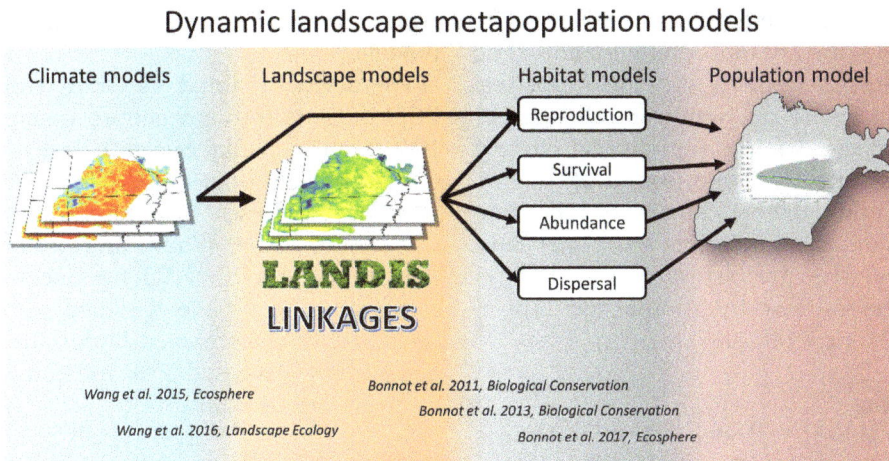

Fig. 2. Overview of dynamic-landscape metapopulation modeling approach used to project the impacts of climate change on landscapes, habitat, and wildlife populations across the Central Hardwoods region.

spatially and temporally explicit representation of habitat and population dynamics and distribution throughout the region.

Future landscape data

We used recent projections of the structure and composition of forests in the Central Hardwoods from 2000 to 2300 under three climate change scenarios (Wang et al. 2015, 2016). Wang et al. used the forest landscape model LANDIS PRO to project forest changes due to succession, harvest, and climate change. LANDIS PRO is a spatial model that operates across grid cells in a landscape, modeling cell-level processes that include species-specific seed dispersal, establishment, growth, competition, and mortality and landscape-level processes such as wind throw and tree harvest. In their scenarios, forest management reflected current patterns in tree harvest throughout the region observed from region-wide Forest Inventory and Analysis data from 1995 to 2005. Wang et al. (2015) directly incorporated changes in climate in LANDIS PRO via the early growth and establishment of different tree species and the maximum allowable tree biomass based on their attributes and cell locations. They estimated these parameters with the ecosystem model LINKAGES III, which integrates temperature and precipitation data with nitrogen availability and soil moisture to model individual tree species growth and mortality at a site (Dijak et al. 2016).

The landscapes were modeled under a current climate scenario and two climate change scenarios based on combinations of general circulation models (GCMs) and emission scenarios from the IPCC (2007). The current climate scenario used temperature, precipitation, and wind speed data for the 30-yr period from 1980 to 2009 observed throughout the region (Wang et al. 2015). The two IPCC-derived climate change scenarios CGCM.T47-A2 and GFDL-A1Fi represented alternative degrees of climate change. The GFDL-A1Fi scenario combined a more substantial and immediate increase in greenhouse gas emissions (A1Fi) with a model that is more sensitive to that increase (GFDL; IPCC 2007). Thus, the GFDL-A1Fi scenario presented more severe changes in climate relative to the CGCM.T47-A2 scenario. For example, by the end of the century the GFDL-A1Fi scenario projects a 4.5°C increase in the mean annual daily maximum temperature as well as twice the number of consecutive summer dry days as has been observed in the region (Fig. 1; Girvetz et al. 2009).

Wang et al. (2015) estimated forest projections from 2000 to 2300 at 10-yr time steps and at a 270-m resolution. The projections comprised cell-based estimates of importance values, basal area, and number and diameter at breast height (dbh) of trees by species and age cohort. Their results suggested a prolonged period (i.e., 300 yr) before substantial shifts in forest composition would occur in response to climate change. When shifts did occur, it was toward more southern and xeric species and lesser northern and mesic species. Although there were no significant changes in overall tree species composition among current climate and climate change scenarios in the region's midterm (100 yr), forests did become more xeric as indicated by lower basal areas and tree densities. The greatest of these changes occurred in the southwest portion of the Central Hardwoods (Wang et al. 2015).

Habitat modeling

We employed habitat models to link landscapes to three demographic processes: the distribution of carrying capacity (K) and abundance, breeding productivity, and dispersal. These models are meta-analytic approaches that integrate published data and findings to quantify habitat and demographic processes (Dijak and Rittenhouse 2009). Although partly conceptual, the flexibility of these approaches allows them to incorporate processes from a range of studies and sites to model local habitat throughout entire regions.

We modeled the distribution of wood thrush and prairie warbler abundance and K using landscape-scale Habitat Suitability Index (HSI) models. Previously developed specifically for the Central Hardwoods, the HSIs indexed the suitability of 30 × 30 m cells based on the habitat attributes of the cell and the surrounding landscape (Tirpak et al. 2009a). Both the prairie warbler and wood thrush models have been independently verified and validated with data from the North American Breeding Bird Survey, a long-term, large-scale bird monitoring program (Tirpak et al. 2009b). The HSI models combined characteristics of land cover (as defined by the National Land Cover Data, NLCD; Fry et al. 2011), landform, and forest seral stage with additional variables that reflected the wood thrush's use of mature hardwood and mixed forests with relatively closed canopies. Prairie warblers inhabit a variety of early successional forest types as well as glades and woodlands. Therefore, an open canopy and shrubby understory were important structural components considered in the prairie warbler model. In addition, HSI models captured each species sensitivity to habitat patch size and the predominance of forest in the surrounding landscape (Tirpak et al. 2009a). See

Appendix S1 for a full description of the HSI models.

We derived habitat variables from LANDIS PRO outputs through geoprocessing in ArcGIS 10.2 (Environmental Systems Research Institute, Redlands, California, USA). We identified NLCD land cover classes by comparing the relative importance values (Smith and Smith 2001) estimated by LANDIS PRO for deciduous versus coniferous species. We classified cells as deciduous forest if the combined importance of deciduous species surpassed 65% and coniferous forest if such species comprised >47% of the cell's importance value. Forested cells not classified as either deciduous or coniferous were assigned to the mixed forest type. We further classified deciduous forest cells as woody wetlands for any cells with this original NLCD class. We used the 65% and 47% thresholds because they produced land cover estimates for the initial (2000) landscape proportional to actual NLCD classes for the same region. We grouped classes into either forest or nonforest to estimate forest patch sizes and the percent forest cover within 1 km and 10 km. We classified forest seral stage as shrub/seedling, sapling, pole, or saw based on quadratic mean tree diameter calculated from LANDIS PRO projections of basal area and tree density according to Tirpak et al. (2009a). We also calculated total tree stocking, from which we estimated canopy closure based on empirical associations between stocking and canopy closure (Johnson et al. 2009, Blizzard et al. 2013). We used the density of all tree species in the 0–10 age cohort output by LANDIS PRO to approximate the density of small stems (<2.54 cm dbh) because most hardwood species take ~10 yr to reach 2.54 cm dbh (Johnson et al. 2009).

We modeled habitat at a 30-m resolution by resampling outputs from Wang et al. (2015) and augmenting gaps in habitat characteristics using spatially explicit, remotely sensed data from ancillary sources. We used 2001 canopy cover estimates from the Multi-Resolution Land Characteristics Consortium (Homer et al. 2004). We obtained data on dbh and small-stem density from efforts integrating Forest Inventory and Analysis data and MODIS imagery (Wilson et al. 2012). For these cells, we held values constant over time.

We followed Bonnot et al.'s (2013) approach to estimating K for cells through a relationship between HSI and densities of prairie warblers and wood thrush found in the literature (Table 1). In the absence of data, we assumed densities of birds reached their maximums at HSI = 1 and declined linearly with HSI. We then scaled density by the area of cells and spatially filtered areas of the landscape that could not support at least one territory given maximum territory sizes for each species. This process more realistically captured the interaction between spatial and resource limitations inherent in estimating K than simply summing K across all cells (see Donovan et al. 2012). Shifts in distribution of habitat over time due to the effects of climate and management were captured by subsequent changes in the distribution of K. We estimated wood thrush and prairie warbler initial distributions in year 2000 throughout the region as a percentage of K (Table 1). We modeled breeding productivity of wood thrush and prairie warblers

throughout the Central Hardwoods over time using a Relative Productivity Index model (RPI; Bonnot et al. 2011). This index of reproductive success of birds ranges 0–1 and is based on the fragmentation paradigm that success is lower in fragmented landscapes and proximate to edge. This concept has a strong basis based on the original studies reporting these effects (Donovan et al. 1995, Robinson et al. 1995, Thompson et al. 2002, Cox et al. 2013) and subsequent reviews and meta-analyses (Chalfoun et al. 2002, Stephens et al. 2004, Lloyd et al. 2005). We estimated RPI for each 30-m cell using the amount of forest cover in a 10 km radius and edge within 200 m. We applied RPIs to the maximum possible productivity identified for each species from the literature to estimate fertility values throughout the region (see Appendix S1 for specific methods).

Finally, we used the dispersal model of Bonnot et al. (2011) to estimate cell-based movements

Table 1. Demographic parameters used in dynamic-landscape metapopulation models for wood thrush and prairie warblers in the Central Hardwoods in the Midwestern United States.

Parameter	Wood thrush		Prairie warbler	
	Estimate	Source	Estimate	Source
Carrying capacity (pairs/ha) at HSI = 1	0.50	Thompson et al. (1992), Roth et al. (1996), Gram et al. (2003), Wallendorf et al. (2007)	1.00	Fink (2003)
Initial abundance (% of carrying capacity)	0.12	Thompson et al. 1992, Gram et al. (2003), Wallendorf et al. (2007)	0.50	Thompson et al. (1992), Fink (2003), Brito-Aguilar (2005), Wallendorf et al. (2007)
Maximum maternity (fem/fem/year)	1.45	Donovan et al. (1995), Anders et al. (1997), Ford et al. (2001)	1.55	Fink (2003), Nolan et al. (2014)
Adult survival	0.61	Conway et al. (1995), Donovan et al. (1995), Powell et al. (2000), Simons et al. (2000)	0.60	Lehnen and Rodewald (2009), Nolan et al. (2014)
Juvenile survival	0.29	Anders et al. (1997)	0.32	Nolan et al. (2014)
Parametric uncertainty (SD)				
Maternity	0.25	Roth et al. (1996)	0.36	Roth et al. (1996)
Adult survival	0.005		0.005	
Juvenile survival	0.005		0.005	
Environmental stochasticity (CI)				
Fertility	0.27	Roth et al. (1996)	0.27	Roth et al. (1996)
Juvenile survival	0.25	Brown and Roth (2004), Schmidt et al. (2008)	0.15	Larson et al. (2004)
Adult survival	0.10	Brown and Roth (2004)	0.10	Brown and Roth (2004)
Demographic stochasticity	Yes		Yes	
Density dependence	Modified ceiling		Modified ceiling	
Percentage of juveniles dispersing annually	90%	Evans et al. (2011)	90%	Nolan et al. (2014)
Percentage of adults dispersing annually	10%	Evans et al. (2011)	20%	Nolan et al. (2014)

Note: Parameter uncertainty and environmental stochasticity are specified by standard deviation (SD) and coefficient of variation (CI), respectively.

of dispersing individuals to the surrounding landscape based on a negative exponential function of distance between cells, weighted by K of the destination cell. Weighting by carrying capacity allowed changes in future dispersal movements to reflect shifts in the distribution of habitat in the region over time. See Appendix S1 for a full description of all habitat modeling.

Population modeling

We modeled regional population growth of wood thrush and prairie warblers through 2100 based on landscapes by treating ecological subsections as subpopulations in a metapopulation model and summarizing their demographics for each subsection over time. The region contained 71 subsections which we delineated into 87 unique subpopulations that ranged in size from 5 to 24,000 km^2 (Fig. 1; Cleland et al. 2007).

For each subpopulation, we summarized results of the habitat models to obtain estimates of initial abundance and K at each decade. We averaged cell fertilities in each subpopulation, weighted by their K, so that estimates of productivity for subpopulations reflected areas where breeding occurred. We derived relative rates of dispersal among subpopulations by combining assumptions about the proportion of birds dispersing with relative estimates of the cell-based movements of those dispersers to the surrounding landscape (Bonnot et al. 2011). We calculated yearly values of demographics by linearly interpolating between decadal estimates because the landscape projections were for 10-yr time steps. Although landscape projections were available through 2300, we only modeled the first 100 yr given the uncertainty associated with predicting population growth.

We developed female-only, Lefkovitch matrix models comprising adult and juvenile stages in R v3.0.1 (R Core Team 2016). We set adult and juvenile survival in prairie warblers and wood thrush at 0.60/0.32 and 0.61/0.29, respectively, and assumed a post-breeding census (Table 1). We redistributed dispersers among the subpopulations according to multinomial distributions with probabilities equal to the relative dispersal rates for that year. We modified the commonly referred to ceiling density dependence (Akçakaya 2000) such that individuals over K in a population were prohibited from breeding but could remain in the

population or disperse (Bonnot et al. 2013), as nonbreeding "floater" adults are relatively common in passerine populations (Smith 1978, Bayne and Hobson 2002).

To quantify viability or risk under the climate scenarios, we used Monte Carlo simulations to induce parameter uncertainty and stochasticity in our population dynamics. We simulated parameter uncertainty by sampling a different survival and fertility rate in each of the 1000 iterations from beta and gamma distributions, respectively, with means equal to their overall estimates and corresponding error, derived from the literature (Table 1; McGowan et al. 2011). In each iteration, the rates drawn were used to construct beta and lognormal distributions, from which annual survival and fertility rates could be drawn. Patterns in annual survival rates were correlated among subpopulations based on a negative exponential relationship with the distances among them (Bonnot et al. 2011). We based variances for these distributions on the amount of temporal variation empirically observed in survival or reproduction (Table 1). In each year, we modeled demographic stochasticity by drawing the number of survivors and the number of young produced in each stage each year from binomial and Poisson distributions, respectively. An example of the R code for these models can be found in Data S1.

RESULTS

Complex shifts in the roles of forest succession and management relative to climate change differentially affected habitats for the two species over the course of the next century. The dominant processes affecting forest change in the first 50 yr were succession and management that produced an aging forest. As a result, wood thrush habitat increased the first 50 yr under all climate scenarios (Fig. 3a). Increases in habitat and, consequently, K leveled off the latter half of the century with K for the two climate change scenarios <5% lower than for current climate (current: 7,581,855 females; CGCM.T47-A2: 7,364,356 females; GFDL-A1FI: 7,266,566 females). The distribution of wood thrush habitat remained mostly constant across subpopulations, with most habitat occurring in the Ozarks subsections of south-central Missouri. Counter to wood thrush, prairie

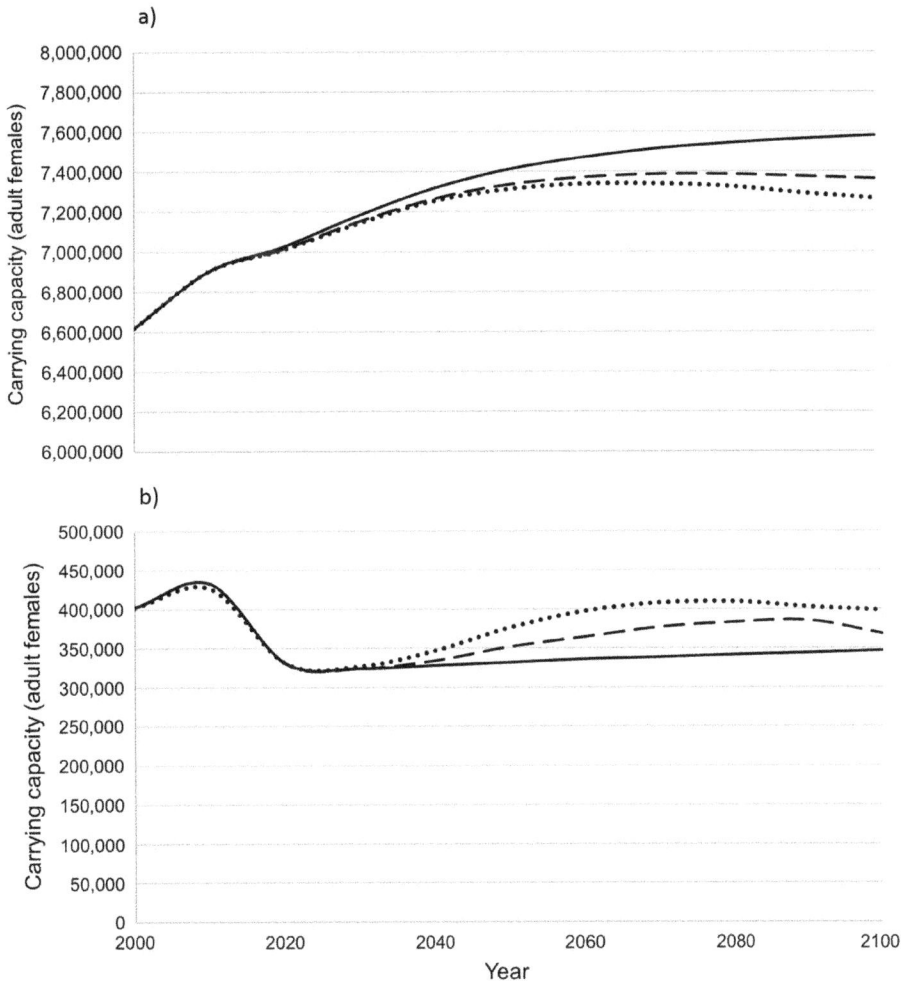

Fig. 3. Impacts of landscape change from climate change on (a) wood thrush and (b) prairie warbler habitat in the Central Hardwoods as indicated by their carrying capacities under a current climate scenario (solid), a moderate CGCM.T47-A2 climate change scenario (dashed), and an extreme climate change GFDL-A1Fi scenario (dotted).

warbler K declined sharply across the region over the first three decades for all scenarios (Fig. 3b). However, while K increased only slightly under the current climate after 2030, K increased more under both climate change scenarios as increasing effects of climate change resulted in more open forests in southwestern subsections. Carrying capacity increased as much as 88% after 2040 in these subsections causing a significant shift in the distribution of habitat under climate change and a >20% rise in K under the GFDL-A1FI scenario than the current climate (Fig. 3b).

The changes in habitat resulted in equally complex effects on population dynamics and

distribution. Wood thrush population dynamics were unaffected by climate change. Despite increasing habitat in the region, we projected >25% declines in wood thrush abundance from the initial estimate of 794,321 adult females under all climate scenarios by 2100 (Fig. 4a). Projected declines averaged <1% per year for all scenarios, but annual dynamics of the regional population ranged between a 3.8% drop to 2.5% growth from year to year (Table 2). These declines were driven by low reproduction in many subsections resulting from habitat fragmentation; however, subpopulations in the Missouri Ozarks grew more than 50%, which concentrated the distribution of wood

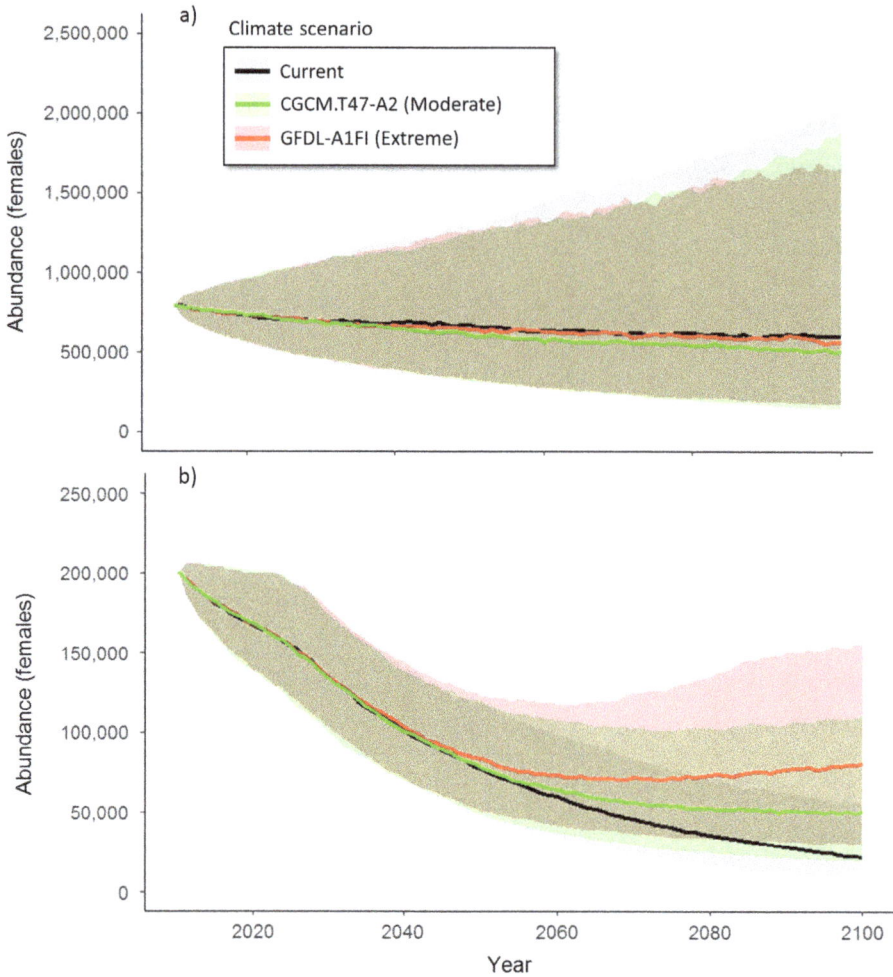

Fig. 4. Projected population dynamics of (a) wood thrush and (b) prairie warblers in the Central Hardwoods based on landscape change under three future climate change scenarios. Shaded regions indicate 85% credible intervals.

Table 2. Predicted dynamics and viability of wood thrush and prairie warbler populations in the Central Hardwoods region of the United States based on future landscapes projected under the current climate and moderate (CGCM.T47-A2) and severe (GFDL-A1Fi) climate change scenarios.

Population parameter	Prairie warbler			Wood thrush		
	Current	CGCM.T47-A2	GFDL-A1Fi	Current	CGCM.T47-A2	GFDL-A1Fi
Initial N†	201,161			794,321		
N in 2100 (median)	19,270	55,490	85,616	579,456	467,395	566,841
Percent change (2000–2100)	−90%	−72%	−57%	−27%	−41%	−29%
Projected average annual trend	−2.32%	−1.28%	−0.85%	−0.31%	−0.53%	−0.34%
Observed BBS trends for Central Hardwoods (1966–2015)‡	−1.98%			−0.62%		
Risk of 50% decline from initial N	65%	56%	48%	21%	23%	22%

† N, abundance of adult females.
‡ Breeding Bird Survey (BBS) results, Sauer et al. (2017)

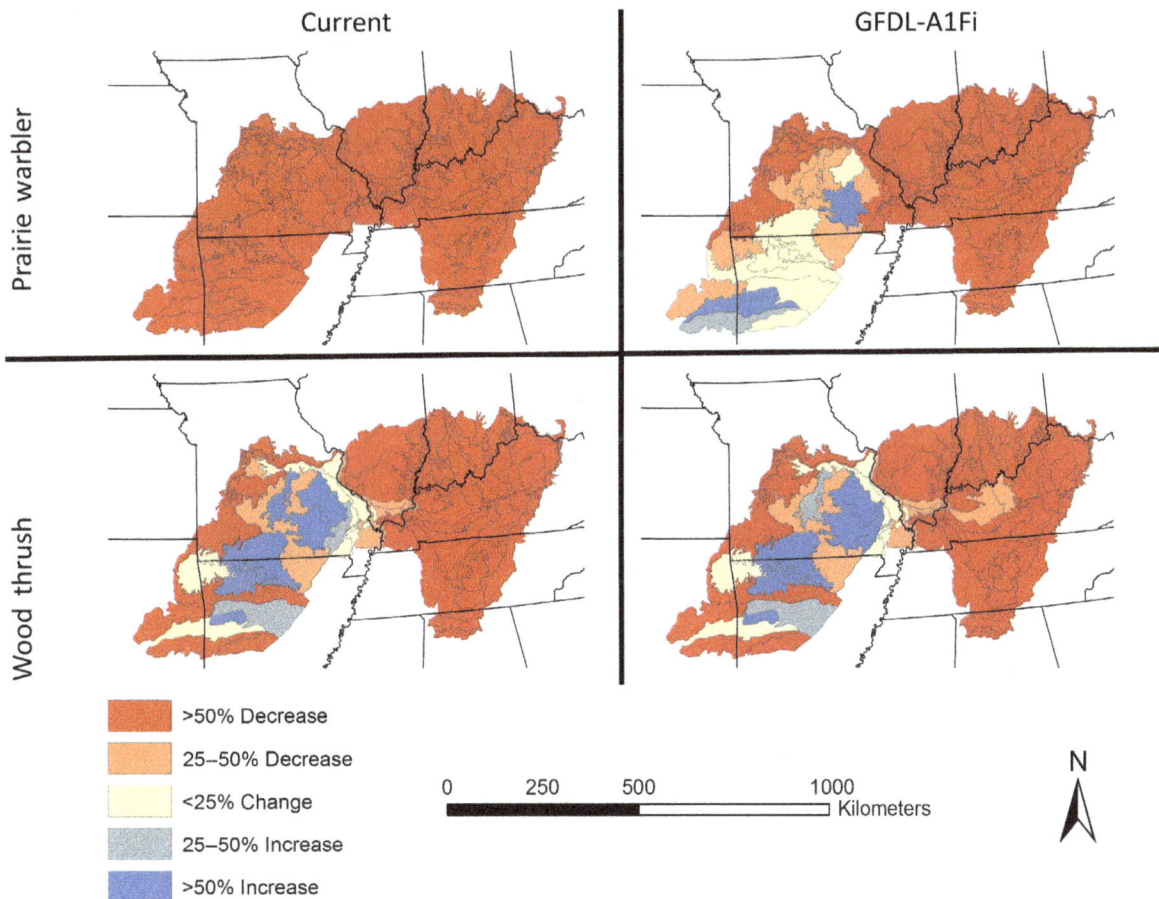

Fig. 5. Predicted changes in distribution of wood thrush and prairie warbler across the Central Hardwoods under a current climate scenario and the extreme (GFDL-A1Fi) climate change scenario. Shifts in distribution are apparent through the projected increases or decreases from initial bird abundances by 2100 across the region's ecological subsections.

thrush in these areas (Fig. 5). Projections for wood thrush were not only similar among scenarios, but they also displayed great uncertainty as the population under any scenario was less than half or more than double the initial abundance (based on 80% confidence intervals). As a result, the risk of decline for wood thrush in the Central Hardwoods was nearly identical under all three future climate scenarios (Fig. 6).

Unlike wood thrush, population dynamics for prairie warblers appeared closely linked to climate-driven increases in habitat over time. We predicted declines exceeding 3% per year through 2050, likely as a result of the decline in K (Table 2). By midcentury, the prairie warbler population was estimated at <50% its initial total of

201,161 females (Fig. 4b). While negative growth continued under the current climate scenario following 2060 (overall 90% loss), the decline slowed under the CGCM.T47-A2 scenario and was ultimately reversed under the GFDL-A1Fi scenario. The positive response of prairie warblers under climate scenarios, however, was primarily seen in the western and southwestern subpopulations (Fig. 5). The shifts in prairie warbler distribution under the two climate change scenarios corresponded with the increase in habitat in these landscapes. No distributional shifts occurred for prairie warblers under the current climate. The beneficial effects of climate change on prairie warbler habitat also translated in improved viability for the regional populations, lowering the risk of

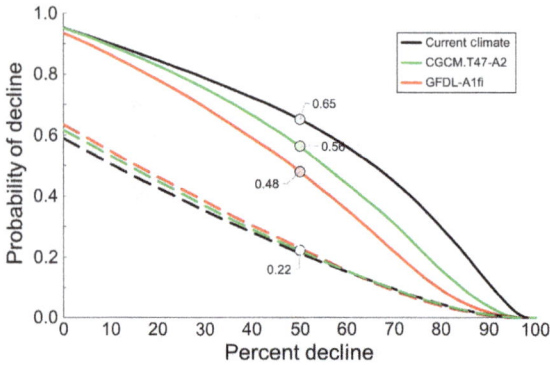

Fig. 6. Projected risk to prairie warbler (solid lines) and wood thrush (dashed lines) populations in the Central Hardwoods from 2000 to 2100 based on dynamic-landscape metapopulation modeling. The probabilities that a population will experience different levels of decline during the simulation are plotted for three future climate scenarios: a current climate scenario, a moderate CGCM.T47-A2 climate change scenario, and an extreme climate change GFDL-A1Fi scenario. Model projections suggest a 17% less chance of prairie warblers declining by half (50% decline) in landscapes under extreme climate change than under a current climate.

a 50% decline sometime during the next century by 17% (Fig. 6).

Discussion

The complex and contrasting responses of prairie warbler and wood thrush populations to climate change demonstrated the importance of mechanistic approaches, such as of DLMPs, that can incorporate important processes. Climate change is predicted to have less effect on the region's forests through 2100 than tree harvest and succession (Wang et al. 2015, 2016). Thus, wood thrush, whose habitat comprises a wider range of forest conditions, saw only slight effects on habitat under climate change and no effect on their population dynamics. Rather, existing landscape-level fragmentation of habitat was responsible for the declines in the population. The shift in distribution to the southwestern subsections stemmed from greater productivity in those less fragmented landscapes, a finding similar to Bonnot et al. (2011), who did not consider climate change. Prairie warbler habitat consists of a

narrow range of forest conditions that include early successional forests or woodland and glade communities that have low canopy closure and high ground and shrub cover. Declines in prairie warbler habitat the first 50 yr under all scenarios were due to forest succession resulting in older, closed-canopy forest, which is consistent with current habitat and population trends in the region (Franzreb et al. 2011). However, by the latter half of the century, reduced precipitation and elevated temperatures in the southwestern portion of the region under the climate change scenarios began to alter forest structure by reducing tree stocking in these areas (Wang et al. 2015). Lower tree stocking resulted in lower canopy cover and more open forest structure that created prairie warbler habitat. These changes occurred in subsections that had poor or droughty soils, which also tended to be areas with a larger proportion of forest land cover because they were less suitable for agricultural land uses. Therefore, climate change created prairie warbler habitat in landscapes with high potential productivity because they had lower levels of fragmentation. Thus, while prairie warblers declined regionally under the current climate, changes in forest structure in Missouri, Arkansas, and Oklahoma from climate change resulted in greater populations because of increased habitat in landscapes with high reproduction. Therefore, while wood thrush were affected little by the effects of climate change on habitat and instead declined from other threats, prairie warblers declined from loss of habitat due to succession but began to recover with the creation of habitat under climate change. The differences between species responses arrived from the interaction of climate, habitat, and demographic process. Without the means to account for these interactive processes, other less mechanistic approaches would have likely produced different predictions that might lead to alternative conservation efforts under climate change.

Our assessment of how two regional songbird populations responded to landscape change from climate warming provides a stark contrast with predictions from less mechanistic approaches. Langham et al. (2015) used SDMs based on climatic variables and projected wood thrush to lose >80% of their summer range, including much of the range in Missouri. While such a prediction seems extreme, it is supported by others who have forecasted major range losses and extinctions of birds and other animals globally, under climate change (Thomas et al. 2004, Warren et al. 2013). The omittance of species habitat and ecology by directly predicting distribution from climate forces SDMs to assume that these processes track climate or that climate is the primary determinant of species range (Keith et al. 2008). We are not the first to question these assumptions (Ralston et al. 2016). It can take long periods for habitat and wildlife populations to respond to climate during which time geological, ecological, and landscape processes could preclude or alter those responses. However, given recent evidence of range expansions and the potential for direct effects of climate change on bird demographics still not accounted for in our models, the true responses likely lie in between. Other SDMs have incorporated habitat with climate and projected similar changes to ours (Matthews et al. 2011). However, the declines we predicted in Wood Thrush were not due to climate, but instead continue the recent trend of studies that show that multiple processes in addition to climate determine the populations' dynamics and distribution (e.g., Swab et al. 2012, Fordham et al. 2013, Franklin et al. 2014). Therefore, because all approaches currently fall short in their ability to fully predict species responses to climate change and other threats, it is wise to base planning on multiple approaches, each with contrasting strengths and weaknesses (Millspaugh et al. 2009, Iverson et al. 2016).

The projected declines of wood thrush and prairie warblers in the Central Hardwoods are good illustrations of the importance of current, anthropogenic, and ecological drivers of change relative to those expected from climate change. The drop in early successional forests in the first three decades that spurred prairie warbler losses occurred because levels of disturbance and timber harvest did not offset habitat losses due to succession. Positive responses of prairie warblers to the formation of open, woodland communities under climate change scenarios further reflect the loss of these natural habitats in the Central Hardwoods from forest management over the past century, which drove much of their declines. The projected declines of both species across most of the region were also due to impaired reproduction resulting

from forest fragmentation/parasitism. Such results highlight the long-held view that anthropogenic habitat loss and fragmentation continue to be a predominant threat to terrestrial species decline (Sala et al. 2000). Nonetheless, even seemingly minor climate impacts on forests created prairie warbler habitat that caused shifts in and reversed declines of an entire regional population. Some have suggested that a key factor in the resiliency of species during past climatic changes has been absence of human-caused impacts (Moritz and Agudo 2013). Indeed, our work suggests that addressing current threats such as habitat loss and fragmentation could be key to resiliency of these species. Ultimately, however, the prairie warbler projections also remind us that impacts from climate change are likely to overwhelm even these processes over the long term; Wang et al. (2015) determined the contribution of climate change to forest landscape change in the region increased substantially from 100 to 300 yr in the future.

Although we achieved more realistic predictions by increasing the number of processes modeled, many processes are still unaccounted for that could change projections. Incorporating the influence of climate change on landscapes and habitat is an important step in modeling future viability of wildlife populations (Fordham et al. 2013). However, the degree to which predictions are improved over other less mechanistic approaches depends on how well our models replicate the actual processes. For example, the landscape projections from Wang et al. (2015) that underlie our results do not incorporate effects of climate on disturbances such as fire, insect outbreaks, and drought, which could be exacerbated by climate change and would provide direct sources of tree mortality. The resiliency of Central Hardwoods forests over the next 100 yr stemmed from the longevity of its trees. Therefore, including mortality from large-scale disturbances would most likely accelerate changes to the forests, species' habitats, and ultimately population dynamics. As a means to isolate the effects of climate change on the landscape, Wang et al. (2015) also maintained current harvest management practices over the duration of their simulations. It is possible that public land managers will adapt forest management under a changing climate, which, given the relative impact of forest management in the near term, could affect predictions. However, much of the current forest management on public lands targets the same conditions brought on by climate change (e.g., opening of forest canopies to restore woodlands). Furthermore, the majority of forests in the Central Hardwoods are privately owned and may not see substantial changes in forest management.

Our models only begin to address the myriad of pathways climate can affect an entire species or population. For example, our DLMPs do not currently incorporate direct effects of climate change on bird demographics, despite knowledge of such relationships (Cox et al. 2013, Bonnot et al., *unpublished manuscript*). While we are currently working to incorporate these demographic processes, preliminary modeling suggests that under future climate change this mechanism could overwhelm the current responses to habitat and drive severe population declines in some species (Bonnot et al., *unpublished manuscript*). Changes in phenology, novel assemblages and invasive species, disease, physiological stress, and food availability are other examples that have been investigated (Thomas et al. 2004, Reed et al. 2013, Selwood et al. 2015, Hache et al. 2016). A century is a long time for animals to evolve in response to environmental change, and adaptations could play a role in how species persist over the long term (Alberti et al. 2017). Finally, our predictions describe impacts to the Central Hardwood's population of these birds and do not account for changes in habitat and other processes outside of the region and throughout the rest or their range. Increasing the comprehensiveness of this approach will require integrating these processes. Therefore, we must continually remain aware and transparent in our uncertainty about predictions, realizing that while our models will always be wrong to some degree, they can still be useful (Box 1979).

Relatively few examples of modeling both distributions and dynamics together under climate change exist. As a result, a dichotomy appears to have evolved where species distribution is the focus at regional and range-wide scales, while a focus on population dynamics occurs at smaller scales. However, the coupling of climate, habitat, and population dynamics is being made easier by advances in modeling and is an emerging area of research that is breaking down this dichotomy.

We showed that the distribution of a population is not solely determined by climatic or even habitat niches but is also the manifestation of the population dynamics that occur throughout its distribution. Likewise, population dynamics are influenced by the climate, habitat, and demographic processes where it is distributed. This link explains both the shifts we projected in bird distributions and the resulting population growth. The inability to address this interdependency risks leaving species threatened by local processes when relying on SDMs or, in the case of population models, larger distributional shifts when focusing on impacts at smaller scales. Therefore, it will be important to build on this study and the works of others (e.g., Hunter et al. 2010, Fordham et al. 2012, 2013, Regan et al. 2012, Franklin et al. 2014) and continue striving for more comprehensive approaches that can model important processes that drive population dynamics and distribution at the scales at which those processes interact with climate.

Because they link local habitat and demographic processes to large-scale population growth, DLMPs address two major conservation planning needs. They combine the ability to predict the impacts of landscape and climate change on populations and simultaneously evaluate the effectiveness of conservation activities to mitigate those impacts. The popularity of DLMPs arises from their ability to incorporate many processes that are important to predicting species responses to climate change (McMahon et al. 2011). As new climate change, habitat, and demographic processes are identified, or current processes are better understood, they are readily integrated to better reflect the complex reality of how climate will affect species. Such adaptability is critical where knowledge of trait-based vulnerabilities of species to climate increasingly exists, but the framework in which to quantify their effects on populations does not (Swab et al. 2012, Fordham et al. 2013, Moritz and Agudo 2013). We have also shown the importance of accounting for threats other than climate, such as land-use change and fragmentation. Further, DLMPs are scalable across species, taxa, and geographies (Jones-Farrand and Bonnot 2014, Bonnot 2016). In DLMPs, planners also have a powerful tool for Strategic Habitat Conservation (National Ecological Assessment Team 2008, Fitzgerald et al. 2009).

By altering projected landscapes or species demographics to simulate habitat restoration or other conservation measures, planners can predict how species will respond to conservation amidst global change when deciding plans. Simultaneously conveying responses of wildlife populations to conservation scenarios and the risk associated with those responses provides managers with a more intuitive and defensible way of comparing plans for species. (Drechsler and Burgman 2004, Bonnot 2016). As DLMPs become increasingly comprehensive, their potential to provide a unifying approach to conserving species in the face of global change grows.

Acknowledgments

We thank J. S. Fraser, H. S. He, W. J. Wang, Todd Jones-Farrand, and W. D. Dijak for data and input on the study. Funding was provided by the Gulf Coastal Plains and Ozarks Landscape Conservation Cooperative, the Department of Interior USGS Northeast Climate Science Center graduate and post-doctoral fellowships, and the U.S.D.A. Forest Service Northern Research Station. The contents of this paper are solely the responsibility of the authors and do not necessarily represent the views of the United States Government.

Literature Cited

Akçakaya, H. R. 2000. Viability analyses with habitat-based metapopulation models. Population Ecology 42:45–53.

Akçakaya, H. R., and B. W. Brook. 2009. Determining viability of wildlife populations in large landscapes. Pages 449–472 in J. J. Millspaugh and F. R. Thompson III, editors. Models for planning wildlife conservation in large landscapes. Elsevier/ Academic Press, Boston, Massachusetts, USA.

Alberti, M., C. Correa, J. M. Marzluff, A. P. Hendry, E. P. Palkovacs, K. M. Gotanda, V. M. Hunt, T. M. Apgar, and Y. Zhou. 2017. Global urban signatures of phenotypic change in animal and plant populations. Proceedings of the National Academy of Sciences. https://doi.org/10.1073/pnas.1606034114

Anders, A. D., D. C. Dearborn, J. Faaborg, and F. R. Thompson III. 1997. Juvenile survival in a population of neotropical migrant birds. Conservation Biology 11:698–707.

Bayne, E. M., and K. A. Hobson. 2002. Annual survival of adult American redstarts and ovenbirds in the southern boreal forest. Wilson Bulletin 114: 358–367.

Bekessy, S. A., B. A. Wintle, A. Gordon, R. A. Chisholm, L. A. Venier, and J. L. Pearce. 2009. Metapopulation models and sustainable forest management. Pages 473–500 in J. J. Millspaugh and F. R. Thompson III, editors. Models for planning wildlife conservation in large landscapes. Elsevier/ Academic Press, Boston, Massachusetts, USA.

Blizzard, E. M., J. M. Kabrick, D. C. Dey, D. R. Larsen, S. G. Pallardy, and D. P. Gwaze. 2013. Light, canopy closure, and overstory retention in upland Ozark forests. Pages 73–79 in J. M. Guldin, editor. Proceedings of the 15th Biennial Southern Silvicultural Research Conference. e-General Technical Report SRS-GTR-175. USDA Forest Service, Southern Research Station, Asheville, North Carolina, USA.

Bonnot, T. W. 2016. Novel approaches to conserving the viability of regional wildlife populations in response to landscape and climate change. Dissertation. University of Missouri, Columbia, Missouri, USA.

Bonnot, T. W., F. R. Thompson III, and J. J. Millspaugh. 2011. Extension of landscape-based population viability models to ecoregional scales for conservation planning. Biological Conservation 144: 2041–2053.

Bonnot, T. W., F. R. Thompson III, J. J. Millspaugh, and D. T. Jones-Farrand. 2013. Landscape-based population viability models demonstrate importance of strategic conservation planning for birds. Biological Conservation 165:104–114.

Box, G. E. P. 1979. Robustness in scientific model building. Pages 201–236 in R. L. Launer and G. N. Wilkinson, editors. Robustness in statistics. Academic Press, New York, New York, USA.

Brito-Aguilar, R. 2005. Effects of even-aged forest management on early successional bird species in Missouri Ozark forest Thesis. University of Missouri, Columbia, Missouri, USA.

Brook, B. W., N. S. Sodhi, and C. J. Bradshaw. 2008. Synergies among extinction drivers under global change. Trends Ecology and Evolution 23:453–460.

Brown, W. P., and R. R. Roth. 2004. Juvenile survival and recruitment of wood thrushes *Hylocichla mustelina* in a forest fragment. Journal of Avian Biology 35:316–326.

Chalfoun, A. D., F. R. Thompson III, and M. J. Ratnaswamy. 2002. Nest predators and fragmentation: a review and meta-analysis. Conservation Biology 16:306–318.

Chen, I. C., J. K. Hill, R. Ohlemuller, D. B. Roy, and C. D. Thomas. 2011. Rapid range shifts of species associated with high levels of climate warming. Science 333:1024–1026.

Cheung, W. W. L., J. L. Sarmiento, J. Dunne, T. L. Frölicher, V. W. Y. Lam, M. L. Deng Palomares, R. Watson, and D. Pauly. 2012. Shrinking of fishes exacerbates impacts of global ocean changes on marine ecosystems. Nature Climate Change 3:254–258.

Cleland, D. T., J. A. Freeouf, J. E. Keys, G. J. Nowacki, C. A. Carpenter, and W. H. McNab. 2007. Ecological subregions: sections and subsections for the conterminous United States. General Technical Report WO-76D. USDA Forest Service, Washington, D.C., USA.

Conway, C. J., G. V. N. Powell, and J. D. Nichols. 1995. Overwinter survival of neotropical migratory birds in early-successional and mature tropical forests. Conservation Biology 9:855–864.

Cox, W. A., F. R. Thompson III, J. L. Reidy, and J. Faaborg. 2013. Temperature can interact with landscape factors to affect songbird productivity. Global Change Biology 19:1064–1074.

Dijak, W. D., B. B. Hanberry, J. S. Fraser, H. S. He, W. J. Wang, and F. R. Thompson III. 2016. Revision and application of the LINKAGES model to simulate forest growth in central hardwood landscapes in response to climate change. Landscape Ecology. https://doi.org/10.1007/s10980-016-0473-8

Dijak, W. D., and C. D. Rittenhouse. 2009. Development and application of habitat suitability models to large landscapes. Pages 367–390 in J. J. Millspaugh and F. R. Thompson III, editors. Models for planning wildlife conservation in large landscapes. Elsevier/Academic Press, Boston, Massachusetts, USA.

Donovan, T. M., R. H. Lamberson, A. Kimber, F. R. Thompson III, and J. Faaborg. 1995. Modeling the effects of habitat fragmentation on source and sink demography of neotropical migrant birds. Conservation Biology 9:1396–1407.

Donovan, T. M., G. S. Warrington, W. S. Schwenk, and J. H. Dinitz. 2012. Estimating landscape carrying capacity through maximum clique analysis. Ecological Applications 22:2265–2276.

Drechsler, M., and M. A. Burgman. 2004. Combining population viability analysis with decision analysis. Biodiversity and Conservation 13:115–139.

Elith, J., and J. R. Leathwick. 2009. Species distribution models: ecological explanation and prediction across space and time. Annual Review of Ecology Evolution and Systematics 40:677–697.

Evans, M., E. Gow, R. R. Roth, M. S. Johnson, and T. J. Underwood. 2011. Wood thrush (*Hylocichla mustelina*). In P. G. Rodewald, editor. The birds of North America. Cornell Lab of Ornithology, Ithaca, New York, USA. https://birdsna.org/Species-Account/bna/species/woothr

Fink, A. D. 2003. Habitat use, demography, and population viability of disturbance-dependent shrubland birds in the Missouri Ozarks. Dissertation. University of Missouri, Columbia, Missouri, USA.

Fischlin, A., G. F. Midgley, J. T. Price, R. Leemans, B. Gopal, C. Turley, M. D. A. Rounsevell, O. P. Dube, J. Tarazona, and A. A. Velichko. 2007. Ecosystems, their properties, goods, and services. Pages 211–272 in M. L. Parry, O. F. Canziani, J. P. Palutikof, P. J. v. d Linden, and C. E. Hanson, editors. Climate Change 2007: impacts, adaptation and vulnerability. Contribution of Working Group II to the Fourth Assessment Report of the Intergovernmental Panel on Climate Change. Cambridge University Press, Cambridge, UK.

Fitzgerald, J. A., W. E. Thogmartin, R. Dettmers, T. Jones, C. Rustay, J. M. Ruth, F. R. Thompson III, and T. Will. 2009. Application of models to conservation planning for terrestrial birds in North America. Pages 593–624 in J. J. Millspaugh and F. R. Thompson III, editors. Models for planning wildlife conservation in large landscapes. Elsevier/Academic Press, Boston, Massachusetts, USA.

Fitzgerald, J. A., C. D. True, D. D. Diamond, T. Ettel, L. Moore, T. Nigh, S. Vorisek, and G. Wathen. 2005. Delineating focus areas for bird conservation in the Central Hardwoods Bird Conservation Region. General Technical Report PSW-GTR-191. USDA Forest Service, Pacific Southwest Research Station, Albany, California, USA.

Ford, T. B., D. E. Winslow, D. R. Whitehead, and M. A. Koukol. 2001. Reproductive success of forest-dependent songbirds near an agricultural corridor in south-central Indiana. Auk 118:864–873.

Fordham, D. A., et al. 2012. Plant extinction risk under climate change: Are forecast range shifts alone a good indicator of species vulnerability to global warming? Global Change Biology 18:1357–1371.

Fordham, D. A., et al. 2013. Population dynamics can be more important than physiological limits for determining range shifts under climate change. Global Change Biology 19:3224–3237.

Franklin, J., H. M. Regan, and A. D. Syphard. 2014. Linking spatially explicit species distribution and population models to plan for the persistence of plant species under global change. Environmental Conservation 41:97–109.

Franzreb, K. E., S. N. Oswalt, and D. A. Buehler. 2011. Population trends for Eastern scrub-shrub birds related to availability of small-diameter upland-hardwood forest. Pages 143–166 in C. H. Greenberg, B. S. Collins, and F. R. Thompson III, editors. Sustaining young forest communities. Springer, Amsterdam, The Netherlands.

Fry, J., G. Xian, S. Jin, J. Dewitz, C. Homer, L. Yang, C. Barnes, N. Herold, and J. Wickham. 2011. Completion of the 2006 National Land Cover Database for the Conterminous United States. Photogrammetric Engineering & Remote Sensing 77:858–864.

Girvetz, E. H., C. Zganjar, G. T. Raber, E. P. Maurer, P. Kareiva, and J. J. Lawler. 2009. Applied climate-change analysis: the climate wizard tool. PLoS ONE 4:e8320.

Gram, W. K., P. A. Porneluzi, R. L. Clawson, J. Faaborg, and S. C. Richter. 2003. Effects of experimental forest management on density and nesting success of bird species in Missouri Ozark Forests. Conservation Biology 17:1324–1337.

Guisan, A., and W. Thuiller. 2005. Predicting species distribution: offering more than simple habitat models. Ecology Letters 8:993–1009.

Hache, S., R. Cameron, M. A. Villard, E. M. Bayne, and D. A. MacLean. 2016. Demographic response of a neotropical migrant songbird to forest management and climate change scenarios. Forest Ecology and Management 359:309–320.

Homer, C., C. Q. Huang, L. M. Yang, B. Wylie, and M. Coan. 2004. Development of a 2001 National Land-Cover Database for the United States. Photogrammetric Engineering and Remote Sensing 70:829–840.

Hunter, C. M., H. Caswell, M. C. Runge, E. V. Regehr, S. C. Amstrup, and I. Stirling. 2010. Climate change threatens polar bear populations: a stochastic demographic analysis. Ecology 91:2883–2897.

Intergovernmental Panel on Climate Change (IPCC). 2007. Climate change 2007: the physical science basis. Contribution of Working Group I to the Fourth Assessment Report of the Intergovernmental Panel on Climate Change. Cambridge University Press, Cambridge, UK.

Iverson, L. R., et al. 2016. Multi-model comparison on the effects of climate change on tree species in the eastern U.S.: results from an enhanced niche model and process-based ecosystem and landscape models. Landscape. https://doi.org/10.1007/s10980-016-0404-8

Johnson, P., S. R. Shifley, and R. Rogers. 2009. The ecology and silviculture of oaks. Second edition. CAB International North America, Cambridge, Massachusetts, USA.

Jones-Farrand, D. T., and T. W. Bonnot. 2014. Assessing the impacts of climate and landscape change on forest interior birds in Indiana's Brown County Hills. Final report to The Nature Conservancy. Blue River Project Office, Columbia, Missouri, USA.

Keith, D. A., H. R. Akcakaya, W. Thuiller, G. F. Midgley, R. G. Pearson, S. J. Phillips, H. M. Regan, M. B. Araujo, and T. G. Rebelo. 2008. Predicting extinction risks under climate change: coupling stochastic population models with dynamic bioclimatic habitat models. Biology Letters 4:560–563.

Langham, G. M., J. G. Schuetz, T. Distler, C. U. Soykan, and C. Wilsey. 2015. Conservation status of North American birds in the face of future climate change. PLoS ONE 10:e0135350.

Larson, M. A., F. R. Thompson III, J. J. Millspaugh, W. D. Dijak, and S. R. Shifley. 2004. Linking population viability, habitat suitability, and landscape simulation models for conservation planning. Ecological Modelling 180:103–118.

LaSorte, F. A., and F. R. Thompson III. 2007. Poleward shifts in winter ranges of North American birds. Ecology 88:1803–1812.

Lehnen, S. E., and A. D. Rodewald. 2009. Investigating area-sensitivity in shrubland birds: responses to patch size in a forested landscape. Forest Ecology and Management 257:2308–2316.

Lloyd, P., T. E. Martin, R. L. Redmond, U. Langner, and M. M. Hart. 2005. Linking demographic effects of habitat fragmentation across landscapes to continental source-sink dynamics. Ecological Applications 15:1504–1514.

Matthews, S. N., L. R. Iverson, A. M. Prasad, and M. P. Peters. 2011. Changes in potential habitat of 147 North American breeding bird species in response to redistribution of trees and climate following predicted climate change. Ecography 34: 933–945.

McGowan, C. P., M. C. Runge, and M. A. Larson. 2011. Incorporating parametric uncertainty into population viability analysis models. Biological Conservation 144:1400–1408.

McMahon, S. M., S. P. Harrison, W. S. Armbruster, P. J. Bartlein, C. M. Beale, M. E. Edwards, J. Kattge, G. Midgley, X. Morin, and I. C. Prentice. 2011. Improving assessment and modelling of climate change impacts on global terrestrial biodiversity. Trends Ecology Evolution 26:249–259.

Millspaugh, J. J., R. A. Gitzen, D. R. Larsen, M. A. Larson, and F. R. Thompson III. 2009. General principles for developing landscape models for wildlife conservation. Pages 1–32 in J. J. Millspaugh and F. R. Thompson III, editors. Models for planning wildlife conservation in large landscapes. Elsevier/Academic Press, Boston, Massachusetts, USA.

Monahan, W. B. 2009. A mechanistic niche model for measuring species' distributional responses to seasonal temperature gradients. PLoS ONE 4:e7921.

Monzón, J., L. Moyer-Horner, and M. B. Palamar. 2011. Climate change and species range dynamics in protected areas. BioScience 61:752–761.

Moritz, C., and R. Agudo. 2013. The future of species under climate change: Resilience or decline? Science 341:504–508.

National Ecological Assessment Team. 2008. Strategic habitat conservation handbook. U.S. Fish and Wildlife Service, Washington D.C., USA.

Nolan, J. V., E. D. Ketterson, and C. A. Buerkle. 2014. Prairie warbler (*Setophaga discolor*). *In* A. D. Rodewald, editor. The birds of North America. Cornell Lab of Ornithology, Ithaca, New York, USA. https://birdsna.org/Species-Account/bna/species/prawar

Panjabi, A. O., et al. 2005. The partners in flight handbook on species assessment. Version 2005. Partners in Flight Technical Series Number 3. Rocky Mountain Bird Observatory, Brighton, Colorado, USA.

Parmesan, C. 2006. Ecological and evolutionary responses to recent climate change. Annual Review of Ecology Evolution and Systematics 37:637–669.

Powell, L. A., J. D. Lang, M. J. Conroy, and D. G. Krementz. 2000. Effects of forest management on density, survival, and population growth of wood thrushes. Journal of Wildlife Management 64:11–23.

R Core Team. 2016. R: a language and environment for statistical computing. R Foundation for Statistical Computing, Vienna, Austria.

Ralston, J., W. V. DeLuca, R. E. Feldman, and D. I. King. 2016. Realized climate niche breadth varies with population trend and distribution in North American birds. Global Ecology and Biogeography 25:1173–1180.

Reed, T. E., V. Grotan, S. Jenouvrier, B. E. Saether, and M. E. Visser. 2013. Population growth in a wild bird is buffered against phenological mismatch. Science 340:488–491.

Regan, H. M., A. D. Syphard, J. Franklin, R. M. Swab, L. Markovchick, A. L. Flint, L. E. Flint, and P. H. Zedler. 2012. Evaluation of assisted colonization strategies under global change for a rare, fire-dependent plant. Global Change Biology 18:936–947.

Robinson, S. K., F. R. Thompson III, T. M. Donovan, D. R. Whitehead, and J. Faaborg. 1995. Regional forest fragmentation and the nesting success of migratory birds. Science 267:1987–1990.

Roth, R. R., M. S. Johnson, and T. J. Underwood. 1996. Wood thrush (*Hylocichla mustelina*). *In* A. Poole, editor. The birds of North America. Cornell Lab of Ornithology, Ithaca, New York, USA.

Sala, O. E., et al. 2000. Global biodiversity scenarios for the year 2100. Science 287:1770–1774.

Sauer, J. R., D. K. Niven, J. E. Hines, D. J. Ziolkowski Jr., K. L. Pardieck, J. E. Fallon, and W. A. Link. 2017. The North American Breeding Bird Survey, results and analysis 1966–2015. Version 2.07.2017. USGS Patuxent Wildlife Research Center, Laurel, Maryland, USA.

Schmidt, K. A., S. A. Rush, and R. S. Ostfeld. 2008. Wood thrush nest success and post-fledging survival across a temporal pulse of small mammal abundance in an oak forest. Journal of Animal Ecology 77:830–837.

Schneider, S. H., and T. L. Root. 2002. Wildlife responses to climate change: North American case studies. Island Press, Washington, D.C., USA.

Selwood, K. E., M. A. McGeoch, and R. Mac Nally. 2015. The effects of climate change and land-use change on demographic rates and population viability. Biological Reviews of the Cambridge Philosophical Society 90:837–853.

Simons, T. R., G. L. Farnsworth, and S. A. Shriner. 2000. Evaluating Great Smoky Mountains National Park as a population source for the wood thrush. Conservation Biology 14:1133–1144.

Smith, S. M. 1978. The "Underworld" in a territorial sparrow: adaptive strategy for floaters. American Naturalist 112:571–582.

Smith, R. L., and T. M. Smith. 2001. Ecology and field biology. Benjamin Cummings, San Francisco, California, USA.

Staudinger, M. D., et al. 2013. Biodiversity in a changing climate: a synthesis of current and projected trends in the US. Frontiers in Ecology and the Environment 11:465–473.

Stephens, S., D. Koons, J. Rotella, and D. Willey. 2004. Effects of habitat fragmentation on avian nesting success: a review of the evidence at multiple spatial scales. Biological Conservation 115:101–110.

Swab, R. M., H. M. Regan, D. A. Keith, T. J. Regan, and M. K. J. Ooi. 2012. Niche models tell half the story: Spatial context and life-history traits influence species responses to global change. Journal of Biogeography 39:1266–1277.

Swab, R. M., H. M. Regan, D. Matthies, U. Becker, and H. H. Bruun. 2015. The role of demography, intra-species variation, and species distribution models in species' projections under climate change. Ecography 38:221–230.

Thomas, C. D., et al. 2004. Extinction risk from climate change. Nature 427:145–148.

Thompson III, F. R., W. D. Dijak, T. G. Kulowiec, and D. A. Hamilton. 1992. Breeding bird populations in Missouri Ozark forests with and without clearcutting. Journal of Wildlife Management 56:23–30.

Thompson III, F. R., T. M. Donovan, R. M. De Graaf, J. Faaborg, and S. K. Robinson. 2002. A multi-scale perspective of the effects of forest fragmentation on birds in eastern forests. Studies in Avian Biology 25:8–19.

Tirpak, J. M., D. T. Jones-Farrand, F. R. Thompson III, D. J. Twedt, C. K. Baxter, J. A. Fitzgerald, and W. B. Uihlein. 2009b. Assessing ecoregional-scale habitat suitability index models for priority landbirds. Journal of Wildlife Management 73: 1307–1315.

Tirpak, J. M., D. T. Jones-Farrand, F. R. Thompson III, D. J. Twedt, and W. B. Uihlein III. 2009a. Multiscale habitat suitability index models for priority landbirds in the Central Hardwoods and West Gulf Coastal Plain/Ouachitas Bird Conservation Regions. General Technical Report WO-NRS-49. USDA Forest Service, Washington, D.C., USA.

U.S. Department of the Interior Fish and Wildlife Service. 2008. Birds of conservation concern 2008. U.S. Department of the Interior, Fish and Wildlife Service, Division of Migratory Bird Management, Arlington, Virginia, USA.

Wallendorf, M. J., P. A. Porneluzi, N. K. Gram, R. L. Clawson, and J. Faaborg. 2007. Bird response to clear cutting in Missouri Ozark Forests. Journal of Wildlife Management 71:1899–1905.

Wang, W. J., H. S. He, R. T. Frank III, J. S. Fraser, B. B. Hanberry, and W. D. Dijak. 2015. Importance of succession, harvest, and climate change in determining future composition in US Central Hardwood Forests. Ecosphere 6:1–18.

Wang, W. J., H. S. He, F. R. Thompson III, J. S. Fraser, and W. D. Dijak. 2016. Landscape- and regional-scale shifts in forest composition under climate change in the Central Hardwood Region of the United States. Landscape Ecology 31:149–163.

Warren, R., et al. 2013. Quantifying the benefit of early climate change mitigation in avoiding biodiversity loss. Nature Climate Change 3:678–682.

Wilson, B. T., A. J. Lister, and R. I. Riemann. 2012. A nearest-neighbor imputation approach to mapping tree species over large areas using forest inventory plots and moderate resolution raster data. Forest Ecology and Management 271:182–198.

Comparative responses of early-successional plants to charcoal soil amendments

Nigel V. Gale,[1,†] Md Abdul Halim,[1,2] Mark Horsburgh,[1] And Sean C. Thomas[1]

[1]*Faculty of Forestry, University of Toronto, 33 Willcocks Street, Toronto, Ontario, Canada*
[2]*Department of Forestry and Environmental Science, School of Agriculture and Mineral Sciences, Shahjalal University of Science and Technology, Sylhet 3114 Bangladesh*

Abstract. Charcoal used as a soil amendment, or "biochar," has received considerable recent research attention as a means to increase plant productivity while mitigating climate change through enhanced carbon sequestration. Interest in biochar for use in the restoration of disturbed sites is growing; however, biochar effects on wild plant species of the early phase of post-disturbance succession have received almost no prior research attention. Physiological adaptations that enable rapid growth in early-successional pioneers (e.g., high rates of photosynthesis) should be advantageous in soils with fresh charcoal since plants with a capacity for expeditious resource capture can capitalize on resource pulses from leachable mineral elements. In a glasshouse study, we tested the effects of biochar applied at two doses (10 and 20 t/ha) to brunisol/juvenile podzol soils, collected from a managed temperate mixed-wood forest, on the growth and physiology of 13 herbaceous old-field pioneers. We measured leaf-level physiology and nutrient supply rates throughout the experiment, and biomass and reproductive performance at experiment completion. Overall, biochar treatments resulted in 30–37% increases in final average aboveground biomass, 13–17% increases in photosynthesis, and an average ~44% increase in leaf-level water-use efficiency (at 10 t/ha), but with a high species-specific variation that included negative responses. We detected weak negative relationships between intrinsic photosynthetic rates (of non-biochar controls) and some biomass responses: Species with high photosynthetic capacities tended to have low or negative biomass responses to biochar. Plants in biochar treatments flowered earlier and on average had double the reproductive biomass overall. Pulses of PO_4^- and K^+ were supplied by biochar in the first four weeks of the experiment, while NO_3^- was significantly immobilized by biochar. These results suggest that by providing a pulse of P and base cations, biochar can improve the restoration of disturbed landscapes by enhancing the physiological and reproductive performance of a subset of pioneers that have moderate photosynthetic rates and nitrogen demand. Biochar has important potential applications to restoration; however, biochar is likely to affect community composition strongly, and careful consideration of the physiological rates and nitrogen requirements of target species will be necessary to maximize the success of biochar-based restoration projects.

Key words: biochar; carbon sequestration; ecophysiology; ecosystem management; photosynthesis; plant nutrients; plant performance; reproduction; restoration; soil nutrients; succession.

† **E-mail:** nigel.gale@mail.utoronto.ca

Introduction

Biochar is the term given to charcoal when used as a soil amendment; it is commonly derived from the pyrolysis (thermal degradation in a low-oxygen atmosphere) of lignocellulosic residues from forestry (sawdust, woodchips, and bark) and agriculture (corn stalks, coconut, rice husks, and manure). Biochar has recently been heralded for its ability to increase plant productivity and ameliorate poor soil conditions across a variety of systems (Biederman and Harpole 2013) while mitigating anthropogenic climate change by enhancing soil carbon sequestration (Lehmann 2007, Woolf et al. 2010). Biochar-enhanced plant growth is attributed to increased nutrient and water availability (Laird et al. 2010, Major et al. 2010), soil pH (Chan et al. 2007), as well as the sorption of growth-inhibitory compounds (Beesley et al. 2010, Thomas et al. 2013).

Biochar effects on plant growth have mainly been investigated in agricultural systems. Average biomass growth responses for crops amended with biochar are 10–30% (Jeffery et al. 2011, Biederman and Harpole 2013, Liu et al. 2013); somewhat larger responses have been estimated in a meta-analysis of tree responses (~40%, across ontogenetic stages; Thomas and Gale 2015). Available data syntheses suggest that biochar enhancement of plant growth is greatest in tropical and boreal systems, with lower responses in temperate systems (Thomas and Gale 2015, Jeffery et al. 2017). Significant increases in plant growth from biochar addition in tropical soils may be primarily due to the combination of liming and increases in nutrient availability and retention, particularly of phosphorus (Park et al. 2011, Lashari et al. 2013). Similar to tropical soils, increases in plant performance to biochar in boreal systems are from liming effects (Hart and Luckai 2013) and phosphorus provisioning (Pluchon et al. 2014); however, the sorption of phenolic compounds is also implicated as an important mechanism (Wardle et al. 1998). Multi-species studies and meta-analyses have also indicated considerable variation in biochar responses among plant species (Pluchon et al. 2014, Thomas and Gale 2015).

Interest in biochar for use in managed systems, such as forests, and in the restoration of disturbed sites is growing; however, biochar effects on wild plant species characteristic of the early phase of post-disturbance succession have received almost no prior research attention. In general, physiological adaptations in pioneer species that facilitate rapid growth, such as high rates of photosynthesis (Bazzaz et al. 1987), should confer an advantage in biochar-amended soils since these plants can capitalize on resource pulses from labile minerals, and utilize high nutrient and water availability. The rate of carbon assimilation is indeed highly dependent on nitrogen (Sage and Pearcy 1987), phosphorus, and, to a lesser extent, cations such as potassium, calcium, and sulfur (Longstreth and Nobel 1980). Leaf-level gas-exchange traits such as stomatal conductance can also be limited by relatively minor deficiencies in soil potassium (Benlloch-González et al. 2008). While biochars typically increases the availability of phosphorus, potassium, and calcium (Clough et al. 2013, Sackett et al. 2015), they are low in available nitrogen and can sorb and immobilize nitrogen, particularly NO_3^- (Rajkovich et al. 2012, Tammeorg et al. 2013, Zheng et al. 2013). Given the high interspecific physiological trait variation and nutrient requirements among early-successional plant species (driven by selection from high environmental heterogeneity), significant species-specific variation is expected in responses of these species to biochar.

Physiological responses to biochar have only been minimally investigated. Increases in photosynthesis in response to biochar have been found in wheat (*Triticum aestivum*; Choi et al. 2009, Akhtar et al. 2015a) and pine (*Pinus densiflora*; Choi et al. 2009), but not in *Abutilon theophrasti* or *Prunella vulgaris* (Thomas et al. 2013). In some prior studies, increased physiological performance in response to biochar was observed only under acute plant stress, including drought stress (Akhtar et al. 2014), and saline soil conditions (Thomas et al. 2013, Akhtar et al. 2015a).

The effects of biochar on plant reproduction are important in a broader ecological context with significant implications to successional trajectories but have also received little research attention to date. Short-lived plants slow or halt vegetative growth to supply reproductive structures with nutrients and photosynthates to increase reproductive success (Bazzaz et al. 1987). In nutrient-limited environments, plants may have a reduced ability to produce reproductive structures, thereby impacting their reproductive potential (Ashman 1994, Petraglia et al. 2014). However, if non-N nutrient availability is enhanced by biochar, then it is possible that plants may be able to produce more and larger reproductive structures than plants growing in nutrient-limited environments (Ashman 1994, Petraglia et al. 2014). Increased reproductive performance in response to biochar is expected to be pronounced for semelparous plants with high photosynthetic capacity in which flowering traits (e.g., flowering time, rate, and inflorescence size) and reproductive performance (e.g., reproductive effort, seed number, and size) are under intense selection pressure (Caruso et al. 2005, Franks et al. 2007, Zeineddine and Jansen 2009). Increases in reproductive performance in response to biochar have been noted in several studies (Tammeorg et al. 2013, Thomas et al. 2013). A recent investigation by Conversa and Bonasia (2015) reported that a low dose of biochar in combination with fertilizer increased the total number of flowers in *Pelargonium zonale*; however, a larger dose of biochar alone depressed flowering traits. Recent work also suggests that hormetic compounds (i.e., compounds that have stimulating and inhibiting effects at low and high concentrations, respectively) leach and evolve from freshly produced biochar (Gale et al. 2016). These compounds include ethylene (Spokas et al. 2010, Fulton et al. 2013), a gaseous plant hormone responsible for regulating growth, reproduction, and senescence (Abeles et al. 1992), and Karrikins, a family of smoke-derived compounds that commonly serve as stimulants of seed germination and seedling growth in fire-adapted species (Kochanek et al. 2016).

To investigate the ecophysiology, growth, and reproductive performance of pioneers to biochar, we grew 13 herbaceous species that commonly occur as colonizers of North American temperate forests and grasslands following disturbance. In a randomized glasshouse experiment, we applied three doses of sugar maple (*Acer saccharum*) sawdust biochar (0, 10, 20 t/ha) applied to brunisol/juvenile podzol soils collected from the mineral layer of a disturbed temperate mixed-wood forest. The following hypotheses are examined: (I) Plants grown in biochar will show increased physiological performance and growth mainly as a result of increased nutrient supply and uptake of major elements—in particular, P, K, Ca, and Mg. (II) Pioneer species with the highest resource requirements and growth rates will respond most positively to biochar. (III) Biochar will increase the reproductive performance of all species, with the largest effects predicted for species with higher resource requirements and growth rates.

MATERIALS AND METHODS

Study species

We selected 13 early-successional pioneer species that commonly colonize disturbed areas in eastern North America: the annuals *Lolium multiflorum* Lam., *Abutilon theophrasti* Medik.; the perennials *Trifolium repens* L., *Tanacetum vulgare* L., *Leonurus cardiaca* L., *Sonchus arvensis* L., *Leucanthemum vulgare* (Vail.) Lam., *Symphyotrichum novae-angliae* (L.) G.L. Nesom, *Mentha arvensis* L., *Solidago nemoralis* Aiton, and *Solidago canadensis* L.; and the biennials *Daucus carota* L., and *Arctium minus* Bernh. (Wieland and Bazzaz 1975, Keeley et al. 1981, Green and Kauffman 1995, Milberg 1995, Bartha 2001). Seeds used in this experiment were derived from three sources: (1) collected in autumn 2013 from disturbed sites in industrial areas in Toronto, Ontario (43°38′ N, 79°19′ W; Toronto Port Lands, Ontario, Canada), (2) acquired from three commercial distributors (Ontario Seed Company, Ontario, Canada; V & J Seed Farms, Woodstock, Illinois, USA; Richters Herbs, Goodwood, Ontario, Canada), and (3) collected from the seed bank of a managed temperate hardwood-dominated forest mineral soil at Haliburton Forest and Wildlife Reserve, Haliburton, Ontario, Canada (hereafter "Haliburton Forest"; 45°15′ N, 78°34′ W). Information regarding species taxonomic affiliations, statuses in North America, and growth forms can be found in Table 1.

Plants were grown in 0.02-m^2 surface area growth containers (3.5 L volume, 18 cm depth, with a fiberglass screen placed at the base to reduce soil loss) in soil amended with three doses of biochar: 0, 10, and 20 t/ha. Biochar was produced at Haliburton Forest (summer 2012) from 80/20% sugar maple/yellow birch (*Acer saccharum/ Betula alleghaniensis*) sawdust, pyrolyzed at a temperature of 525°C for ~0.5 h in a long feed-screw pyrolysis unit with the feedstock tightly packed to reduce oxygen access. To alleviate potential phytotoxicity from mobile organic compounds leaching and evolving from this biochar, we heat-treated it at 100°C for 48 h (Gale et al. 2016). Because wildfire-generated char is incorporated primarily in the top 10 cm of soil in temperate and boreal systems (Hart and Luckai 2013), we mixed biochar in the top 10 cm of soil in the pots (i.e., ~the top half). Biochar mass equivalents were 20 g (~0.4% w/w of the total pot) and 40 g (~0.8% w/w of the total

Table 1. List of species used in this study, including information on their taxonomic affiliation, growth form, status in North America, and source of seed.

Latin name	Family	Growth form	Status	Seed source
Lolium multiflorum Lam.	Poaceae	Annual grass	Introduced	Ontario Seed Company
Abutilon theophrasti Medik.	Malvaceae	Annual forb	Introduced	V & J seed farms
Tanacetum vulgare L.	Asteraceae	Perennial forb	Introduced	Toronto field collections
Leonurus cardiaca L.	Lamiaceae	Perennial forb	Introduced	Richters Herbs
Sonchus arvensis L.	Asteraceae	Perennial forb	Introduced	Toronto field collections
Leucanthemum vulgare (Vail.) Lam.	Asteraceae	Perennial forb	Introduced	Haliburton Forest field collections
Symphyotrichum novae-angliae (L.) G.L. Nesom	Asteraceae	Perennial forb	Native	Ontario Seed Company
Mentha arvensis L.	Lamiaceae	Perennial forb	Native	Toronto field collections
Solidago nemoralis Aiton	Asteraceae	Perennial forb	Native	Haliburton Forest field collections
Solidago canadensis L.	Asteraceae	Perennial forb	Native	Haliburton Forest field collections
Daucus carota L.	Apiaceae	Biennial forb	Introduced	Toronto field collections
Arctium minus Bernh.	Asteraceae	Biennial forb	Introduced	Toronto field collections
Trifolium repens L.	Fabaceae	N-fixing, perennial forb	Introduced	Ontario Seed Company

pot) of biochar added to 10 and 20 t/ha pots containing 3 L of soil, respectively. The soil used in the experiment was a mixture of 2/3 a composite brunisol/juvenile podzol soils from glacial origins that was sandy loam in texture, collected in the summer of 2013 from the uppermost mineral layer (<10 cm containing A and B horizons) of a managed stand at Haliburton Forest (Gradowski and Thomas 2006). The other 1/3 of the mixture was granitic sand from an abandoned quarry collected summer 2013 from Haliburton Forest to replicate coarsely textured temperate disturbed soils (Parker et al. 2001). Biochar and soil chemical properties are reported in Table 2. Biochar addition rates selected for the present study corresponded to those used in restoration trials in the region (Sackett et al. 2015, Mitchell et al. 2016, Gale et al. 2016, Kuttner and Thomas 2017). Additionally, these doses are representative of those most commonly used to increase plant performance in previous field trials (Biederman and Harpole 2013, Liu et al. 2013, Thomas and Gale 2015). Seedlings were germinated in flats of vermiculite and transplanted at the cotyledon stage into unfertilized pre-filled pots. One plant per pot with six replicates of each treatment was used (6 reps × 13 species × 3 doses, $n = 234$ plants). Fertilizer was applied eight days following planting: 100 mL of diluted fertilizer (12 g of 20-20-20 NPK fertilizer in 4000 mL of H_2O) approximating 40 $kg \cdot ha^{-1} \cdot yr^{-1}$ N. Plants were grown in a glasshouse (Toronto, Ontario, Canada) from 23

December 2013 to 17 May 2014—a total of 146 d. Pots were randomly placed on benches at the beginning of the experiment and shuffled three times throughout. Glasshouse conditions reflect average growing conditions found in Toronto

Table 2. Properties of the maple/birch biochar and of the 2:1 composite brunisol/juvenile podzol:granitic sand used in the experiment.

Attribute	Maple flow-through BC (±SE)	Brunisol–juvenile podzol soil
Moisture (%)	2.7 (0.0)	2.72
Ash (%)	2.74 (0.06)	12.3
Volatile matter (%)	29.87 (3.07)	n/a
Fixed carbon (%)	64.68 (3.07)	n/a
pH	7.39 (0.23)	6.6
EC (µS/cm)	105 (4)	n/a
Elemental composition		
C (%)	85.01 (2.3)	n/a
N (%)	0.71	n/a
P ($NaHCO_3$, mg/L)	n/a	15
K (mg/L)	n/a	89
Mg ($NH_4C_2H_3O_2$, mg/L)	n/a	150
NH_4 (KCl-NH_4, mg/kg)	n/a	3.2
NO_3 (KCl-NH_4, mg/kg)	n/a	138

Notes: EC, electrical conductivity; SE, standard error. Biochar was produced in a flow-through screw-fed pyrolyzer with a temperature of 500–600°C (Gale et al. 2016), and soil was collected from the uppermost mineral layer in a temperate forest, Haliburton, Ontario, Canada. Soil analysis was performed at the Agriculture and Food Laboratory (AFL), University of Guelph, Guelph, Ontario, Canada. Test procedures and/or units are reported in parentheses. Three replicates were used for biochar analysis and physically pooled for soil analysis at AFL.

according to the Government of Canada's Canadian Climate Normals for May to September, 1961–1990: Average daily temperature in the glasshouse was 17.5°C with a mean daily minimum and maximum temperature of 16.5° and 24°C, respectively. Supplemental lighting was applied to maintain a 16-h photoperiod. All plants were watered every other day to soil field capacity and allowed to drain.

Growth, physiological, and reproductive measurements

To examine physiological responses to biochar, we measured the leaf-level gas-exchange parameters light-saturated photosynthetic rate (A_{max}) and stomatal conductance (g_s) between 8:00 and 11:00 hours local time at days 42–44, and at days 87–88 for species that were too small at day 42 of the experiment (*L. cardiaca*, *M. arvensis*, *L. vulgare*, *D. carota*, *S. canadensis*, *S. nemoralis*) using an LI-6400 XT Portable Photosynthesis System (Li-Cor, Lincoln, Nebraska, USA) on leaves large enough to fill the 6-cm^2 cuvette either completely or by half. Triplicate measurements were made on the most recently developed fully expanded leaf. The flow rate was set to 400 mmol/s, and CO_2 concentration of the sample set to 400 ppm; relative humidity in the chamber was kept at approximately 50%. For leaves that filled the chamber by half, chamber area was reduced to 3 cm^2 (either 3 × 1 cm or 1.5 × 2 cm). A red-blue light source (Li-Cor SI-355 red LED) with a photosynthetic photon flux density of 1000 mmol·m^{-2}·s^{-1} was used for measurements. We also calculated instantaneous leaf-level water-use efficiency (WUE), expressed as the ratio of photosynthetic rate to transpiration rate (A_{max}/E), where E is the product of g_s and the leaf-to-air vapor pressure deficit held constant during gas-exchange measurements (1.8–2.0 KPa). Gas flux rates were monitored during measurements to ensure steady-state values before recording data.

Non-destructive allometric estimates of leaf area were used to measure growth and development three times throughout the growing period (days 25, 43, and 72). The length of every leaf on each plant was measured to the nearest centimeter. Allometric relationships between leaf length and area were developed for each species using an additional subset of plants grown in non-biochar control soils. Approximately 30–60 leaves of each

species were randomly harvested from at least three replicates. Leaf length was measured, and the leaf area was determined using a Li-3100C leaf area scanner with a resolution of 1 mm^2 (Li-Cor Biosciences). Allometric equations for leaf area were mostly described by $A = aL^b$, where a and b are constants, and L is leaf length, fit with type I regression of log–log-transformed data; only R^2 values >0.90 were considered (Thomas et al. 1999). Leaf area growth of *L. vulgare* and *S. novae-angliae* was not included due to a large variation in leaf shape. Leaf area allometric equations for each species are listed in Appendix S1: Table S2. Leaf area at harvest was determined using a Li-3100C leaf area scanner. Aboveground biomass and belowground biomass were separated at the soil surface at harvest. Belowground biomass was separated from soil and biochar media by dry sieving followed by gentle washing of roots with water. Aboveground biomass and belowground biomass, including scanned leaves, were dried for 48 h at 65°C, then weighed.

To assess whether biochar affected reproductive potential of plants, we measured the development of reproductive structures, specifically inflorescences in *A. theophrasti*, *T. repens*, *M. arvensis*, *T. vulgare*, *S. canadensis*, and *S. nemoralis*. See Appendix S2 for methods used to assess reproduction performance of different species.

Biochar and soil characterization, nutrient supply rates, and leaf mineral nutrition

Biochar was characterized as follows: Electrical conductivity (EC) and pH were measured in a 1:20 (w:v) biochar to H_2O solution using pH and EC probes (Rajkovich et al. 2012). Total moisture content (%), ash content (%), volatile matter content, and fixed carbon content were determined by oven drying at 105°C, by muffle furnace combustion at 750°C, by heating at 950°C in sealed containers, and through loss on ignition for 0.5 h in a muffle furnace at 500°C, following ASTMD1762-84 (ASTM 2013), respectively. Total carbon and nitrogen was quantified using combustion analysis by an Elementar VarioMax (Elementar Analysensysteme GmbH, Hanau, Germany).

Soil was characterized as follows: Soil pH was measured on a saturated paste using an Orion 4STAR meter (Thermo Fisher Scientific, Waltham, Massachusetts, USA). For cation (K^+, Mg^{2+}) determination, soil samples were extracted using 1.0 N ammonium acetate and the concentration determined using ICP–OES on a Varian Vista Pro instrument (Varian Inc, Palo Alto, California, USA). Olsen P was determined by extracting samples with 0.5 mol/L sodium carbonate solution and determining the concentration of P colorimetrically using a SEAL AutoAnalyzer 3 (SEAL Analytical, Fareham, Hampshire, UK).

Plant root simulator probes (PRS probes; Western Ag Innovations, Saskatoon, Saskatchewan, Canada) were used to measure soil nutrient (cation and anion) supply rates at 10, 20, and 30 d in pot incubations of control and biochar treatments. Each PRS probe consisted of either a positively or negatively charged ion exchange membrane (17.5 cm^2 surface area), which adsorbed ions from the soil solution over the burial period. We were primarily interested in biochar effects on the plant macronutrients NO_3^-, NH_4^+, PO_4^-, K^+, S and the micronutrients B, Mn^{2+}, Zn^{2+}, Ca^{2+} and Mg^{2+}. Plant root simulator probe pairs (one anion and one cation) were installed at a depth of 0–5 cm which encompasses most of the biochar-treated area and the rhizosphere. Three PRS probe pairs were inserted into one non-planted pot, and treatments were replicated three times (i.e., 6 probes/pot × 3 treatments × 3 replicates; $n = 9$ pots, 54 probes). One pair of PRS probes was retrieved from each pot at 10, 20, and 30 d, and was washed with deionized water. At each incubation date, PRS probe pairs were physically pooled from each treatment and shipped to Western Ag Innovations for analysis by ion chromatography.

Aboveground foliar nutrient content (N, P, K, Ca, and Mg) was analyzed for the two species that demonstrated the most positive (*M. arvensis*, *T. vulgare*) and negative (*S. arvensis*, *A. minus*) growth and biomass responses to biochar. Analyses were based on 10 g of dried ground plant tissue (2 g from five replicates pooled into one sample). Plant and soil analyses (pH, major nutrients) were conducted at the Agriculture and Food Laboratory, University of Guelph, Guelph, Ontario.

Foliar nutrient analysis is a useful tool in measuring plant responses to stress, and foliar nutrition analysis is increasingly used to indicate biochar effects on plant nutrient use and performance (Major et al. 2010, Jones et al. 2012, Bai et al. 2015). Because plant nutrition is a function of nutrient concentration and biomass, plant growth is driven by changes in biomass

production and nutrient uptake. Vector diagrams demonstrate leaf nutrient content (amount; the bottom horizontal x-axis) which is calculated from the tissue nutrient concentration (vertical y-axis) multiplied by its dry biomass (the top horizontal z-axis). Plant responses to biochar are expressed relative to the control plants (normalized to 100) to compare the different biochar doses and nutrients. Responses are depicted by vectors where length represents the magnitude of responses, and the direction identifies specific nutritional responses. For example, if biochar increases nutrient uptake (content) and dry biomass production, but decreases nutrient concentration, then dilution of nutrients is suggested to occur with biochar application. If a similar response occurs, but without changing the nutrient concentration, then nutrients are likely sufficiently provisioned by biochars since concentration kept up with increased growth. Enriched responses to biochar occur when all variables increase, and antagonistic responses happen when all variables decrease (further interpretations can found in Isaac and Kimaro 2011).

We plotted vector nutrient diagrams to assess the nutritional responses of the two species that demonstrated the strongest positive and negative growth responses to biochar. Foliar N, P, K, Ca, and Mg concentrations (percentage of dry mass) were multiplied by the average dry foliar mass of that species to calculate nutrient contents (g) for each treatment. Relative nutrient concentration, nutrient content, and unit dry mass were calculated from:

$$R = \left(\frac{\mu_{char}}{\mu_{con}}\right) \times 100$$

where μ_{char} represents the mean of the biochar treatment and μ_{con} is the mean of the non-biochar controls. To construct the vector diagrams, we plotted the data points (x, y, z) on a relative nutrient content (x), relative nutrient concentration (y), and relative unit dry weight (z) space.

Statistical analysis

To test the effects of biochar dose on the growth and physiological performance of herbaceous early-successional pioneers, we used a two-way analysis of variance (ANOVA) with dose and species as independent factors, and biomass and physiological traits as dependent variables. Post hoc comparisons were performed using Tukey's honestly significant difference test. Because plant biomass and growth responses to biochar have commonly been reported as effect sizes in investigations involving many species (Biederman and Harpole 2013, Thomas and Gale 2015), we present biomass and physiological response ratios calculated from

$$RR = \ln(B/C)$$

in Fig. 2, where RR is the response ratio metric, B is the mean value of plants in biochar at a given dose, and C is the mean value for control plants.

Although not an original objective of this research, comparing the responses of native vs. non-native species to biochar is important to evaluate its restoration potential (e.g., the unintended proliferation of non-native species). Our work provides the opportunity to conduct a preliminary assessment that can facilitate further study on this comparison. Since the data did not meet the assumptions of an ANOVA, we used a general linearized model fit with a gamma distribution, to test the influence of species status (native vs. non-native), biochar treatment, and their interaction on performance traits using the "glm" function in the R statistical software. Post hoc analysis was conducted to compare differences between treatments using least square means adjusted with Tukey's method using the package "lsmeans."

To test the influence of biochar on leaf area growth, we separated exponential early growth (days 25, 43, and 72) from asymptotic late growth (days 72 and 146) since plant growth was sigmoidal (Appendix S3: Fig. S1), and since responses to biochar are expected to differ with ontogeny. Because data were not normally distributed, we used a generalized linear mixed-effects model (GLMM) to describe the influence of biochar on early growth. Trends were fit using penalized quasi-likelihood with a Gaussian distribution of leaf area, with biochar treatment and species as the fixed effects, and growth day (time) as the random effect, using the function "glmmPQL" in the R statistical software. We log-transformed dependent variates in the GLMM to get the best fit for the model using the link = "log" function. Estimates for the fixed effects (biochar treatments and species) are reported as the differences in response slopes

compared to non-amended controls, and P-values indicate whether these fixed effects influenced leaf area growth using two-sided hypothesis tests.

For late leaf area growth, we used a three-way ANOVA with treatment, time (day 72 and 146), and species (and their interactions) as explanatory variables, and leaf area as the response variable. Assumptions of homoscedasticity and normality were met.

Linear models were used to test the effects of biochar on plant reproductive structures: see Appendix S2.

The response of plant performance to biochar is expected to depend on the dose (Biederman and Harpole 2013). We performed regression analysis to develop biochar dose–response relationships, testing the influence of biochar dose on growth and biomass traits—aboveground biomass, belowground biomass, and final leaf area using the lm function in the R statistical software. The slope of the relationship indicates the magnitude and direction of response to biochar (e.g., Δ biomass/t $\times$ ha^{-1} biochar). To test whether physiological traits are responsible for biomass and growth responses to biochar, we performed additional regression analyses testing (I) the influence of species mean $A_{\max}$ response ratio on biomass effect sizes to biochar and (II) the influence of $A_{\max}$ species mean values in the non-biochar control on biomass biochar dose responses (slope of previous regression), again using the lm function.

Soil nutrient supply concentrations were analyzed with repeated measures analysis using a linear mixed-effects model. Biochar treatment was the fixed factor with growth day as the random effect in the model using the R package "nlme" (Pinheiro et al. 2011). Tukey's multiple comparison tests were performed using the glht function in the R package "multcomp" (Hothorn et al. 2008). All analyses were performed in the R statistical software version 3.2.3 (R Core Team 2016). The Meta data used for all analyses can be found in Data S1 and in Gale et al. 2017. R Code is provided in Data S2 to perform the statistical analyses used.

Results

Plant growth and physiological responses

On average, biochar additions increased aboveground biomass by 37% at 10 t/ha and 30% at 20 t/ha ($P < 0.0001$); however, species responses varied considerably and included neutral and negative responses (Table 3, Figs. 1, 2A; Appendix S1: Table S1). Biochar addition, regardless of dose, did not influence belowground biomass. Similarly, biochar did not affect final leaf area, although a trend toward a treatment × species interaction was detected (Table 3). Indeed, leaf area growth throughout the experiment was not influenced by biochar treatments in either the early or asymptotic stages of growth (Tables 4, 5; Appendix S3: Fig. S1).

Dose response analyses revealed that aboveground biomass was positively influenced by increasing biochar dose in six species and negatively influenced in three species (Appendix S1: Table S4). Photosynthetic capacity (A_{max}) was also positively affected by increasing biochar dose in three species but negatively influenced by

Table 3. Two-way analysis of variance outputs for growth, physiological, and reproductive traits responses to biochar soil treatment (Trt), across species (Spec).

Source of variation	df	Type III SS	MS	F	P
Aboveground biomass					
Trt	2	185	92.73	87.64	**<0.0001**
Spec	12	3220	268.33	253.62	**<0.0001**
Trt × Spec	24	523	21.78	20.58	**<0.0001**
Error	190	201	1.06		
Leaf area					
Trt	2	4582	2291	0.14	0.86
Spec	9	3,075,238	341,693	21.56	**<0.0001**
Trt × Spec	18	471,133	26,174	1.65	**0.054**
Error	146	2,313,598	15,847		
Stomatal conductance					
Trt	2	0.0136	0.0068	1.49	0.228
Spec	10	0.889	0.088	13.49	**<0.0001**
Trt × Spec	20	0.217	0.0108	2.375	**0.0016**
Error	152	0.6940	0.00457		
Reproductive biomass					
Trt	2	2649	1324	43.90	**<0.0001**
Spec	8	115,940	14,492	480.43	**<0.0001**
Trt × Spec	16	722	45	1.49	0.114
Error	107	3228	30		
Belowground biomass					
Trt	2	176	88.0	1.63	0.198
Spec	12	25,186	2098.9	38.95	**<0.0001**
Trt × Spec	24	1953	81.4	1.51	0.067
Error	188	10,132	53.9		
Photosynthetic rate					
Trt	2	166.8	83.41	13.84	**<0.0001**
Spec	10	1137.4	113.74	18.87	**<0.0001**
Trt × Spec	20	452.0	22.60	3.75	**<0.0001**
Error	152	915.9	6.03		
Flowering time					
Trt	2	2649	1324	43.90	**<0.0001**
Spec	8	115,940	114,492	480.43	**<0.0001**
Trt × Spec	16	722	45	1.49	0.114
Error	107	3228	30		
Water-use efficiency					
Trt	2	0.00017	8.66e-05	9.994	**<0.0001**
Spec	10	0.00069	6.96e-05	6.421	**<0.0001**
Trt × Spec	10	0.00469	2.34e-05	2.166	**0.0046**
Error	150	0.00162	1.04e-05		

Notes: Statistically significant relationships ($P < 0.05$) are in boldface type. SS = Sums of Squares, MS = Mean Square.

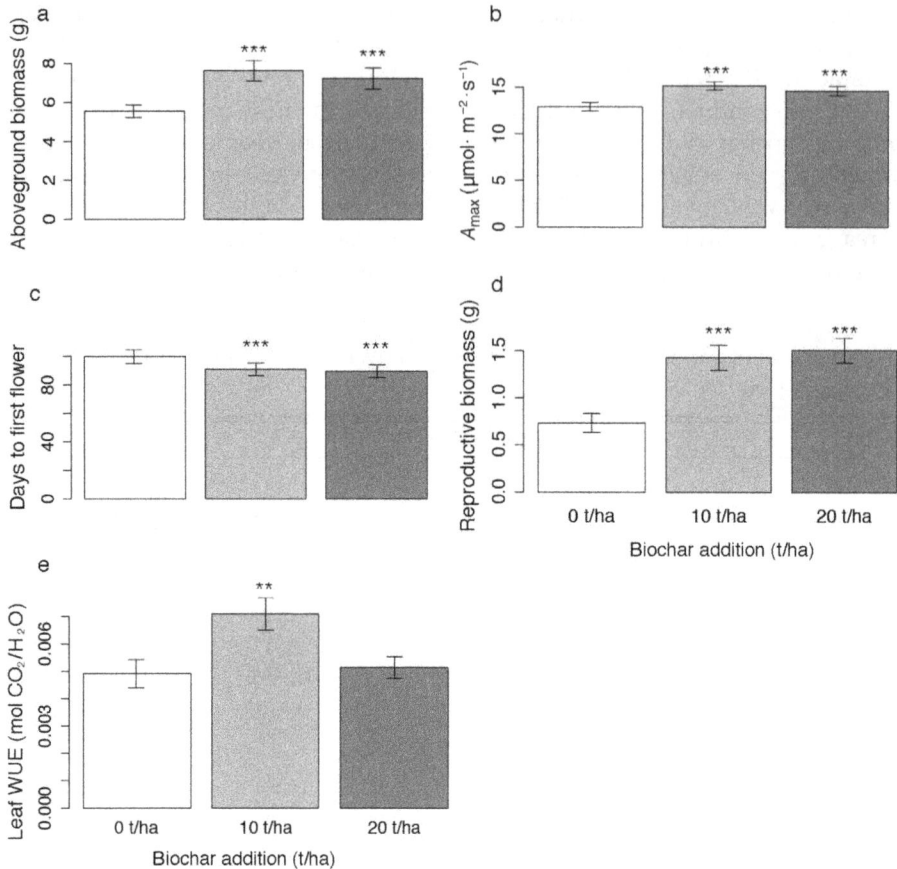

Fig. 1. Mean aboveground biomass (a), photosynthetic rate (b), flowering time (c), reproductive biomass (d), and water-use efficiency (e) for all species with (gray) and without (white) biochar. Asterisks indicate significant differences between means of biochar treatments and the non-biochar controls revealed by Tukey's HSD post hoc tests (P = <0.05*, < 0.001**, <0.0001***). Error bars depict standard error. WUE, water-use efficiency.

increasing dose in *Arctium minus*. Belowground biomass and stomatal conductance showed significant dose–response relationships in three species. In three of the nine species that flowered, reproductive biomass was positively associated with increasing biochar dose (Appendix S1: Table S4).

Biochar increased the maximum rate of photosynthesis (A_{max}) by an average of 15% overall—13% and 17% for 10 and 20 t/ha, respectively (Fig. 1B)—with species differing considerably in A_{max} responses to biochar, indicated by a significant treatment × species interaction (Table 3). Biochar did not influence stomatal conductance (g_s) overall; however, stomatal conductance responses to biochar did differ by species (e.g., a significant treatment × species interaction; Table 3). When amended with biochar, plants

had significantly increased average WUE efficiency of ~44% overall for 10 t/ha (P = 0.0011), but not for 20 t/ha (Fig. 1E, Table 3; Appendix S1: Table S1).

Species-level A_{max} responses to biochar showed a correlation with aboveground biomass responses at 20, but not 10 t/ha (10 t/ha: t_9 = 1.33; 20 t/ha: t_9 = 4.53; Fig. 3A). Species-level mean A_{max} in non-biochar soils did not significantly predict aboveground biomass biochar dose-response (t_9 = −1.637); however, a suggestive negative trend is apparent—that is, plants with high photosynthetic rates tended to have null and negative biomass responses to biochar (Fig. 3B). A_{max} did marginally predict leaf area biochar dose–response slopes (t_9 = −2.221, P = 0.053; Fig. 3C), and belowground biomass dose responses (t_9 = −1.86, P = 0.09, not shown), although a negative

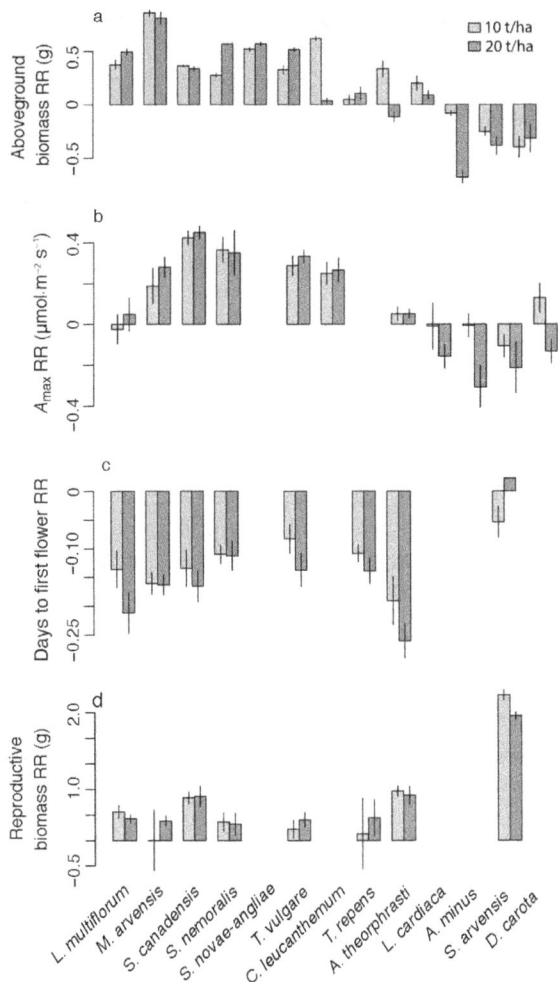

Fig. 2. Mean aboveground biomass (a), photosynthetic rate (b), flowering time (c), and reproductive biomass (d) response ratios (RR) = $(\ln(\mu_{char}/\mu_{control}))$ of early-successional pioneers to maple biochars. Error bars depict standard error.

trend was detected. Mean biomass, physiological, and reproductive responses by species are reported in Appendix S1: Table S1.

Native species in 20 t/ha biochar had 60% greater biomass than the controls ($P = 0.01$); while introduced species biomass had null responses to biochar. Similarly, the photosynthetic rate of native, but not introduced species, was 41% and 46% greater in 10 ($P = 0.002$) and 20 t/ha ($P = 0.0004$) biochar compared to the controls, and these responses significantly influenced overall photosynthetic responses ($t_{10} = 2.35$, $P = 0.019$; $t_{20} = 3.34$, $P = <0.001$). No other significant

differences between native and non-native species were detected for any other performance traits.

Reproductive performance

For plants that flowered, biochar doubled reproductive biomass with 20 t/ha increasing reproductive biomass the most (105%; Fig. 1D). Biochar influenced reproductive biomass allocation similarly across species (Table 3). Plants in biochar flowered 10 d earlier on average, with no differences between dose or treatment × species interaction (Figs. 1C, 2C; Appendix S1: Table S1).

Nutrient supply and foliar mineral nutrition

Soil nutrient supply PRS probes revealed mixed support for the hypothesis that biochar improves nutrient availability. Biochar immobilized total soil nitrogen, significantly reducing supply rates at 20 t/ha ($t_4 = -5.32$, $P = 0.006$), and marginally reducing supply at 10 t/ha in the first 30 d. Immobilization of total N by biochar was so strong that the supply of total N at 20 t/ha was significantly lower than at 10 t/ha ($P = 0.0059$; data not shown). The immobilization of N was mainly NO_3-N: The supply of NO_3^- was significantly reduced at 20 t/ha (Appendix S1: Table S3, Appendix S3: Fig. S2), and this reduction was significantly lower than at 10 t/ha ($P = 0.02$). The supply of NH_4^+, however, was not influenced by

Table 4. Generalized linear mixed models for early leaf area growth.

Early leaf area growth (days 25, 43, 72)					
Parameter	Est.	SE	df	t	P
Control 0 t/ha	5.025	0.431	84	11.639	**<0.0001**
Biochar 10 t/ha	0.010	0.039	84	0.266	0.790
Biochar 20 t/ha	0.041	0.390	84	1.051	0.295
Arctium minus	−0.162	0.119	84	−1.355	0.1779
Daucus carota	0.169	0.106	84	0.445	0.6568
Leonurus cardiaca	−0.722	0.154	84	−3.53	**<0.0001**
Lolium multiflorum	0.023	0.104	84	0.924	0.3577
Mentha arvensis	−1.432	0.337	84	−4.297	**<0.0001**
Solidago canadensis	0.350	0.094	84	3.7355	**<0.0001**
Solidago nemoralis	0.649	0.087	84	7.454	**<0.0001**
Sonchus arvensis	1.116	0.081	84	13.125	**<0.0001**
Tanacetum vulgare	0.342	0.095	84	3.32	**0.0013**
Trifolium repens	0.505	0.148	84	5.882	**<0.0001**

Notes: SE, standard error. Generalized linear mixed-effects model was fit using penalized quasi-likelihood with a Gaussian-log distribution (A); estimates of fixed effects are given, and P-values indicate whether biochar dose or species influenced leaf area growth. All signification relationships ($P < 0.05$) are in boldface type.

Table 5. Three-way analysis of variance outputs for late leaf area growth across species (Spec), time (growth day), biochar treatment (Trt) (B).

Source of variation	df	Late leaf area growth (day 72–146)			
		Type II SS	MS	F	P
Treatment (Trt)	2	16,040	8020	0.267	0.766
Species (Spec)	10	10,745,399	1,074,540	35.820	**<0.0001**
Time (d)	1	853,909	8,353,909	28.465	**<0.0001**
Trt × Spec	20	712,920	35,646	1.188	0.262
Spec × Day	10	3,949,464	394,946	13.166	**<0.0001**
Trt × Day	2	12,950	6475	0.216	0.806
Trt × Day × Spec	20	277,613	13,881	0.463	0.978
Residuals	319	9,569,475	29,998		

Notes: All signification relationships (*P* < 0.05) are in boldface type. SS = Sums of Squares, MS = Mean Square.

biochar. The supply of cations Ca^{2+}, K^+, Mg^{2+}, and Fe^{2+} was not affected by biochar significantly but did show a trend toward positive increases in supply over the first 30 d (Appendix S1: Table S3, Appendix S3: Fig. S2).

Foliar vector diagrams reveal dilutions of nitrogen in foliar tissues of *T. vulgare* and *M. arvensis* in response to biochar, with the increase in biomass from biochar associated with a decrease in tissue nitrogen concentration (Fig. 4). However, the positive biomass responses to biochar in these two species were associated with increases in foliar concentrations of P, and to a lesser extent K and Mg, indicating possible limitations of these elements in the non-biochar control. In contrast, vector diagrams of *S. arvensis* and *A. minus*, the two species showing the most pronounced negative responses to biochar, show signals of excessive P when biochar is applied, in that the decrease in biomass from biochar application is associated with an increase in foliar P concentration and a decrease in content. Antagonistic responses of biochar application are revealed in *S. arvensis and A. minus*, where decreased biomass is associated with reduced nutrient concentration and content of Mg, N, Ca, and K (Fig. 4).

DISCUSSION

Positive plant growth responses in early-successional pioneers

Strong positive responses to biochar additions were observed in aboveground biomass, photosynthesis rate, and reproductive performance traits in most species in the set of early-successional plants examined; however, there was significant variation in responses for all traits across species, including null and negative responses. These results somewhat exceed the mean ranges of crop responses to biochar reported in recent meta-analyses of ~10–30% (Biederman and Harpole 2013, Liu et al. 2013). For example, we found greater increases in biomass than did Rajkovich et al. (2012) where corn was grown in temperate agricultural soils across low (2.6 t/ha) to high (91 t/ha) doses of pine- and oak-derived biochars produced at moderate temperatures. Greater increases in biomass responses observed here may be due to a combination of P limitation in the forest soil examined (Gradowski and Thomas 2006, 2008), the alleviation of potential phytotoxicity through thermal pre-treatment of biochar (Gale et al. 2016), and the wide selection of species and life histories compared which represents a large range of possible responses. We did not detect any influence of biochar on species leaf area growth throughout the experiment. We speculate that enhanced carbon gains through increased photosynthesis reduced the need to make additional photosynthetic structures in species showing positive responses and that nitrogen immobilization likely limited the production of N-costly leaf tissues.

Contrary to the enhanced biomass responses of old-field species to biochar observed here, other recent work by Biederman et al. (2017) found no influence of hardwood biochar on the biomass of native perennials in a five-year temperate agricultural field restoration trial. One potential reason for higher responses observed in the present study is likely the inclusion of annual/biennial species characterized by rapid resource acquisition and growth rates. Indeed, the meta-analysis by Biederman and Harpole (2013) reports positive

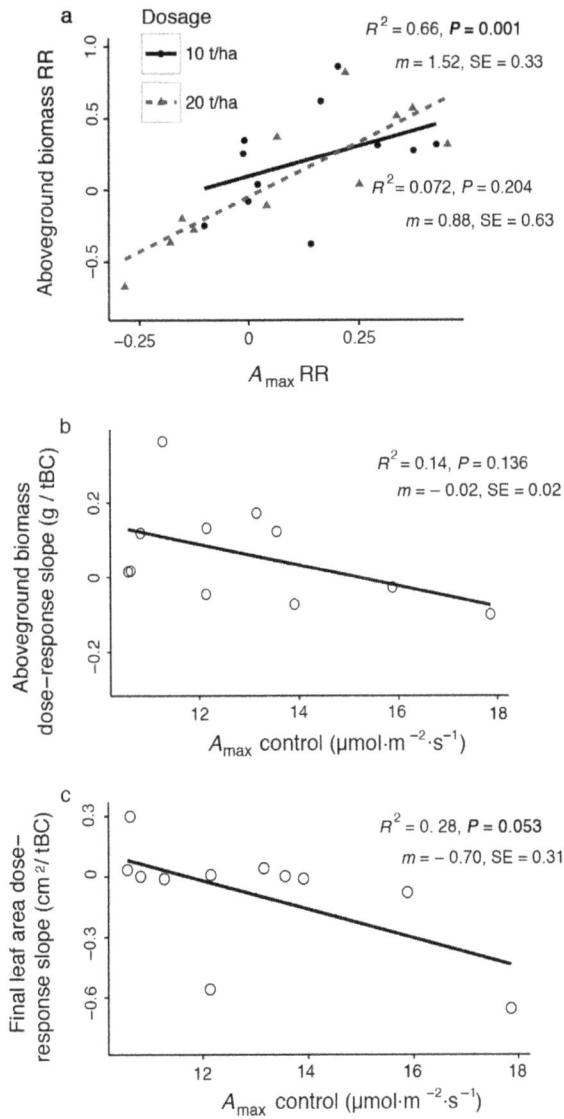

Fig. 3. Influence of photosynthetic rate (A_{max}) response ratio ($\ln(\mu_{char}/\mu_{control})$) on aboveground biomass response ratio (a); and the influence of mean photosynthetic rate of non-biochar controls on the species-wise biochar dose–response slopes of aboveground biomass (b) and final leaf area (c). Slope (m), standard error (SE), R-squared, and the P-values of the models are presented. Significant relationships are in boldface type.

responses of annuals, but not perennials, to biochars. Higher responses observed here (in comparison with Biederman et al. [2017]) could also be due to a combination of several other factors including the short duration of the experiment, as plant responses to biochar are generally greatest

in the short term (Thomas and Gale 2015), and/or the relative infertility of the forest mineral soils used. While plant responses to biochar in pot trials, including glasshouse studies, are comparable to those in the field (Thomas and Gale 2015), much more research on the influence of biochar on the growth of plants in the field is needed to evaluate biochar suitability for restoration. Further comparative tests of native vs. non-native species responses to biochar are also necessary to assess potential unintended effects of biochar-based restoration (e.g., the proliferation of non-natives).

Comparative ecophysiological responses to biochar: potential physiological mechanisms

Ecophysiological responses to biochar exposure may provide insight into the mechanisms responsible for growth and biomass responses, and species-specific variation observed in the literature, but have received surprisingly little investigation. Increases in leaf-level carbon gain through photosynthetic rate improvements should directly translate into increases in biomass and leaf area (Evans 1989). Biochar significantly increased the light-saturated photosynthetic rate by 15–17%, but did not influence stomatal conductance, providing some support for our second hypothesis that biochar will increase physiological performance in early-successional pioneers. These responses are higher than the few other studies to date that demonstrate little to no influence of biochar on the photosynthetic rates of fast-growing herbaceous annuals (Thomas et al. 2013, Akhtar et al. 2014, 2015b). Our findings contrast with Akhtar et al. (2015b) who report increased stomatal conductance, density, and aperture when biochar was applied to potato (*Solanum tuberosum*). Similarly, biochar significantly improved stomatal conductance and photosynthesis in wheat (*Triticum aestivum*; Akhtar et al. 2015a). In the present study, the pattern of increasing photosynthetic rate without a corresponding increase in stomatal conductance also increased leaf-level WUE and may be of importance in a restoration context, providing a mechanism for enhanced growth under water-limited conditions. Species biomass responses to biochar were positively correlated with A_{max} responses (Fig. 3A), suggesting that increased carbon gain associated with high photosynthetic responses to biochar is one mechanism

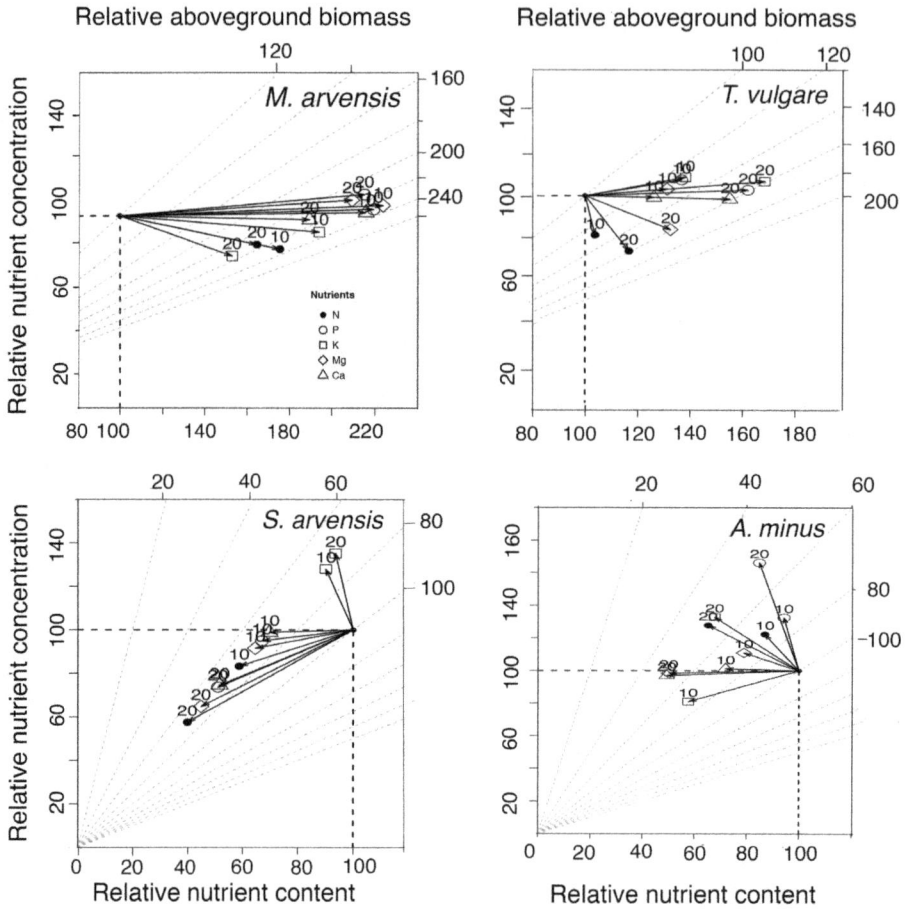

Fig. 4. Vector diagrams of changes in foliar nutrients with biochar. Changes in foliar nutrient content (the bottom horizontal x-axis) are expressed as the nutrient concentration (vertical y-axis) multiplied by dry biomass (the top horizontal z-axis). Vector length represents the magnitude of responses, and the direction identifies the specific nutritional responses to biochar. Nutrient enrichment from biochar occurs when all variables increase, while a decrease in all variables represents antagonistic responses. An increase in nutrient content and biomass growth, without a change in nutrient concentration, signifies nutrient sufficiency. Nutrient dilution is when a similar response occurs, except with a decrease in nutrient concentration. Foliar nutrient responses to biochar are expressed relative to the non-biochar control plants: (char/con) $\times$ 100. Gray dashed lines indicate the z-axis response by referencing to the zero point on the z- and y-axis.

to explain large biomass responses to biochar reported in the literature. Further study of the eco-physiological responses of non-agricultural crops is necessary to assess biochar's broader use in ecosystem restoration.

Intrinsic species physiological rates—specifically A_{max} of the controls—did not positively predict biomass responses to biochar. Instead, species with high photosynthesis in control soils had null, and even negative, biomass dose–response slopes. The highest biochar dose responses for biomass and leaf areas were for plants with low-to-

moderate photosynthetic rates. One possible explanation for the low responses of species with high photosynthesis is N-immobilization. Photosynthesis is highly nitrogen dependent—high N consumption rates are thus potentially maladaptive in N-limited soils (Chapin et al. 1987). Species with moderate photosynthesis can benefit from biochar without N deficiency. Indeed, *M. arvensis* and *T. vulgare* had the greatest positive biomass responses to biochar accompanied by moderate biochar dose–response slopes and moderate photosynthetic rates. These species also demonstrated dilution of N when biochar was applied, meaning they were trending toward deficiency. In contrast, *A. minus* and *S. arvensis* had the strongest negative responses to biochar, highly negative biochar dose–response slopes, and relatively high photosynthetic rates. Foliar tissue elemental analysis revealed nitrogen limitation for these plants that were grown in biochar treatments. Whether high anthropogenic N deposition under field conditions would offset reduced growth from N-immobilization by biochar remains to be tested.

Reproductive performance is greatly enhanced by biochar

Biochar effects on reproductive traits other than agricultural yields have received minimal investigation (Thomas et al. 2013, Conversa and Bonasia 2015). Here, we report substantial decreases in time to reproduction (days to flowering) and significant increases in reproductive biomass from biochar across multiple species (e.g., 265% in *S. canadensis* and 157% in *T. vulgare*). The doubling of reproductive biomass from biochar observed in the present study greatly exceeds the increases shown in Conversa and Bonasia (2015) who report 30% increases in floral clusters in *Pelargonium zonale* in response to biochar additions. Thomas et al. (2013) showed a 10-fold increase in flower production in the temperate pioneer *Prunella vulgaris* when a comparable wood biochar was applied at 50 t/ha to a non-soil medium. Recent work demonstrated substantially increased flower and fruit production in *A. theophrasti*, and early onset of flowering up to two weeks when a high-temperature mixed-wood biochar was applied to a non-soil medium (Seehausen et al. 2017).

Increased reproductive output and reduced time to reproductive onset are likely to be particularly strongly favored in early-successional semelparous plants. We detected limitation of P, and possibly K, in the non-biochar temperate forest mineral soil; these nutrients are essential for floral development. Furthermore, increases in carbon gain allow greater investment into reproduction. Interestingly, the iteroparous species *S. arvensis* had the highest reproductive biomass response, but the lowest aboveground biomass response, and signs of nitrogen limitation when amended with biochar. When nutrient limited, iteroparous perennials sometimes initiate reproduction given

the high future costs associated with growing vegetatively (Reekie and Bazzaz 2011). Similarly, semelparous plants under nitrogen limitation may flower earlier due to high future costs of reproduction. Further work should investigate the adaptive potential (i.e., fitness benefit) of increased reproductive performance of pioneer species in response to charcoals produced naturally and engineered biochars since reproductive success is fundamental to succession following disturbance.

Mineral nutrition influenced by biochar: supply and foliar uptake

We found mixed support for the hypothesis that biochar increases nutrient availability. Biochar immobilized nitrogen, specifically NO_3-N, while improving the short-term soil supply and foliar uptake of P, K. In crop plants, similar reductions in nitrogen uptake were found in corn by Rajkovich et al. (2012) where plants were grown in temperate soils with wood biochars at doses >6.5 t/ha. Tammeorg et al. (2013) found that nitrogen content in biomass of wheat and turnip, but not the leguminous faba bean, was reduced when biochar was applied in a three-year field experiment. Wood biochars have demonstrated effectiveness as nitrogen-sorbing substrates, reducing nitrogen leaching in several studies (Ventura et al. 2013, Sika and Hardie 2014). Biochar has mostly been noted for its ability to adsorb NH_4^+-N, with most investigations reporting the minimal ability of biochar to sorb NO_3-N (Hollister et al. 2013, Gai et al. 2014). On the contrary, and similar to Ventura et al. (2013), biochar in the present study strongly immobilized NO_3^- while having little influence on NH_4^+. Feedstock properties and pyrolysis conditions have shown considerable influence on ammonium and nitrate sorption by biochars (Gai et al. 2014), and thus, immobilization here might be biochar-specific. The mechanisms responsible for nitrogen sorption and immobilization require further exploration.

Biochar increased foliar P and K concentrations which resulted in increased aboveground biomass in *M. arvensis* and *T. vulgare*—the species with the highest biochar growth response investigated here. Phosphorus in nine wood biochars was associated with the enhanced growth of northern tree seedlings (Pluchon et al. 2014) and has shown rapid mineralization from wood-derived biochars (Mukherjee and Zimmerman 2013, Sackett et al. 2015). In a temperate grassland restoration study using biochar, van de Voorde et al. (2014) found significantly increased available P and K five months following application. However, in a five-year field trial testing the influence of wood biochars on prairie development, Biederman et al. (2017) found no effect of biochars on soil nutrient status (except total C and N). While we did not test the extractable elements of the biochar used in the present study, we note that a comparable maple sawdust biochar used in Sackett et al. (2015), and Gale et al. (2016), had appreciable amounts of extractable P (197 mg/kg), Ca (6015 mg/kg), K (3443 mg/kg), and Mg (619 mg/kg), and we thus expect the biochar used here to release pulses of these constituents. Indeed, in the present study, biochar tended toward increasing the supply of P and K in soil; however, these effects were not significant. Importantly, Gradowski and Thomas (2006, 2008) found strong co-limitations of Ca, Mg, K, and P in saplings and mature *Acer saccharum* in the managed forest from which our soil was collected. One limitation of this work is that although the PRS probe pairs were replicated (three times), they were physically pooled before analyses. Future tests should include additional replicates of physically pooled probes.

Vector diagrams offer a visual interpretation of plant nutritional responses to soil amendments and provide a useful tool in biochar research (Headlee et al. 2013, Omil et al. 2013). Here, we examined responses of the four species with the greatest biomass responses to biochar (two negatives and two positives). Results indicate that positive responses are associated with provision of P, and to a lesser extent Ca and Mg, while negative responses correspond to an over-supply of P. While this provides some information on the overall nutritional response of early-successional pioneers, additional work is necessary to fully quantify biochar effects on plant nutrient use and uptake in other species and through all ontogenetic stages. In particular, there is evidence for an initial flux of P and K from freshly applied biochar, followed by the more gradual release of base cations (Sackett et al. 2015). The capacity for early plant uptake and storage of P and K may thus be a major correlate of species response to biochar.

Conclusions

Our results support the general conclusion that biochar improves plant ecophysiological performance, but with high species-specific variation. Contrary to the hypothesis that fast-growing species will benefit the most from biochar fertilization effects, we detected weak negative relationships between species intrinsic photosynthetic rate (in control soils) and biomass responses to biochar, potentially due to nitrogen immobilization. In managed systems, biochar has the potential to enhance the growth of early-successional pioneers when thermally treated and mixed into the soil. Biochar has important potential applications for restoration; however, careful consideration of the physiological rates and nitrogen requirements of target species will be necessary to maximize the success of biochar-based restoration projects. Given the additional interest and potential global importance of biochar applications for enhancing carbon sequestration, longer-term field studies in a variety of systems are called for to evaluate the broader ecological consequences.

Acknowledgments

We thank Haliburton Forest and Wildlife Reserve Ltd, Haliburton, Ontario for research support and provision of biochar and soil. Special thanks are due to Katerina Eyre, Sam Dehdashti, Jad Murtada, Maria Al Zayat, William Merritt, Maha Mansoor, Lutchmee Seejeun, Janise Herridge, and James Hall for help with experimentation and data collection. Much thanks to Marney Isaac, Adam Martin, Nathan Basiliko, and Tattersall Smith for comments on various stages of this work. We additionally thank three anonymous reviewers, and the subject-matter editor, Michael Perring, for helpful comments on the manuscript. Research funding was provided by the Natural Science and Engineering Research Council, including funding through the NSERC Industrial Research Chair Program with contributions from the Ontario Mining Association and Haliburton Forest and Wildlife Reserve Ltd.

Literature Cited

Abeles, F. B., P. W. Morgan, and M. E. Saltveit Jr. 1992. Ethylene in plant biology. Academic Press, San Diego, California, USA.

Akhtar, S. S., M. N. Andersen, and F. Liu. 2015*a*. Residual effects of biochar on improving growth, physiology and yield of wheat under salt stress. Agricultural Water Management 158:61–68.

Akhtar, S. S., M. N. Andersen, and F. Liu. 2015b. Biochar mitigates salinity stress in potato. Journal of Agronomy and Crop Science 201:368–378.

Akhtar, S. S., G. Li, M. N. Andersen, and F. Liu. 2014. Biochar enhances yield and quality of tomato under reduced irrigation. Agricultural Water Management 138:37–44.

Ashman, T. L. 1994. A dynamic perspective on the physiological cost of reproduction in plants. American Naturalist 144:300–316.

ASTM. 2013. Standard test method for chemical analysis of wood charcoal. ASTM standard 1762-84. Annual book of ASTM standards. ASTM International, West Conshohocken, Pennsylvania, USA.

Bai, S. H., C. Y. Xu, Z. Xu, T. J. Blumfield, H. Zhao, H. Wallace, F. Reverchon, and L. Van Zwieten. 2015. Soil and foliar nutrient and nitrogen isotope composition ($\delta15N$) at 5 years after poultry litter and green waste biochar amendment in a macadamia orchard. Environmental Science and Pollution Research 22:3803–3809.

Bartha, S. 2001. Spatial relationships between plant litter, gopher disturbance and vegetation at different stages of old-field succession. Applied Vegetation Science 4:53–62.

Bazzaz, F. A., N. R. Chiariello, P. D. Coley, and L. F. Pitelka. 1987. Allocating resources to reproduction and defense. BioScience 37:58–67.

Beesley, L., E. Moreno-Jiménez, and J. L. Gomez-Eyles. 2010. Effects of biochar and greenwaste compost amendments on mobility, bioavailability and toxicity of inorganic and organic contaminants in a multi-element polluted soil. Environmental Pollution 158:2282–2287.

Benlloch-González, M., O. Arquero, J. M. Fournier, D. Barranco, and M. Benlloch. 2008. K+ starvation inhibits water-stress-induced stomatal closure. Journal of Plant Physiology 165:623–630.

Biederman, L. A., and W. S. Harpole. 2013. Biochar and its effects on plant productivity and nutrient cycling: a meta-analysis. GCB Bioenergy 5:202–214.

Biederman, L. A., J. Phelps, B. J. Ross, M. Polzin, and W. S. Harpole. 2017. Biochar and manure alter few aspects of prairie development: a field test. Agriculture, Ecosystems & Environment 236:78–87.

Caruso, C. M., D. L. D. Remington, and K. E. Ostergren. 2005. Variation in resource limitation of plant reproduction influences natural selection on floral traits of Asclepias syriaca. Oecologia 146:68–76.

Chan, K. Y., L. Van Zwieten, I. Meszaros, A. Downie, and S. Joseph. 2007. Agronomic values of greenwaste biochar as a soil amendment. Soil Research 45:629–634.

Chapin III, F. S., A. J. Bloom, C. B. Field, and R. H. Waring. 1987. Plant responses to multiple environmental factors. BioScience 37:49–57.

Choi, D., K. Makoto, A. M. Quoreshi, and L. Qu. 2009. Seed germination and seedling physiology of Larix kaempferi and Pinus densiflora in seedbeds with charcoal and elevated CO_2. Landscape and Ecological Engineering 5:107–113.

Clough, T., L. Condron, C. Kammann, and C. Müller. 2013. A review of biochar and soil nitrogen dynamics. Agronomy 3:275–293.

Conversa, G., and A. Bonasia. 2015. Influence of biochar, mycorrhizal inoculation and fertilizer rate on growth and flowering of pelargonium (Pelargonium zonale L.) plants. Frontiers in Plant Science 6:429.

Evans, J. R. 1989. Photosynthesis and nitrogen relationships in leaves of C3 plants. Oecologia 78:9–19.

Franks, S. J., S. Sim, and A. E. Weis. 2007. Rapid evolution of flowering time by an annual plant in response to a climate fluctuation. Proceedings of the National Academy of Sciences USA 104:1278–1282.

Fulton, W., M. Gray, F. Prahl, and M. Kleber. 2013. A simple technique to eliminate ethylene emissions from biochar amendment in agriculture. Agronomy for Sustainable Development 33:469–474.

Gai, X., H. Wang, J. Liu, L. Zhai, S. Liu, T. Ren, and H. Liu. 2014. Effects of feedstock and pyrolysis temperature on biochar adsorption of ammonium and nitrate. PLoS ONE 9:e113888.

Gale, N. V., T. E. Sackett, and S. C. Thomas. 2016. Thermal treatment and leaching of biochar alleviates plant growth inhibition from mobile organic compounds. PeerJ 4:e2385.

Gradowski, T., and S. C. Thomas. 2006. Phosphorus limitation of sugar maple growth in central Ontario. Forest Ecology and Management 226:104–109.

Gradowski, T., and S. C. Thomas. 2008. Responses of Acer saccharum canopy trees and saplings to P, K and lime additions under high N deposition. Tree Physiology 28:173–185.

Gale, N. V., M. A. Halim, M. Horsburgh, and S. C. Thomas. 2017. Data from: Gale et al., Comparative responses of early-successional plants to charcoal soil amendments. Ecosphere. https://doi.org/10.6084/m9.figshare.5371033.v1

Green, D. M., and J. B. Kauffman. 1995. Succession and livestock grazing in a northeastern Oregon riparian ecosystem. Journal of Range Management 48:307.

Hart, S., and N. Luckai. 2013. Charcoal function and management in boreal ecosystems. Journal of Applied Ecology 50:1197–1206.

Headlee, W. L., C. E. Brewer, and R. B. Hall. 2013. Biochar as a substitute for vermiculite in potting mix for hybrid poplar. Bioenergy Research 7:120–131.

Hollister, C. C., J. J. Bisogni, and J. Lehmann. 2013. Ammonium, nitrate, and phosphate sorption to and solute leaching from biochars prepared from corn stover (*Zea mays* L.) and oak wood (*Quercus* spp.). Journal of Environmental Quality 42:137–144.

Hothorn, T., F. Bretz, and P. Westfall. 2008. Simultaneous inference in general parametric models. Biometrical Journal 50:346–363.

Isaac, M. E., and A. A. Kimaro. 2011. Diagnosis of nutrient imbalances with vector analysis in agroforestry systems. Journal of Environmental Quality 40:860–866.

Jeffery, S., D. Abalos, M. Prodana, A. C. Bastos, J. W. van Groenigen, B. A. Hungate, and F. Verheijen. 2017. Biochar boosts tropical but not temperate crop yields. Environmental Research Letters 12:053001.

Jeffery, S., F. G. A. Verheijen, M. Van Der Velde, and A. C. Bastos. 2011. A quantitative review of the effects of biochar application to soils on crop productivity using meta-analysis. Agriculture, Ecosystems & Environment 144:175–187.

Jones, D. L., J. Rousk, G. Edwards-Jones, T. H. DeLuca, and D. V. Murphy. 2012. Biochar-mediated changes in soil quality and plant growth in a three-year field trial. Soil Biology and Biochemistry 45:113–124.

Keeley, S. C., J. E. Keeley, S. M. Hutchinson, and A. W. Johnson. 1981. Postfire succession of the herbaceous flora in southern California chaparral. Ecology 62:1608–1621.

Kochanek, J., R. L. Long, A. T. Lisle, and G. R. Flematti. 2016. Karrikins identified in biochars indicate postfire chemical cues can influence community diversity and plant development. PLoS ONE 11:e0161234.

Kuttner, B. G., and S. C. Thomas. 2017. Interactive effects of biochar and an organic dust suppressant for revegetation and erosion control with herbaceous seed mixtures and willow cuttings. Restoration Ecology 25:367–375.

Laird, D., P. Fleming, B. Wang, R. Horton, and D. Karlen. 2010. Biochar impact on nutrient leaching from a Midwestern agricultural soil. Geoderma 158:436–442.

Lashari, M. S., Y. Liu, L. Li, W. Pan, J. Fu, G. Pan, J. Zheng, J. Zheng, X. Zhang, and X. Yu. 2013. Effects of amendment of biochar-manure compost in conjunction with pyroligneous solution on soil quality and wheat yield of a salt-stressed cropland from central China great plain. Field Crops Research 144:113–118.

Lehmann, J. 2007. A handful of carbon. Nature 447:143–144.

Liu, X., A. Zhang, C. Ji, S. Joseph, R. Bian, L. Li, G. Pan, and J. Paz-Ferreiro. 2013. Biochar's effect on crop productivity and the dependence on experimental conditions—a meta-analysis of literature data. Plant and Soil 373:583–594.

Longstreth, D. J., and P. S. Nobel. 1980. Nutrient influences on leaf photosynthesis effects of nitrogen, phosphorus, and potassium for *Gossypium hirsutum* L. Plant Physiology 65:541–543.

Major, J., M. Rondon, D. Molina, S. J. Riha, and J. Lehmann. 2010. Maize yield and nutrition during 4 years after biochar application to a Colombian savanna oxisol. Plant and Soil 333:117–128.

Milberg, P. 1995. Soil seed bank after eighteen years of succession from grassland to forest. Oikos 72:3–13.

Mitchell, P. J., A. J. Simpson, R. Soong, J. S. Schurman, S. C. Thomas, and M. J. Simpson. 2016. Biochar amendment and phosphorus fertilization altered forest soil microbial community and native soil organic matter molecular composition. Soil Biochemistry 130:227–245.

Mukherjee, A., and A. R. Zimmerman. 2013. Organic carbon and nutrient release from a range of laboratory-produced biochars and biochar–soil mixtures. Geoderma 194:122–130.

Omil, B., V. Piñeiro, and A. Merino. 2013. Soil and tree responses to the application of wood ash containing charcoal in two soils with contrasting properties. Forest Ecology and Management 295:199–212.

Park, J. H., G. K. Choppala, N. S. Bolan, J. W. Chung, and T. Chuasavathi. 2011. Biochar reduces the bioavailability and phytotoxicity of heavy metals. Plant and Soil 348:439–451.

Parker, W. C., K. A. Elliott, D. C. Dey, E. Boysen, S. G. Newmaster, W. C. Parker, K. A. Elliott, D. C. Dey, E. Boysen, and S. G. Newmaster. 2001. Managing succession in conifer plantations: converting young red pine (*Pinus resinosa* Ait.) plantations to native forest types by thinning and underpainting. Forestry Chronicle 77:721–734.

Petraglia, A., M. Tomaselli, A. Mondoni, L. Brancaleoni, and M. Carbognani. 2014. Effects of nitrogen and phosphorus supply on growth and flowering phenology of the snowbed forb *Gnaphalium supinum* L. Flora 209:271–278.

Pinheiro, J., D. Bates, S. DebRoy, and D. Sarkar. 2011. Package 'nlme'. Linear and nonlinear mixed effects models. http://cran.r-project.org/web/packages/nlme/nlme.pdf

Pluchon, N., M. J. Gundale, M.-C. Nilsson, P. Kardol, and D. A. Wardle. 2014. Stimulation of boreal tree seedling growth by wood-derived charcoal: effects of charcoal properties, seedling species and soil fertility. Functional Ecology 28:766–775.

R Core Team. 2016. R: a language and environment for statistical computing. R Foundation for Statistical Computing, Vienna, Austria.

Rajkovich, S., A. Enders, K. Hanley, C. Hyland, A. R. Zimmerman, and J. Lehmann. 2012. Corn growth and nitrogen nutrition after additions of biochars with varying properties to a temperate soil. Biology and Fertility of Soils 48:271–284.

Reekie, E. and F. A. Bazzaz, editors. 2011. Reproductive allocation in plants. Academic Press, Burlington, Massachusetts, USA.

Sackett, T. E., N. Basiliko, G. L. Noyce, C. Winsborough, J. Schurman, C. Ikeda, and S. C. Thomas. 2015. Soil and greenhouse gas responses to biochar additions in a temperate hardwood forest. Global Change Biology: Bioenergy 7:1062–1074.

Sage, R. F., and R. W. Pearcy. 1987. The nitrogen use efficiency of C3 and C4 plants I. Leaf nitrogen, growth, and biomass partitioning in *Chenopodium album* (L.) and *Amaranthus retroflexus* (L.). Plant Physiology 84:954–958.

Seehausen, M. L., N. V. Gale, S. Dranga, V. Hudson, N. Liu, J. Michener, E. Thurston, C. Williams, S. M. Smith, and S. C. Thomas. 2017. Is there a positive synergistic effect of biochar and compost soil amendments on plant growth and physiological performance? Agronomy 7. https://doi.org/10.3390/agronomy7010013

Sika, M. P., and A. G. Hardie. 2014. Effect of pine wood biochar on ammonium nitrate leaching and availability in a South African sandy soil. European Journal of Soil Science 65:113–119.

Spokas, K. A., J. M. Baker, and D. C. Reicosky. 2010. Ethylene: potential key for biochar amendment impacts. Plant and Soil 333:443–452.

Tammeorg, P., A. Simojoki, P. Mäkelä, F. L. Stoddard, L. Alakukku, and J. Helenius. 2013. Biochar application to a fertile sandy clay loam in boreal conditions: effects on soil properties and yield formation of wheat, turnip rape and faba bean. Plant and Soil 374:89–107.

Thomas, S. C., S. Frye, N. Gale, M. Garmon, R. Launchbury, N. Machado, S. Melamed, J. Murray, A. Petroff, and C. Winsborough. 2013. Biochar mitigates negative effects of salt additions on two herbaceous plant species. Journal of Environmental Management 129:62–68.

Thomas, S. C., and N. Gale. 2015. Biochar and forest restoration: a review and meta-analysis of tree growth responses. New Forests 46:931–946.

Thomas, S. C., M. Jasienski, and F. A. Bazzaz. 1999. Early vs asymptotic growth responses of herbaceous plants to elevated CO_2. Ecology 80:1552–1567.

van de Voorde, T. F., T. M. Bezemer, J. W. Van Groenigen, S. Jeffery, and L. Mommer. 2014. Soil biochar amendment in a nature restoration area: effects on plant productivity and community composition. Ecological Applications 24:1167–1177.

Ventura, M., G. Sorrenti, P. Panzacchi, E. George, and G. Tonon. 2013. Biochar reduces short-term nitrate leaching from a horizon in an apple orchard. Journal of Environmental Quality 42:76–82.

Wardle, D. A., O. Zackrisson, and M. C. Nilsson. 1998. The charcoal effect in boreal forests: mechanisms and ecological consequences. Oecologia 115:419–426.

Wieland, N. K., and F. A. Bazzaz. 1975. Physiological ecology of three codominant successional annuals. Ecology 56:681–688.

Woolf, D., J. E. Amonette, F. A. Street-Perrott, J. Lehmann, and S. Joseph. 2010. Sustainable biochar to mitigate global climate change. Nature Communications 1:56.

Zeineddine, M., and V. A. A. Jansen. 2009. To age, to die: parity, evolutionary tracking and Cole's paradox. Evolution 63:1498–1507.

Zheng, H., Z. Wang, X. Deng, S. Herbert, and B. Xing. 2013. Impacts of adding biochar on nitrogen retention and bioavailability in agricultural soil. Geoderma 206:32–39.

Is initial post-disturbance regeneration indicative of longer-term trajectories?

Nathan S. Gill [iD],[1,2,]† Daniel Jarvis,[1,3] Thomas T. Veblen,[4] Steward T. A. Pickett [iD],[5] and Dominik Kulakowski[1]

[1]Graduate School of Geography, Clark University, 950 Main Street, Worcester, Massachusetts 01610 USA
[2]Pacific Island Ecosystems Research Center, 344 Crater Rim Drive, Volcano, Hawaii 96718 USA
[3]Vermont Technical College, 124 Admin Drive, Randolph Center, Vermont 05061 USA
[4]Geography Department, University of Colorado-Boulder, Guggenheim 110, 260 UCB, Boulder, Colorado 80309 USA
[5]Cary Institute of Ecosystem Studies, Box AB, 2801 Sharon Turnpike, Millbrook, New York 12545 USA

Abstract. The ability to estimate and model future vegetation dynamics is a central focus of contemporary ecology and is essential for understanding future ecological trajectories. It is therefore critical to understand when the influence of initial post-disturbance regeneration versus stochastic processes dominates long-term post-disturbance ecological processes. Often, conclusions about post-disturbance dynamics are based upon initial regeneration in the years immediately after disturbances. However, the degree to which initial post-disturbance regeneration indicates longer-term trends is likely to be contingent on the types, intensities, and combinations of disturbances, as well as pre-disturbance ecosystem structure and composition. Our relatively limited understanding of why initial post-disturbance regeneration is sometimes a poor predictor of future ecosystem trajectories represents a critical gap in post-disturbance ecological forecasting. We studied the composition and density of regeneration of tree species following wind blowdown in 1997, wildfire in 2002, and compounded disturbances by blowdown and wildfire in subalpine forests of Colorado. We examined regeneration of *Picea engelmannii*, *Abies lasiocarpa*, *Pinus contorta*, and *Populus tremuloides* in 180 permanent plots across 12 sites (classified by pre-disturbance age and composition) in 2003, 2010, and 2015. At sites that were blown down but not burned, regeneration was dense and dominated by *Picea* and *Abies*. At these sites, regeneration observed from 2003 to 2005 (hereafter *initial regeneration*) was also highly predictive of regeneration 5–10 yr later. In contrast, at sites that were burned and sites that were blown down and burned, regeneration was less dense and dominated by a mix of species. At these sites, initial regeneration was a poor predictor of longer-term trends as species dominance and overall density fluctuated over the 13-yr period. These findings call into question our ability to confidently predict ecosystem trajectories based upon observations made in the years immediately after large, severe disturbances such as wildfires and compounded disturbances. As compounded disturbances become more common under climatically driven changes in disturbance regimes, post-disturbance ecosystem trajectories may become increasingly stochastic and unpredictable.

Key words: blowdown; compounded disturbance; ecosystem trajectories; legacy; linked disturbance; post-fire regeneration; resilience.

† E-mail: ngill@clarku.edu

INTRODUCTION

Although climate change can affect forest ecosystems gradually due to changes in atmospheric and climatic conditions, the most dramatic changes are likely to be abrupt and modulated by climatically driven disturbances (Frelich and Reich 2009, Turner 2010). Disturbances strongly influence the structure and composition of most ecosystems, and understanding the ways in which disturbances accelerate or introduce change is key to predicting future ecosystem trajectories (Sousa 1984, Pickett and White 1985). Disturbances cause mortality and often provide opportunity for new establishment or promote reorganization of surviving vegetation. The first few years of regeneration after a disturbance can be the best available indicator of future trajectories and therefore are relied upon for management decisions and individual-based ecological modeling (Egler 1954, Peet 1981, Veblen et al. 1991, Van Mantgem and Stephenson 2005, Zald et al. 2008, Donato et al. 2009b, Baker et al. 2013, Collins and Roller 2013, Grimm and Railsback 2013, Xiang et al. 2013). However, initial post-disturbance regeneration may not always be indicative of longer-term trends (Pickett et al. 2001), and it is critical to understand when initial post-disturbance regeneration versus stochastic processes dominate post-disturbance ecosystem dynamics. The lack of long-term monitoring of forest regeneration following large, severe disturbances has been stressed as one of the major deficiencies facing restoration efforts (Holl and Cairns 2002) and is seen as an urgent need in ecological science (Van Leeuwen et al. 2010), especially in the context of multiple interacting disturbances. Our relatively limited understanding of when and why initial post-disturbance regeneration is at times a poor predictor of future ecosystem trajectories represents a critical gap in post-disturbance ecological forecasting, with important implications for the resilience and vulnerability of ecosystem services (Turner et al. 2013).

For decades, succession theory was strongly influenced by Egler's (1954) Initial Floristic Composition Hypothesis (Finegan 1984), which predicts that successional pathways depend on the communities that are present from the start of succession (i.e., complete initial floristics; Wilson et al. 1992). However, understanding of succession and species composition evolved with the presentation of the intermediate disturbance hypothesis (Connell 1978), which states that maximum species richness is achieved not immediately following disturbance, but when sufficient time between disturbances allows for a balance of pioneer and late-successional species. Although this hypothesis spurred much research into the role of disturbances in determining biological diversity, this theory has been called into question because of lack of supporting theory and empirical evidence (Fox 2013). While contemporary ecology recognizes a more nuanced relationship between disturbance regime attributes and species richness, the fact remains that ecologists and managers often operate under the assumption that initial plant communities following disturbances are indicative of long-term successional pathways. This assumption should be adequately examined, especially as disturbances become larger, more frequent, and more severe, and ecosystems are increasingly affected by multiple disturbances over shorter periods of time.

Current understanding of regeneration following multiple disturbances suggests that an interacting effect may emerge in two ways—through the effect of a first disturbance on the occurrence, intensity, severity, or other attributes of a subsequent disturbance, or through a compounded effect that alters the nature of post-disturbance development (Donato et al. 2009a, Simard et al. 2011, Kulakowski and Veblen 2015). In extreme cases, compounded disturbances may push a community to an alternate stable state (Paine et al. 1998). For example, a short interval between two disturbances may lead to immaturity risk whereby seed may be less available at the time of the second disturbance, causing a shift to non-forest (Enright et al. 2015). Promoting ecosystem resilience through a deeper understanding of natural disturbance regimes is commonly a central management objective (DeRose and Long 2009, Nagel et al. 2014), but questions remain as to how resilience can best be achieved in the context of compounded disturbances or how to recognize cases in which compound effects are most likely. Such effects may not be immediately evident after disturbances, but emerge many years later in post-disturbance development through the presence (or absence) of biotic legacies (Pickett et al. 2005, Turner 2010, Royo et al. 2016) and may be sensitive to varying spatial scales of disturbance (Svoboda et al. 2014, Jogiste et al. 2017).

The subalpine forests of the Colorado Rocky Mountains are influenced by a variety of natural disturbances including wildfires, native beetle outbreaks, and wind blowdowns (Veblen et al. 1994, Kulakowski and Veblen 2003). These disturbance agents have played key roles in forest dynamics for thousands of years, but recent climatic warming has accelerated their frequency, enhanced their intensity, and increased their size (Westerling et al. 2006, Bentz et al. 2010, Evangelista et al. 2011). Colorado subalpine forests are composed primarily of Engelmann spruce (*Picea engelmannii*), subalpine fir (*Abies lasiocarpa*), lodgepole pine (*Pinus contorta*), and quaking aspen (*Populus tremuloides* Michx.), each of which utilizes a different regeneration strategy and fills a unique ecological niche.

Picea engelmannii and *A. lasiocarpa* are *avoiders* (Rowe 1983, Veblen 1986a, Buma and Wessman 2012), which are not particularly adapted to disturbances, but generally have a competitive advantage in the shadier conditions prevalent in the years after regeneration has established. *Picea engelmannii* and *A. lasiocarpa* both produce high seedling densities but differ in that *P. engelmannii* tend to have fewer seedlings but a higher rate of survivorship, while *A. lasiocarpa* regenerate in greater abundance initially but thin out as seedlings become saplings and saplings eventually become mature adult trees (Veblen 1986a, Maher et al. 2005). *Pinus contorta* exhibit an *evader* strategy (Rowe 1983) through the dispersal of seed by serotinous cones. Adult *P. contorta* are consumed in severe wildfires such as the Mount Zirkel Fire Complex of 2002, but many *P. contorta* are serotinous, and therefore, high fire temperatures allow cones to release plentiful seed, potentially leading to very dense regeneration. This is especially effective in dense stands of sexually mature and serotinous *P. contorta* (Tinker et al. 1994). *Populus tremuloides* can take an *endurer* approach (Rowe 1983) by regenerating asexually through vegetative resprouting. *Populus tremuloides* are also capable of regenerating from seed, although there is disagreement over the frequency of this occurrence in the subalpine forests of the Rockies (Kay 1993, Howard 1996, Romme et al. 1997, Quinn

and Wu 2001). A majority of regenerating aspen in the study site in the years immediately after the 2002 wildfires resprouted vegetatively (Kulakowski et al. 2013), but increased aspen seedling regeneration in more recent years has been documented in stands that were previously dominated by spruce and fir (Buma and Wessman 2012). Both *P. tremuloides* and especially *P. contorta* have a competitive advantage over *P. engelmannii* and *A. lasiocarpa* in full sunlight, as is typical following a stand-replacing disturbance (Parker and Parker 1983). The general strategies of these four species are common among tree species worldwide, and are well-studied following single disturbances (Rowe 1983), but the efficacy of these strategies is highly complex following compounded disturbances and not well understood (Buma and Wessman 2012, Kulakowski et al. 2013). Here, we examine regeneration over a 13-yr period following individual and compounded disturbances in four major forest types to identify scenarios under which initial post-disturbance regeneration is indicative of longer-term trends and under which it is more stochastic.

MATERIALS AND METHODS

Study area

The study area lies within Routt National Forest and the Mt. Zirkel Wilderness in northern Colorado and is defined as the sum of the areas affected by severe blowdown in 1997 and fire in 2002. The elevation of the study area ranges from 2400 to 3600 m above sea level. The climate is continental, and mean monthly temperatures since 1893 range from a minimum of −17.1°C in January to a maximum of 28.1°C in July. Mean annual precipitation is 60.2 cm of rain and 423 cm of snowfall (Western Regional Climate Center, http://www.wrcc.dri.edu/). Upland forests in this region are underlain by coarse-textured soils consisting of glacial deposits and Precambrian crystalline parent material, while low-lying valleys are derived from poorly drained alluvial deposits (Snyder et al. 1987). Forests are dominated by *Pinus contorta*, *Populus tremuloides*, *Picea engelmannii*, and *Abies lasiocarpa*. Portions of the study area were burned by stand-replacing fires in 1879 and 1880 (Kulakowski and Veblen 2002).

In October 1997, ~10,340 ha of the National Forest was blown down in a severe windstorm. Five years later, 12,354 ha of the forest burned in the Mount Zirkel Fire complex, and an additional 1724 ha burned in the Green Creek Fire that same year. Approximately 40% of the blown-down forest was burned in 2002.

Data and analysis

In 2003, we established 540 2 × 1 m micro-plots nested in 180 10 × 2 m macroplots across 24 sites within the study area. Sites were located according to a stratified random sampling scheme across classes of species composition (>90% *P. engelmannii/A. lasiocarpa*, >40% *P. contorta*, or >50% *P. tremuloides*) as determined by Kulakowski and Veblen (2002), stand structural stage (understory reinitiation stands that originated after fires in the 1880s or old-growth stands >200 yr old; Oliver 1981), and a combination of recent disturbances (severe 1997 blowdown, 2002 stand-replacing wildfire, or both). Other differences among sites such as elevation and precipitation were not included in the analysis because these factors were fairly consistent among sites by design.

We measured and recorded the regeneration density of all tree species in each plot in 2003, 2010, and 2014–2015. Sites that burned were also measured in 2004 and 2005. All saplings (individuals >140 cm in height but <4 cm dbh) were counted within each macroplot. We sampled the density of seedlings (conifer individuals <140 cm in height) and *P. tremuloides* ramets <140 cm in height in each microplot. Ramets were counted as separate stems if they were not adjoined above the soil. The average density from three nested microplots was then averaged with the single macroplot density of saplings and sapling-sized ramets to yield an estimate of regeneration density (juveniles per hectare) for each species within each macroplot (equal weight given to seedling density and sapling density to yield regeneration density).

Linear regression models were created in R (R Development Core Team 2008) using total initial regeneration density (2003 after blowdown or 2005 after wildfire) as the independent variable and subsequent regeneration densities (both 2010 and 2014/15 for all stands) as the dependent variable. One model was created for each category based upon pre-disturbance stand structural stage (Stage III understory reinitiation or Stage IV old-growth; Oliver 1981), pre-disturbance composition, (spruce/fir dominated or lodgepole dominated), and disturbance type(s) (blowdown, fire, or both). Adjusted R^2 and the F statistic were used to measure significance.

Results

In stands that were affected by the 1997 blowdown but not the 2002 wildfire, initial regeneration density was generally a strong indicator of the dominant species (Table 1) and densities (Table 2) of regeneration both 7 and 11 yr later, with the exception of old-growth *Pinus contorta* stands (Table 2). In all cases, blowdown-only stands were typified by early *Abies lasiocarpa* dominance followed by gradual increases in the densities of *A. lasiocarpa*, *Picea engelmannii* (Fig. 1), and some *P. contorta* regeneration in the case of stands that were in Stage III of structural development at the time of blowdown (Fig. 1a, b). Rankings of regenerating species by density in later years were very similar to 2003 for all stand categories that were blown down and not burned (Table 1, Fig. 1).

In stands that burned but were not blown down, initial regeneration tended to be sparse in most cases (Fig. 2) and exhibited numerous changes in ranking of species by density (Table 3, Fig. 2). Spruce/fir stands of either stage and old-growth lodgepole pine stands exhibited no more

Table 1. Rankings of post-1997 blowdown regeneration of *Abies lasiocarpa* (AL), *Picea engelmannii* (PE), *Pinus contorta* (PC), and *Populus tremuloides* (PT) for stands of different pre-blowdown structural stage and composition.

Year	Stage III PE/AL	Stage III PC	Stage IV PE/AL	Stage IV PC
2003	AL, PE, PC, PT	AL, PC, PE, PT	AL, PE, PC-PT	AL, PE, PC-PT
2010	AL, PC, PE, PT	AL, PE, PC, PT	AL, PE, PC, PT	AL, PE, PC-PT
2014	AL, PC, PE, PT	AL, PC, PE, PT	AL, PE, PC, PT	AL, PE, PC-PT

Notes: Species are ranked from highest to lowest density. None of the sites presented in this table were burned in 2002.

Table 2. Dynamics of regeneration for stands that were blown down but not burned.

Pre-disturbance composition/ structural stage	Subsequent year	Adjusted R^2	F	P
Stage III PE/AL	2010	0.3937	58.8*	<0.001
Stage III PE/AL	2014	0.4582	76.3*	<0.001
Stage IV PE/AL	2010	0.1505	16.76*	<0.001
Stage IV PE/AL	2014	0.1512	16.85*	<0.001
Stage III PC	2010	0.5573	113.0*	<0.001
Stage III PC	2014	0.3832	56.3*	<0.001
Stage IV PC	2010	0.0000	0.9	0.336
Stage IV PC	2014	0.0497	5.7	0.020

Notes: Results of linear regression models using total initial (2003) regeneration density as the independent variable and total subsequent regeneration density as the dependent variable. Data are separated by pre-disturbance composition (PE/AL, *Picea engelmannii/Abies lasiocarpa* or PC, *Pinus contorta*) and stage of structural development. Asterisk denotes significance at the 1% level.

than 600 trees/ha of any one species of regeneration, with most species regenerating under 200 trees/ha. Even in cases where regeneration density was near 600 trees/ha, the most abundant species had been absent in previous years

(Fig. 2a,c). These stands are characterized by low regeneration densities and numerous shifts in density rankings (Table 3, Fig. 2). Stage III *P. contorta* stands exhibited consistent dominance by very dense *P. contorta* regeneration (Fig. 2b), but older *P. contorta* stands and old and young *P. engelmannii/A. lasiocarpa* stands saw shifts in the most dominant species of regeneration (Table 3, Fig. 2). Early regeneration density in Stage III *P. contorta* stands was an effective predictor of later regeneration density through 2014, while early regeneration in Stage IV *P. contorta* stands was a good predictor for 2010 but not 2014 (Table 4).

Initial regeneration in stands affected by both disturbances exhibited high predictive power of later trajectories in some stands while others exhibited great fluctuations in regeneration density and species ranks. Regeneration of Stage III stands (of both *P. engelmanni/A. lasiocarpa* and *P. contorta*) was for the first two years dominated by *P. contorta*, but by 2005 *Populus tremuloides* regeneration became more dense and persisted as the dominant species of regeneration (Fig. 3a, b,

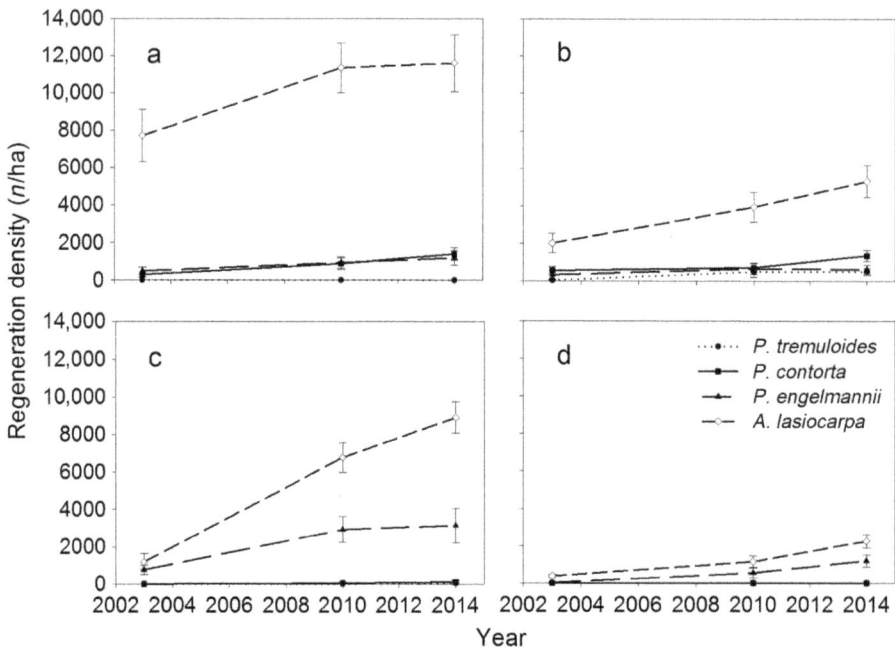

Fig. 1. Regeneration density (mean ± SE) of four species in stands affected only by 1997 blowdown, organized by pre-disturbance stand dominance and structural stage. Number of 20-m² plots = 30 per graph. (a) Understory reinitiation *Picea engelmannii/Abies lasiocarpa* stands. (b) Understory reinitiation *P. contorta* stands. (c) Old-growth *P. engelmannii/A. lasiocarpa* stands. (d) Old-growth *P. contorta* stands.

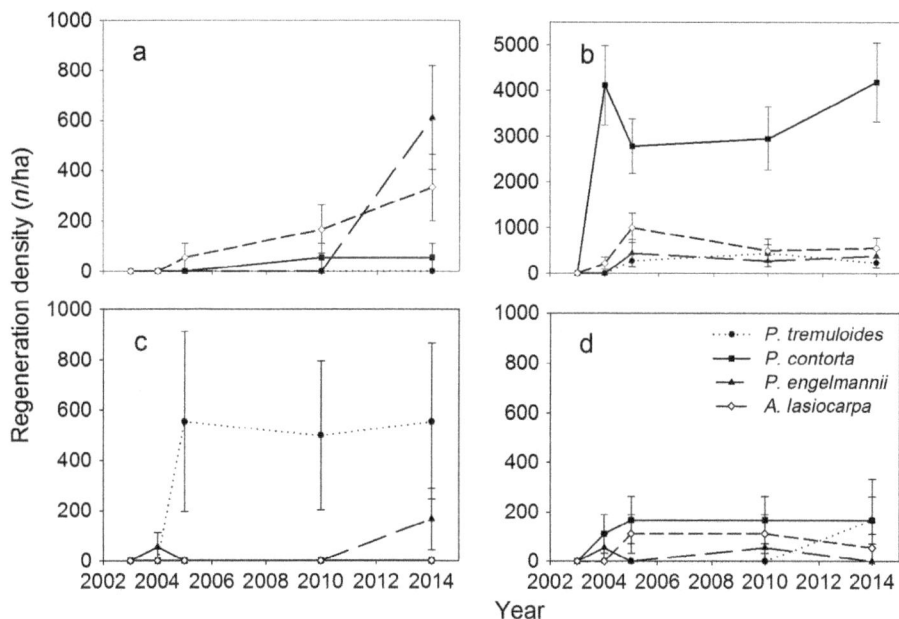

Fig. 2. Regeneration density (mean ± SE) of four species in stands affected only by 2002 fire, organized by pre-disturbance stand dominance and structural stage. Number of 20-m^2 plots = 30 per graph. (a) Understory reinitiation *Picea engelmannii/Abies* lasiocarpa stands. (b) Understory reinitiation *P. contorta* stands. (c) Old-growth *P. engelmannii/A. lasiocarpa* stands. (d) Old-growth *P. contorta* stands. Note the different scale for the *y*-axis of panel B.

Table 5). Similarly, in old-growth *P. engelmannii/ A. lasiocarpa* stands that were blown down and burned, *P. contorta* is initially the only species regenerating, but from 2005 to 2010, *P. tremuloides* and *A. lasiocarpa* are ascending together and *P. contorta* becomes completely absent across these sites. Between 2010 and 2014, *P. tremuloides* also fails to persist, leaving *A. lasiocarpa* as the dominant regeneration species, while *P. contorta* establishes once again (Fig. 3c, Table 5). Old-growth *P. contorta* stands that experienced both disturbances are dominated by dense

P. tremuloides regeneration from 2004 onward (Fig. 3d). Following compounded disturbances, Stage III and Stage IV *P. contorta* stands had early regeneration patterns which effectively predicted trends for 2010 and 2014, while patterns in Stage IV spruce/fir stands were effective predictors only through 2010, and patterns in Stage III spruce/fir stands predicted 2014 densities effectively but not 2010 densities (Table 6).

When regeneration plots were categorized by initial regeneration composition and density rather than by pre-disturbance stand structure,

Table 3. Rankings of post-2002 wildfire regeneration of *Abies lasiocarpa* (AL), *Picea engelmannii* (PE), *Pinus contorta* (PC), and *Populus tremuloides* (PT) for stands of different pre-fire structural stage and composition.

Year	Stage III PE/AL	Stage III PC	Stage IV PE/AL	Stage IV PC
2003	...	...	...	...
2004	...	PC, AL, PT, PE	PE, AL-PC-PT	PC, PE, AL-PT
2005	AL, PC, PE-PT	PC, AL, PE, PT	PT, AL-PC-PE	PC, AL, PE-PT
2010	AL, PC, PE-PT	PC, AL, PT, PE	PT, AL-PC-PE	PC, AL, PE, PT
2014	PE, AL, PC, PT	PC, AL, PE, PT	PT, PE, AL-PC	PT-PC, AL, PE

Notes: Species are ranked from highest to lowest density. None of the sites presented in this table were blown down in 1997. An ellipsis indicates complete absence of regeneration.

Table 4. Dynamics of regeneration for stands that were burned but not blown down.

Pre-disturbance composition/ structural stage	Subsequent year	Adjusted R^2	F	P
Stage III PE/AL	2010	0.0000	0.1	0.831
Stage III PE/AL	2014	0.0242	3.2	0.077
Stage IV PE/AL	2010	0.0000	0.1	0.783
Stage IV PE/AL	2014	0.0000	0.4	0.528
Stage III PC	2010	0.2097	24.6*	<0.001
Stage III PC	2014	0.3302	44.9*	<0.001
Stage IV PC	2010	0.0949	10.33*	0.002
Stage IV PC	2014	0.0333	4.1	0.047

Notes: Results of linear regression models using total initial (2005) regeneration density as the independent variable and total subsequent regeneration density as the dependent variable. Data are separated by pre-disturbance composition (PE/AL, *Picea engelmannii/Abies lasiocarpa* or PC, *Pinus contorta*) and stage of structural development. An asterisk denotes significance at the 1% level.

composition, and disturbance agent(s), early observations in plots which exhibited regeneration densities of any species >1000 trees/ha were strong predictors of trajectories through 2010 and 2014. Plots with low early regeneration

densities (<1000 but >0 trees/ha) had very little predictive power of future trends (Table 7). Plots with absolutely no regeneration through the first three years after disturbance often remained with little to no regeneration through 2014.

Discussion

The current study shows that disturbance type and (secondarily) pre-disturbance stand structure and composition influence the ability to predict regeneration densities from the initial years of regeneration, even through the first 1–2 decades. The predictive power of models of regeneration for stands that were affected by a single disturbance agent varied strongly based on disturbance agent. In all cases, stands that were blown down but not burned were dominated by the same species in 2014 as they were in all prior years since the blowdown, across classes of structural stage and pre-disturbance species composition. Linear models of these regeneration trajectories effectively predict regeneration 13–17 yr following disturbance based on regeneration patterns

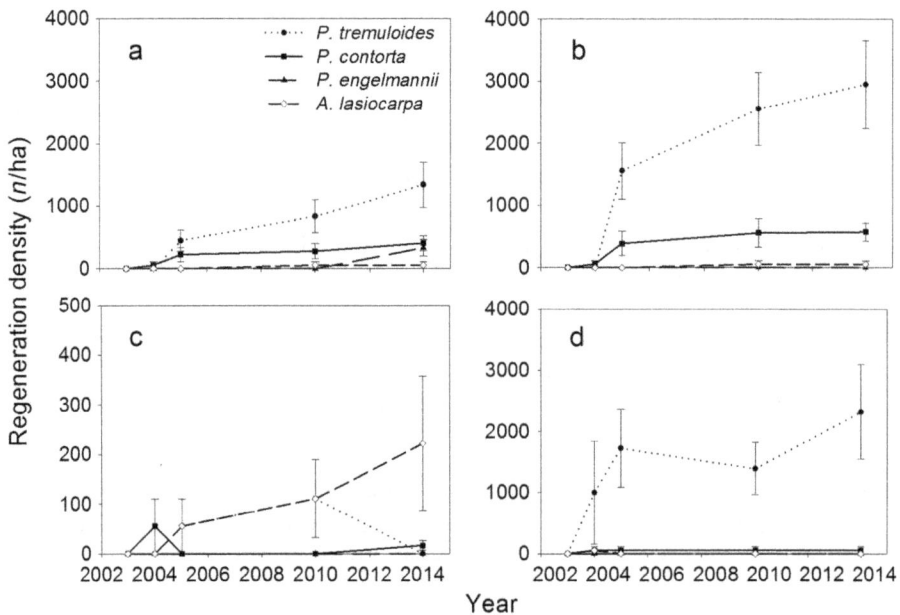

Fig. 3. Regeneration density (mean ± SE) of four species in stands affected by 1997 blowdown and 2002 fire, organized by pre-disturbance stand dominance and structural stage. Number of 20-m² plots = 30 per graph. (a) Understory reinitiation *Picea engelmannii/Abies lasiocarpa* stands. (b) Understory reinitiation *P. contorta* stands. (c) Old-growth *P. engelmannii/A. lasiocarpa* stands. (d) Old-growth *P. contorta* stands. Note the different scale for the *y*-axis of panel C.

Table 5. Rankings of regenerating *Abies lasiocarpa* (AL), *Picea engelmannii* (PE), *Pinus contorta* (PC), and *Populus tremuloides* (PT) following blowdown and wildfire for stands of different pre-blowdown structural stage and composition.

Year	Stage III PE/AL	Stage III PC	Stage IV PE/AL	Stage IV PC
2003	...	...	...	...
2004	PC, AL-PE-PT	PC, AL-PE-PT	PC, AL-PE-PT	PT, AL-PC, PE
2005	PT, PC, AL-PE	PT, PC, AL-PE	AL-PT, PE-PC	PT, PC, AL-PE
2010	PT, PC, AL, PE	PT, PC, AL, PE	AL-PT, PE-PC	PT, PC, AL-PE
2014	PT, PC, PE, AL	PT, PC, AL, PE	AL, PC, PE-PT	PT, PC, AL-PE

Notes: Species are ranked from highest to lowest density. An ellipsis indicates complete absence of regeneration.

observed 5 yr after the blowdown. In stands that were severely burned in 2002 and not previously blown down, most of the temporal trends in the composition and density of regeneration were volatile as the abundance and composition of regeneration fluctuated substantially. The exception is the highly serotinous younger *Pinus contorta* category, for which the model exhibited strong predictive power through 2014. Models of less-serotinous (though still partly serotinous) old-growth *P. contorta* stands demonstrated high predictive power for regeneration density through 2010, but not 2014. In this way, the strength of post-fire regeneration models appears to be associated with stand serotiny, although future research should analyze this relationship with greater specificity (i.e., measure percentages of serotinous cones and trees; Tinker et al. 1994).

In the current study, regeneration composition and density during the first 5 yr following compounded blowdown and fire were strongly indicative of future trajectories in some cases, especially when early post-fire regeneration was abundant, but not indicative in other cases (despite the relatively short study period). Kulakowski et al. (2013) found that *Populus tremuloides* regeneration was no less dense following compounded blowdown and fire disturbances than following fire only, unlike conifer regeneration, which was largely inhibited by compounded disturbances. Likewise, models of regeneration density in the current study

Table 6. Regeneration dynamics for stands that were blown down in 1997 and burned in 2002.

Pre-disturbance composition/ structural stage	Subsequent year	Adjusted R^2	F	P
Stage III PE/AL	2010	0.1595	17.9	<0.001
Stage III PE/AL	2014	0.1711	19.37*	<0.001
Stage IV PE/AL	2010	0.1010	11.0*	0.001
Stage IV PE/AL	2014	0.0613	6.8	0.011
Stage III PC	2010	0.4626	75.8*	<0.001
Stage III PC	2014	0.2710	34.1*	<0.001
Stage IV PC	2010	0.5021	90.8*	<0.001
Stage IV PC	2014	0.8055	369.7*	<0.001

Notes: Results of linear regression models using total initial (2005) regeneration density as the independent variable and total subsequent regeneration density as the dependent variable. Data are separated by pre-disturbance composition (PE/AL, *Picea engelmannii/Abies lasiocarpa* or PC, *Pinus contorta*) and stage of structural development. An asterisk denotes significance at the 1% level.

Table 7. Regeneration dynamics for stands of varying patterns of initial regeneration of *Abies lasiocarpa* (AL), *Picea engelmannii* (PE), *Pinus contorta* (PC), and *Populus tremuloides* (PT), with results of four linear regression models using total initial regeneration density as the independent variable and total subsequent regeneration density as the dependent variable.

Composition of early regeneration (2003–2005)	Subsequent year	Adjusted R^2	F	P
PE/AL ≥1000/ha	2010	0.3527	72.4*	<0.001
PE/AL ≥1000/ha	2014	0.3138	60.9*	<0.001
PC ≥1000/ha	2010	0.266	18.48*	<0.001
PC ≥1000/ha	2014	0.1723	11.8*	0.001
PT ≥1000/ha	2010	0.1854	10.33*	0.003
PT ≥1000/ha	2014	0.3957	27.84*	<0.001
Total regeneration <1000/ha	2010	0.000	0.3	0.568
Total regeneration <1000/ha	2014	0.000	0.0	0.892

Note: Model data are categorized by plot initial regeneration species, regardless of pre-disturbance stand dominance and disturbance agent(s). An asterisk denotes significance at the 1% level.

exhibited high predictive power when initial regeneration density was high, frequently due to *P. tremuloides* after compounded disturbance. In contrast, models were weak when initial regeneration was sparse, a condition associated with compounded disturbance in this landscape (Kulakowski et al. 2013).

The findings of the current study partly reflect differences between stand-replacing and less severe disturbance agents. While even a high-severity blowdown typically will leave surviving individuals, a stand-replacing wildfire by definition consumes all vegetation. The role of legacies is critical to the trajectory of ecosystem processes and should receive special attention as management strategies are developed (Grumbine 1994, Buma and Wessman 2011, Sturtevant et al. 2014). Considering that all stands in this study were >120 yr old in the late 20th century, the presence of understory juveniles would be expected. Wind blowdown events tend to cause mortality among the tallest trees, while understory vegetation is typically protected (Veblen et al. 2001, Kulakowski and Veblen 2003). With the elimination of competing taller trees, the understory individuals are able to access more resources and are effectively released from competition, resulting in sustained increases in annual growth (Marks 1974, Kulakowski and Veblen 2003). These individuals would have established over the years preceding disturbance and have a competitive advantage over post-disturbance regeneration. The presence of legacies brings a steadiness to the trajectory of post-disturbance ecosystem development.

Regeneration following high-severity fires is free of competition from above-ground legacies but is more sensitive to environmental conditions. This fact may in part explain the observed fluctuations in regeneration density, shifts in species dominance, and the failure of some cohorts to persist (see St Clair et al. 2013). Cyclic wetting and drying and fluctuating temperatures are known inhibitors of early growth in aspen (McDonough 1985). Fluctuating temperatures may also influence seed production and availability—for example, current warming temperatures have been found to favor increased seed production of *Abies lasiocarpa* (Buechling et al. 2016). Post-disturbance climate appears to affect new seedlings more than established individuals

(Turner 2010, Dodson and Root 2015), and vegetative versus sexual regeneration may be differentially sensitive to these factors. Because of monotonic shifts in temperature and moisture conditions, plant community reshuffling is expected to occur more frequently under future climate change (Dodson and Root 2015). The results of the current study may demonstrate exactly such a case and emphasize the important role of legacies in influencing the predictability of ecosystem trajectories.

The persistence of initial regeneration varies across species based upon competing regeneration strategies that are found not only in North America, but in forests around the globe. *Abies lasiocarpa* in this study appear to be least sensitive to environmental fluctuations, consistently persisting after establishment, while *Picea engelmannii*, *P. contorta*, and most notably *P. tremuloides* each exhibited sudden establishment, decline, and failure to persist in some cases. In stands affected only by blowdown, this may be primarily a result of the fact that *A. lasiocarpa* are typically more abundant among the advance regeneration, creating a competitive advantage for this species following a disturbance that was not stand-replacing (Schmid and Hinds 1974). *Abies lasiocarpa* regeneration was consistent from year to year, even in burned stands, and even while other regeneration fluctuated at the same sites. *Abies lasiocarpa* is a shade-tolerant species, and establishes at much lower densities immediately after stand-replacing disturbances, if at all (Veblen 1986b, Bigler and Veblen 2009), while *P. contorta* and *P. tremuloides* thrive in open sunlight and are sensitive to changes in light availability (Calder 2009). Otherwise, *A. lasiocarpa* tend to recruit high numbers of individuals in early years of regeneration, but few of these reach maturity due to high juvenile mortality rates (Veblen 1986a, Antos et al. 2000).

Interacting disturbances may have compound effects, even causing a shift to an alternate stable state (Paine et al. 1998), but drivers of forest trajectories following multiple disturbances may include nuanced differences in stand attributes, species-specific regeneration strategies, and a fluctuating microenvironment. Initially, prolific regeneration of *P. tremuloides* in Routt National Forest suggested that favorability for *P. tremuloides* may provide a negative feedback to the frequency of fire and beetle disturbances (Kulakowski et al. 2013). However, in some stand types, initial indications of trajectories have already shifted in a brief 13-yr period of post-fire stand development, highlighting the complexity of estimating trajectories in an environment with highly interactive disturbance regimes. The need to understand compound effects of multiple disturbances is of utmost importance (Turner 2010). The effects of multiple disturbances on regeneration are not necessarily synergistic, but depend on species regeneration strategies and the interactions of multiple species (Buma and Wessman 2012). To this, we add that the response varies not only by species, but also with pre-disturbance structural stage (White et al. 2015) and potentially with fluctuating environmental conditions.

Conclusions

Often, the regeneration density in the years immediately after a disturbance is indicative of the future trajectory of an ecosystem, but in some cases, particularly following stand-replacing or compounded disturbances, initial regeneration density and composition can be poor indicators of future trends. Competitive advantages in an environment void of advance regeneration, such as that following stand-replacing disturbance, may shift with subtle changes in resource availability caused by weather patterns, climate, or other factors. In contrast, regeneration in stands that have legacies that survive a disturbance tends to be consistent from the years immediately after the event into the next decades. Future research should measure the degree to which changing climate conditions may affect trends in post-disturbance regeneration, as these trends have important implications for forest management and ecosystem modeling. Patterns of regeneration after fire can directly influence patterns of future fire severity. Additionally, drier climatic conditions expected with climate change may alter fire regimes such that intervals between fires become shorter, decreasing the window of opportunity for the recruitment of fire-intolerant woody plants (Westerling et al. 2011). Often, management plans and models of future carbon stocks and other resources are based solely upon as little as one year of post-disturbance stand development, and rarely more than five years. As high-severity and compounded disturbances become more common under climatically driven changes in disturbance regimes, post-disturbance ecosystem trajectories may become increasingly stochastic and unpredictable, particularly when densities are low during initial years of regeneration.

Acknowledgments

This work was supported by the National Science Foundation under grants 1262691 and 1262687 and by a National Science Foundation Graduate Research Fellowship. NG, DK, and TV conceived the ideas and designed methodology; NG, DK, TV, and DJ collected the data; NG, SP, and DK analyzed the data; NG led the writing of the manuscript. All authors contributed critically to the drafts and gave final approval for publication.

Literature Cited

Antos, J., R. Parish, and K. Conley. 2000. Age structure and growth of the tree-seedling bank in subalpine spruce-fir forests of south-central British Columbia. American Midland Naturalist 143:342–354.

Baker, S. C., M. Garandel, M. Deltombe, and M. G. Neyland. 2013. Factors influencing initial vascular plant seedling composition following either aggregated retention harvesting and regeneration burning or burning of unharvested forest. Forest Ecology and Management 306:192–201.

Bentz, B. J., J. Régnière, C. J. Fettig, E. M. Hansen, J. L. Hayes, J. A. Hicke, R. G. Kelsey, J. F. Negron, and S. J. Seybold. 2010. Climate change and bark beetles of the western United States and Canada: direct and indirect effects. BioScience 60:602–613.

Bigler, C., and T. T. Veblen. 2009. Increased early growth rates decrease longevities of conifers in subalpine forests. Oikos 118:1130–1138.

Buechling, A., P. H. Martin, C. D. Canham, W. D. Shepperd, and M. A. Battaglia. 2016. Climate drivers of seed production in *Picea engelmannii* and response to warming temperatures in the southern Rocky Mountains. Journal of Ecology 104:1051–1062.

Buma, B., and C. A. Wessman. 2011. Disturbance interactions can impact resilience mechanisms of forests. Ecosphere 2:64.

Buma, B., and C. A. Wessman. 2012. Differential species responses to compounded perturbations and implications for landscape heterogeneity and resilience. Forest Ecology and Management 266:25–33.

Calder, W. J. 2009. Ecophysiological mechanisms underlying aspen to conifer succession. Brigham Young University Scholars Archive: All Theses and

Dissertations. Paper 2307. Brigham Young University, Provo, Utah, USA.

Collins, B. M., and G. B. Roller. 2013. Early forest dynamics in stand-replacing fire patches in the northern Sierra Nevada, California, USA. Landscape Ecology 28:1801–1813.

Connell, J. H. 1978. Diversity in tropical rain forests and coral reefs. Science 199:1302–1310.

DeRose, R. J., and J. N. Long. 2009. Wildfire and spruce beetle outbreak: simulation of interacting disturbances in the central Rocky Mountains. Ecoscience 16:28–38.

Dodson, E. K., and H. T. Root. 2015. Native and exotic plant cover vary inversely along a climate gradient 11 years following stand-replacing wildfire in a dry coniferous forest, Oregon, USA. Global Change Biology 21:666–675.

Donato, D. C., J. B. Fontaine, J. L. Campbell, W. D. Robinson, J. B. Kauffman, and B. E. Law. 2009b. Conifer regeneration in stand-replacement portions of a large mixed-severity wildfire in the Klamath-Siskiyou Mountains. Canadian Journal of Forest Research 39:823–838.

Donato, D. C., J. B. Fontaine, W. D. Robinson, J. B. Kauffman, and B. E. Law. 2009a. Vegetation response to a short interval between high-severity wildfires in a mixed-evergreen forest. Journal of Ecology 97:142–154.

Egler, F. E. 1954. Vegetation science concepts I. Initial floristic composition, a factor in old-field vegetation development with 2 figs. Vegetatio 4:412–417.

Enright, N. J., J. B. Fontaine, D. M. J. S. Bowman, R. A. Bradstock, and R. J. Williams. 2015. Interval squeeze: Altered fire regimes and demographic responses interact to threaten woody species persistence as climate changes. Frontiers in Ecology and the Environment 13:265–272.

Evangelista, P. H., S. Kumar, T. J. Stohlgren, and N. E. Young. 2011. Assessing forest vulnerability and the potential distribution of pine beetles under current and future climate scenarios in the Interior West of the US. Forest Ecology and Management 262:307–316.

Finegan, B. 1984. Forest succession. Nature 312:109–114.

Fox, J. W. 2013. The intermediate disturbance hypothesis should be abandoned. Trends in Ecology & Evolution 28:86–92.

Frelich, L. E., and P. B. Reich. 2009. Will environmental changes reinforce the impact of global warming on the prairie–forest border of central North America? Frontiers in Ecology and the Environment 8:371–378.

Grimm, V., and S. F. Railsback. 2013. Individual-based model and ecology. Princeton University Press, Princeton, New Jersey, USA.

Grumbine, E. R. 1994. What is ecosystem management? Conservation Biology 8:27–38.

Holl, K. D., and J. J. Cairns. 2002. Monitoring and appraisal. Pages 411–432 in M. R. Perrow and A. J. Davy, editors. Handbook of ecological restoration. Cambridge University Press, Cambridge, UK.

Howard, J. L. 1996. *Populus tremuloides*. Fire effects information system. USDA Forest Service, Rocky Mountain Research Station, Fire Sciences Laboratory, Missoula, Montana, USA. https://www.fs.fed.us/database/feis/plants/tree/poptre/all.html

Jogiste, K., et al. 2017. Hemiboreal forest: natural disturbances and the importance of ecosystem legacies to management. Ecosphere 8:e01706.

Kay, C. E. 1993. Aspen seedlings in recently burned areas of Grand Teton and Yellowstone National Parks. Northwest Science 67:94–104.

Kulakowski, D., C. Matthews, D. Jarvis, and T. T. Veblen. 2013. Compounded disturbances in subalpine forests in western Colorado favor future dominance by quaking aspen (*Populus tremuloides*). Journal of Vegetation Science 24:168–176.

Kulakowski, D., and T. T. Veblen. 2002. Influences of fire history and topography on the pattern of a severe wind blowdown in a Colorado subalpine forest. Journal of Ecology 90:806–819.

Kulakowski, D., and T. T. Veblen. 2003. Subalpine forest development following a blowdown in the Mount Zirkel Wilderness, Colorado. Journal of Vegetation Science 14:653–660.

Kulakowski, D., and T. T. Veblen. 2015. Bark beetles and high-severity fires in Rocky Mountain subalpine forests. Pages 149–174 in A. Dellasalla and C. T. Hanson, editors. Mixed-high severity fires: ecosystem processes and biodiversity. Elsevier, Amsterdam, The Netherlands.

Maher, E. L., M. J. Germino, and N. J. Hasselquist. 2005. Interactive effects of tree and herb cover on survivorship, physiology, and microclimate of conifer seedlings at the alpine tree-line ecotone. Canadian Journal of Forest Research 35:567–574.

Marks, P. L. 1974. The role of pin cherry (*Prunus pensylvanica* L.) in the maintenance of stability in northern hardwood ecosystems. Ecological Monographs 44:73–88.

McDonough, W. T. 1985. Sexual reproduction, seeds, and seedlings. Pages 25–28 in N. V. DeByle and R. P. Winokur, editors. Aspen: ecology and management in the western United State. USDA Forest Service General Technical Report RM-119. Rocky Mountain Forest and Range Experiment Station, Fort Collins, Colorado, USA.

Nagel, T. A., M. Svoboda, and M. Kobal. 2014. Disturbance, life history traits, and dynamics in an old-growth forest landscape of southeastern Europe. Ecological Applications 24:663–679.

Oliver, C. D. 1981. Forest development in North America following major disturbances. Forest Ecology and Management 3:153–168.

Paine, R. T., M. J. Tegner, and E. A. Johnson. 1998. Compounded perturbations yield ecological surprises. Ecosystems 1:535–545.

Parker, A. J., and K. C. Parker. 1983. Comparative successional roles of trembling aspen and lodgepole pine in the southern Rocky Mountains. Great Basin Naturalist 43:447–455.

Peet, R. K. 1981. Forest vegetation of the Colorado Front Range. Vegetatio 45:3–75.

Pickett, S. T. A., M. L. Cadenasso, and S. Bartha. 2001. Implications from the Buell-Small Succession Study for vegetation restoration. Applied Vegetation Science 4:41–52.

Pickett, S. T. A., M. L. Cadenasso, and S. J. Meiners. 2005. Vegetation dynamics. Pages 172–198 in E. van der Maarel, editor. Vegetation ecology. Blackwell Science, Malden, Massachusetts, USA.

Pickett, S. T. A., and P. S. White. 1985. The ecology of natural disturbances and patch dynamics. Academic Press, Cambridge, Massachusetts, USA.

Quinn, R. O., and L. Wu. 2001. Quaking aspen reproduce from seed after wildfire in the mountains of southeastern Arizona. Pages 369–376 in W. D. Shepperd, D. Binkley, D. L. Bartos, T. J. Stohlgren, and L. G. Eskew, editors. Sustaining aspen in western landscapes: Symposium proceedings, 13–15 June 2000, Grand Junction, CO. Proceedings RMRS-P-18. USDA Forest Service, Rocky Mountain Research Station, Fort Collins, Colorado, USA.

R Development Core Team. 2008. R: a language and environment for statistical computing. R Foundation for Statistical Computing, Vienna, Austria.

Romme, W. H., M. G. Turner, R. H. Gardner, W. W. Hargrove, G. A. Tuskan, D. G. Despain, and R. A. Renkin. 1997. A rare episode of sexual reproduction in aspen. Natural Areas 17:17–25.

Rowe, J. S. 1983. Concepts of fire effects on plant individuals and species. Pages 135–151 in R. Wein and D. A. MacLean, editors. The role of fire in northern circumpolar ecosystems. John Wiley and Sons, Hoboken, New Jersey, USA.

Royo, A. A., C. J. Peterson, J. S. Stanovick, and W. P. Carson. 2016. Evaluating the ecological impacts of salvage logging: Can natural and anthropogenic disturbances promote coexistence? Ecology 97:1566–1582.

Schmid, J. M., and T. E. Hinds. 1974. Development of spruce-fir stands following spruce beetle outbreaks. USDA Forest Service Research Paper RM-13. Rocky Mountain Forest and Range Experiment Station, Fort Collins, Colorado, USA.

Simard, M., W. H. Romme, J. M. Griffin, and M. G. Turner. 2011. Do mountain pine beetle outbreaks change the probability of active crown fire in lodgepole pine forests? Ecological Monographs 81:3–24.

Snyder, G. L., L. L. Patten, and J. J. Daniels. 1987. Mineral resources of the Mount Zirkel Wilderness and northern Park Range vicinity, Jackson and Routt counties, Colorado. Bulletin 1554. U.S. Geological Survey, Washington, D.C., USA.

Sousa, W. P. 1984. The role of disturbance in natural communities. Annual Review of Ecology and Systematics 15:353–391.

St Clair, S. B., X. Cavard, and Y. Bergeron. 2013. The role of facilitation and competition in the development and resilience of aspen forests. Forest Ecology and Management 299:91–99.

Sturtevant, B. R., B. R. Miranda, P. T. Wolter, P. M. A. James, M. J. Fortin, and P. A. Townsend. 2014. Forest recovery patterns in response to divergent disturbance regimes in the Border Lakes region of Minnesota (USA) and Ontario (Canada). Forest Ecology and Management 313:199–211.

Svoboda, M., et al. 2014. Landscape-level variability in historical disturbance in primary Picea abies mountain forests of the Eastern Carpathians, Romania. Journal of Vegetation Science 25:386–401.

Tinker, D. B., W. H. Romme, W. W. Hargrove, R. H. Gardner, and M. G. Turner. 1994. Landscape-scale heterogeneity in lodgepole pine serotiny. Canadian Journal of Forest Research 24:897–903.

Turner, M. G. 2010. Disturbance and landscape dynamics in a changing world. Ecology 91:2833–2849.

Turner, M. G., D. C. Donato, and W. H. Romme. 2013. Consequences of spatial heterogeneity for ecosystem services in changing forest landscapes: priorities for future research. Landscape Ecology 28:1081.

Van Leeuwen, W. J. D., G. M. Casady, D. G. Neary, S. Bautista, J. A. Alloza, Y. Carmel, L. Wittenberg, D. Malkinson, and B. J. Orr. 2010. Monitoring post-wildfire vegetation response with remotely sensed time-series data in Spain, USA and Israel. International Journal of Wildland Fire 19:75–93.

Van Mantgem, P. J., and N. L. Stephenson. 2005. The accuracy of matrix population model projections for coniferous trees in the Sierra Nevada, California. Journal of Ecology 93:737–747.

Veblen, T. T. 1986a. Treefalls and the coexistence of conifers in subalpine forests of the central Rockies. Ecology 67:644–649.

Veblen, T. T. 1986b. Age and size structure of subalpine forests in the Colorado Front Range. Bulletin of the Torrey Botanical Club 113:225–240.

Veblen, T. T., K. S. Hadley, E. M. Nel, T. Kitzberger, M. Reid, and R. Villalba. 1994. Disturbance regime and disturbance interactions in a Rocky Mountain subalpine forest. Journal of Ecology 82: 125–135.

Veblen, T. T., K. S. Hadley, M. S. Reid, and A. J. Rebertus. 1991. The response of subalpine forests to spruce beetle outbreak in Colorado. Ecology 72:213–231.

Veblen, T. T., D. Kulakowski, K. S. Eisenhart, and W. L. Baker. 2001. Subalpine forest damage from a severe windstorm in northern Colorado. Canadian Journal of Forest Research 31:2089–2097.

Westerling, A. L., H. G. Hidalgo, D. R. Cayan, and T. W. Swetnam. 2006. Warming and earlier spring increase western U.S. forest wildfire activity. Science 313:940–943.

Westerling, A. L., M. G. Turner, E. A. H. Smithwick, W. H. Romme, and M. G. Ryan. 2011. Continued warming could transform Greater Yellowstone fire regimes by mid-21st century. Proceedings of the National Academy of Sciences 108:13165–13170.

White, S. D., J. L. Hart, C. J. Schweitzer, and D. C. Dey. 2015. Altered structural development and accelerated succession from intermediate-scale wind disturbance in Quercus stands on the Cumberland Plateau, USA. Forest Ecology and Management 336:52–64.

Wilson, J., H. Gitay, S. Roxburgh, W. M. King, and R. Tangney. 1992. Egler's concept of 'Initial Floristic Composition' in succession: Ecologists citing it don't agree what it means. Oikos 64:591–593.

Xiang, W., S. Liu, S. C. Frank, D. Tian, G. Wang, and X. Deng. 2013. Secondary forest floristic composition, structure, and spatial pattern in subtropical China. Journal of Forest Research 18:111–120.

Zald, H. S. J., A. N. Gray, M. North, and R. A. Kern. 2008. Initial tree regeneration responses to fire and thinning treatments in a Sierra Nevada mixed-conifer forest, USA. Forest Ecology and Management 256:168–179.

4

Biogeomorphic impact of oligochaetes (Annelida) on sediment properties and *Salicornia* spp. seedling establishment

M. van Regteren [iD],[1,2,†] R. ten Boer,[1] E. H. Meesters,[1] and A. V. de Groot[1]

[1]*Wageningen Marine Research, Wageningen University & Research, Ankerpark 27, 1781 AG Den Helder, The Netherlands*
[2]*Environmental Sciences Group, Wageningen University & Research, Postbus 47, 6700 AA Wageningen, The Netherlands*

Abstract. Oligochaetes (Annelida) are active bioturbators that can be present in high densities in the transition zone between intertidal flats and salt marshes, though their occurrence and functional role remain understudied. This study aimed to clarify the biogeomorphic role of oligochaete bioturbation in facilitating or hindering vegetation establishment. Two microcosm experiments were performed to assess the effect of oligochaete bioturbation on sediment properties, oxidation depth, algal biomass, seed distribution, and germination success of pioneer species *Salicornia* spp. Oligochaetes created burrow networks in the sediment matrix, which, together with upward conveyor belt feeding, lead to substrate mixing. Sediment reworking rates of oligochaetes were compared with those of polychaete macrofauna. Bioturbation and bio-irrigation of burrows can stimulate resource flows into the sediment. Oxidation depth increased almost tenfold in the presence of oligochaetes. Their bioturbation did not seem to affect sediment properties such as dry bulk density, porosity, and organic matter content. Sediment reworking, however, significantly reduced algal biomass at the surface with possible cascading effects on sediment stability and erodibility. Oligochaete conveyor belt feeding buried *Salicornia* spp. seeds until below the critical germination depth, thus negatively affecting *Salicornia* spp. germination and seedling establishment. Our study indicates that small, though numerous, oligochaete bioturbators may reduce lateral expansion potential of salt marshes by hindering the establishment of pioneer vegetation in the transition zone. Additionally, in dynamic fine-grained habitats, these oligochaetes have the feature to quickly oxygenate the sediment top layer.

Key words: bioturbation; oligochaetes; oxidation depth; pioneer vegetation; *Salicornia;* salt marsh; intertidal flat.

† **E-mail:** marin.vanregteren@wur.nl

Introduction

Salt marshes and their adjoining intertidal flats are vital areas in coastal systems. They harbor unique plant communities, host migratory and breeding birds and also function as coastal protection through wave attenuation (Bouma et al. 2005, Reise et al. 2010, Temmerman et al. 2013, Schuerch et al. 2014). Sea-level rise and increased inclemency call for improved coastal defenses (Möller 2006, Bouma et al. 2014). Ecosystem-based coastal defense, such as salt marshes, provides a sustainable and cost-effective improvement (Temmerman et al. 2013). Salt marshes are often fixed on their landward side by hard structures, such as dikes, which cause salt-marsh expansion to be possible only on the seaward side (Elias et al. 2012). Lateral expansion is critical for the preservation of salt marshes (Balke et al. 2016, Bouma et al. 2016). Most studies, however, have focused on vertical marsh accretion (Kirwan and Guntenspergen 2010, Mariotti and Fagherazzi 2010, Fagherazzi et al. 2012, Marani et al. 2013). Lateral salt-marsh expansion is broadly considered to be a dynamic process with alternating periods of erosion and re-establishment of vegetation (van de Koppel et al. 2005, Schuerch et al. 2014).

Lateral salt-marsh expansion is governed by sediment dynamics and vegetation establishment in the transition zone (Friedrichs and Perry 2001, Meysman et al. 2006, Van Wesenbeeck et al. 2007, Schuerch et al. 2014). Salt-marsh vegetation increases sedimentation rates by impeding water flow, and increases soil stability by binding sediment with their roots, aiding vertical as well as lateral expansion (Gerdol and Hughes 1993, Van Wesenbeeck et al. 2008b, van der Wal and Herman 2012). In many salt-marsh transition zones, the main pioneer plant species is the strong ecosystem engineer *Spartina anglica*. Once established, its tussocks induce a local positive feedback, initiating bed-level elevation, tussock expansion, and eventually, marsh succession (Bouma et al. 2005, 2013, Van Wesenbeeck et al. 2008a). In many other transition zones, the first colonizer of bare intertidal flats is the annual species *Salicornia* spp., which can also facilitate the accretion of fine sediment (Boorman et al. 2001, Wolters et al. 2005). However, sedimentation can also be detrimental for its emergence: *Salicornia* cannot germinate when burrowed deeper than 1 cm below the sediment surface (Gerdol and Hughes 1993, Wirtz 1994). A reduction in *Salicornia* development as a result of high accretion or burrowing rates may inhibit lateral salt-marsh expansion.

Both vegetation establishment and sediment dynamics are affected by interactions between vegetation and benthos (Fagherazzi et al. 2012). For example, *Corophium volutator* can inhibit *Salicornia* development directly through burial of seeds, but also indirectly by disturbing the sediment matrix, which prevents root anchorage (Gerdol and Hughes 1993). This, in turn, reduces sediment stability and deposition. Similarly, the lugworm *Arenicola marina* destabilizes the sediment, which impairs root growth and anchorage of *S. anglica* (Van Wesenbeeck et al. 2007). On the other hand, *S. anglica* patches are high in silt content and contain a dense root network, which is unfavorable for the bioturbating lugworm. This results in a clear distinction between areas dominated by *S. anglica* and those by *A. marina*, because both species modify their habitat to their own benefit, negatively affecting the other (Van Wesenbeeck et al. 2007). *Hediste diversicolor*

buries cordgrass seeds and consumes them once sprouted (Zhu et al. 2016). When not consumed, though, these buried seeds are retained in the soil and thereby protected from hydrodynamics (Zhu et al. 2016). *Tubifex tubifex*, a limnetic oligochaete, has been shown to exert a positive influence on macrophyte growth through oxygen stress amelioration (Mermillod-Blondin and Lemoine 2010). As pioneer vegetation establishment on the intertidal flats can be limited by macrofauna bioturbation (Gerdol and Hughes 1993, Van Wesenbeeck et al. 2007), it might similarly be obstructed by bioturbation of small, though numerous, oligochaetes.

Sediment reworking rates by oligochaetes range from 0.003 (Ravera 1955) to 0.49 (Matisoff et al. 1999) $cm \cdot d^{-1} \cdot (100,000 \text{ individuals})^{-1} \cdot m^{-2}$, and abundance can range up to a million per m^2 (Bagheri and McLusky 1982, Giere 2006). Although oligochaetes are significant bioturbators, their abundance and ecological role are hardly ever studied (Evans et al. 1979, McCann 1989, Giere 1993, Seys et al. 1999). A likely cause is that these worms belong to meiofauna that are regularly not included in research (Seys et al. 1999, Chen et al. 2016). Most studies performed on their bioturbation capacities are of the freshwater model species *T. tubifex* (Fisher et al. 1980, Reible et al. 1996, Mermillod-Blondin et al. 2004, Dafoe et al. 2011). Sediment reworking, bioturbation, and bio-irrigation cause cascading effects on resource flows and microbial activity in the sediment (Mermillod-Blondin et al. 2005, Kristensen et al. 2012, Pigneret et al. 2016). Understanding the interaction effects between sediment reworking and aboveground vegetation helps clarify processes that govern lateral marsh expansion (Meysman et al. 2006).

We studied the role of bioturbating oligochaetes in facilitating or hindering vegetation establishment in the transition zone from salt marsh to intertidal flats. The salt marsh at Westhoek, Friesland (The Netherlands), is a naturally expanding marsh and therefore an excellent location to study expansion processes in the transition zone. A pilot study in the transition zone at Westhoek revealed that a major component of the species abundance and diversity consisted of marine or brackish oligochaetes. Using microcosm laboratory experiments, this study examined whether oligochaetes play a biogeomorphic role in pioneer seedling establishment. For this purpose, we tested whether oligochaete bioturbation affects algal biomass, chemical and bulk sediment properties, and whether it affects pioneer *Salicornia* seed germination and seedling establishment.

METHODS

The study consisted of two microcosm experiments. The first experiment was conducted to test the effects of oligochaete bioturbation on abiotic sediment characteristics. The second experiment consisted of a full-factorial design to determine the role of oligochaetes in *Salicornia* germination and seedling establishment. Algal biomass measurements were non-invasive and therefore performed in both experiments.

Sediment properties: experimental set-up

Sediment was collected at the intertidal flats at Westhoek, Province of Friesland (53°16.31 N, 5°33.14 E, The Netherlands, 2016). A naturally expanding salt marsh in its early stages is located at Westhoek, predominated by pioneer, low and middle marsh species, and inundated on average once per day. The salt marsh is located in the Wadden Sea area and water is at times brackish through freshwater discharge at Kornwerderzand sluices and harbor of Harlingen. Sediment was sieved through a 1-mm sieve to remove all large particles and fauna. Sediment, 70% mud (<63 μm), was mechanically homogenized and left at 3°C until further use. Anoxic sediment was transferred into 28 transparent microcosm cores of 4.4 cm diameter and 20 cm long. Microcosms with anoxic sediment were left overnight to oxidize the top layer, after which oxidation depth was measured from four fixed points. Subsequently, the microcosms were transferred into a tidal basin and subjected to a simulated tidal regime of 2 h of submergence every 24 h and a 12-h light/dark period, corresponding to spring conditions at the source. Oligochaetes were collected at Westhoek by gathering clumps of sediment with worms visible. Extraction of oligochaetes was performed by diluting sediment clumps with brackish water (15 PSU), after which suspended sediment was transferred into a transparent tray and placed on a light table. Live oligochaetes were selected from the tray, counted, and introduced into the treatment microcosms the

same day ($n = 14$, 200 oligochaetes per microcosm: resembling a density of 131.493 individuals/m^2). To the control microcosms ($n = 14$), no worms were added. It was in this stage not possible to discriminate between species of oligochaetes (van Haaren 2016). Species composition was, afterwards, determined to be approximately 60% *Heterochaeta costata* and 40% *Enchytraideae* spp. Visual inspection revealed that all worms had burrowed within a few hours. The microcosms were systematically placed, control and treatment microcosms alternating, into a tidal basin in a 15°C climate chamber. Water salinity was 16.4 ± 0.7‰ and water temperature was 14.0° ± 0.2°C during the experiment. After 36 d, the experiment was terminated. Burrows were clearly visible by brown oxidized sediment in the otherwise black anoxic matrix (Mermillod-Blondin et al. 2008, Mermillod-Blondin and Lemoine 2010, De Lucas Pardo et al. 2013). Oxidation depth was measured from the same four fixed points and split up into solid oxidation and deep burrow oxidation (Fig. 1a). Solid oxidation refers to the uninterrupted brown oxidized layer, measured from the surface. Deep burrow oxidation was the depth of the interrupted brown oxidized burrows at four fixed points (Fig. 1a). Solid and deep burrow oxidation had the same value when burrows were absent. Algal biomass, consisting of diatoms, cyanobacteria, and green algae, was measured at the surface with a BenthoTorch fluorometer (bbe Moldaenke GmBH, Schwentinental, Germany). Subsequently, all microcosms were frozen to halt ammonium formation and breakdown processes and stored at −20°C until further processing.

Measurements

From the 28 experimental units, 10 control and 10 treatment cores were sliced to determine bulk sediment properties. Slice intervals were 0–1, 1–2, 2–3, 3–4, 4–5, 5–6, 6–7, and 7–8 cm. The remaining sediment in the microcosm (8–20 cm) was not analyzed because the oligochaetes concentrate mainly in the upper 5 cm (Appleby and Brinkhurst 1970, McCann 1989, Seys et al. 1999). The outside of the frozen cores was first briefly thawed with hot water; then with a piston, we gently extruded the sediment bottom upward until the specific interval was reached and sawn off. Slice thickness was measured at four points to calculate the volume of each individual slice (V). Each slice was weighed to

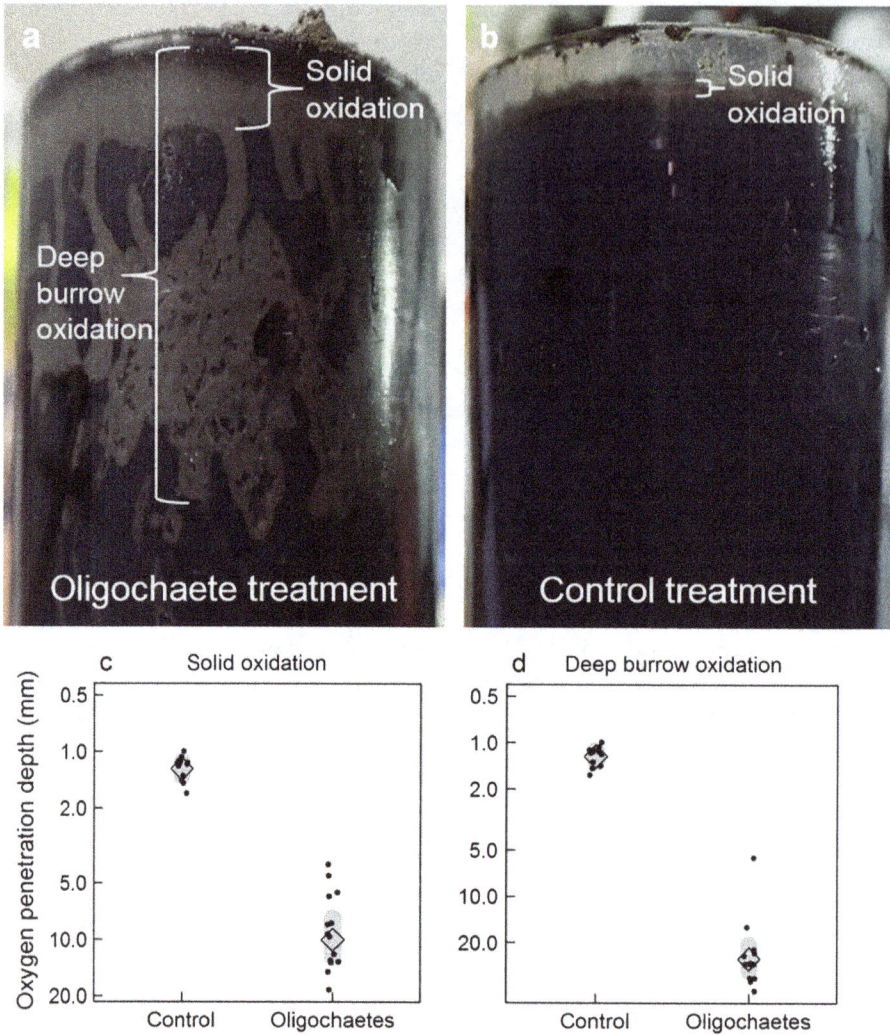

Fig. 1. Examples of oxidation depth, solid oxidation, and deep burrow oxidation are indicated, in a core (a) with oligochaetes and (b) control treatment without oligochaetes after 36 d. Oxidation depth (log-scale) after 36 d for (c) solid oxidation and (d) deep burrow oxidation, $n = 14$ per treatment. Treatments are control, that is, without oligochaetes, and oligochaetes with on average 200 worms per microcosm. Diamonds indicate the mean, gray bars indicate the 95% confidence limits, and black dots depict individual measurements.

obtain the wet weight (WW) and samples were dried for 24 h at 105°C to obtain dry weight (DW; Compton et al. 2013). Subsequently, they were incinerated for 6 h at 560°C to obtain the ash-free dry weight (AFDW; Compton et al. 2013). Mass water content and dry bulk density were calculated as follows (Flemming and Delafontaine 2000):

$$w_a\,(\%) = 100 \times \left(\frac{WW - DW}{WW}\right) \qquad (1)$$

$$BD_{dry} = \frac{DW}{V} \qquad (2)$$

The density of the solids (ρ_s) in natural sediment mixtures is assumed to be 2.65 g/cm^3 (De Groot et al. 2009). Porosity was calculated as follows (De Groot et al. 2009):

$$\varepsilon = \frac{\rho_s - BD_{dry}}{\rho_s} \qquad (3)$$

Organic matter content in weight percent was calculated with the loss on ignition (LOI) method as follows (Heiri et al. 2001):

$$LOI = 100 \times \left(\frac{DW - AFDW}{DW} \right) \qquad (4)$$

Water and organic matter content values were more accurate than those of bulk density and porosity as they were not influenced by possible inaccuracy in volumetric measurements.

Pore-water ammonium

The remaining four control and four oligochaete treatment microcosms were used for pore-water ammonium (NH_4^+) analysis. Cores were sliced into intervals at 0–1, 1–2, 2–3, 3–4, and 4–5 cm. Sediment from each slice was transferred into a tube and centrifuged for 60 min at 1937 g. Pore water was drained, filtered, and diluted threefold. Analysis of NH_4^+ was performed with a Spectroquant Merck kit (Ammonium Cell Test 1.14558.0001, Darmstadt, Germany). Analysis results from one control microcosm were discarded because of a dilution error.

Salicornia germination: experimental set-up

Twenty-eight microcosm sediment cores were collected at Westhoek. Microcosms were defaunated by freezing them at $-20°C$ for 15 d. The sediment was purposely not homogenized to resemble the field situation. Oligochaetes were collected in the same way as described in the previous experiment. *Salicornia* seeds were collected in October 2015. The experiment was conducted in a full-factorial design with seven microcosms per treatment and two factors: oligochaetes and *Salicornia* seeds (Table 1). Seventy-five seeds were added to the seeds (S) and seeds and oligochaetes (SO) treatments, and 200 oligochaetes were added to the oligochaete (O and SO) treatments, respectively. The control treatment (C) contained neither seeds nor oligochaetes. The microcosms were exposed to the same conditions as the previous experiment. Water salinity was $15.9 \pm 0.8‰$ and water temperature $13.9° \pm 0.3°C$ during the experiment. After three days, oligochaetes were introduced into the O and SO treatment microcosms. On day 4, *Salicornia* seeds were added to the S and SO treatments. Studies indicate that a window of

Table 1. Amount of oligochaetes and *Salicornia* seeds retrieved from the microcosms after experiment termination.

Treatments	Oligochaetes		*Salicornia* seeds	
	Added	Retrieved	Added	Retrieved
Seeds and oligochaetes (SO)	1400	1194	525	345
Oligochaetes (O)	1400	1338	0	0
Control (C)	0	16	0	0
Seeds (S)	0	20	525	426
Outside microcosms	—	85	—	41
Total	2800	2653	1050	812

Note: Outside microcosms were seeds and worms that have drifted out of the microcosms with the simulated tide.

opportunity without flooding is necessary for the establishment of pioneer vegetation on an intertidal flat (Balke et al. 2011). To simulate this, the cores were not submersed for the first five days after seed addition, and to initiate germination, 3 mL of freshwater was added to each core during the first three days (Wirtz 1994, Hu et al. 2015). Subsequently, the tidal cycle with 2 h submergence every 24 h continued for the remainder of the experiment and 1 mL freshwater was added to each core three times a week to simulate rainfall. The experiment was terminated 33 d after seed addition, and algal biomass was measured. To terminate oligochaete activity and fix their position in the sediment matrix, microcosms were frozen at $-80°C$ and stored at $-20°C$ until further processing. The microcosms were sectioned into slices of 0–0.5, 0.5–1, 1–2, 2–3, 3–4, 4–5, 5–6, and 6–12 cm. Each layer was sieved through a 300-µm sieve and placed in a transparent tray on a light table. Oligochaetes and *Salicornia* seeds were counted. *Salicornia* seeds were classified into three separate groups: intact seeds, germinated seeds, and seedlings. Seeds were classified as germinated when the radicle emerged (Khan et al. 2000, Carter and Ungar 2003), and seeds were considered a seedling when the capsule was shed and the cotyledons were visible (Davy et al. 2001).

Data analyses

Data were analyzed taking into account several possible sources of correlation between the data. Spatial correlation may exist between slices at different depths in the same microcosm. To model these sources of variation, we used linear and additive mixed- effect models (Pinheiro and Bates 2000, Wood 2006). Additionally, to allow for heterogeneity between different treatments, control and oligochaetes, in these models, a variance factor was included. Oxidation depth and algal biomass data were analyzed using generalized least squares (GLS) to allow for unequal variances (heteroscedasticity) in the treatments (Pinheiro and Bates 2000). Likelihood ratio (LR) tests were used for model selection and to assess significance of individual factors. All mean values are accompanied by their 95% confidence limits (CL). The protocol by Zuur et al. (2010) was used for data exploration and assessing the assumptions of the analyses. Distribution of oligochaetes is known to be non-linear over depth (Fisher et al. 1980, Seys et al. 1999, Mermillod-Blondin and Lemoine 2010); therefore, their effect on the sediment matrix might also be non-linear. To allow for non-linearity, models were constructed using generalized additive mixed models for the analyses of mass water content and ammonium. Microcosm was added as a random effect to allow for correlation between samples from the same microcosm. Mean slice depth was used to construct an autocorrelation structure to model spatial (i.e., sediment depth) correlation among observations (continuous autoregressive model, corCAR1 class, Pinheiro and Bates 2000). Essentially, this means that the correlation between samples decreases exponentially as a function of the distance between them. Model assumptions (e.g., homogeneity of variances and normality of residuals) were assessed graphically. Exploratory analyses gave no indication for non-linearity in the remaining sediment properties (bulk density, porosity, and organic matter content) which were therefore analyzed using linear mixed-effects models. All statistical analyses were performed with the statistical program R (R Core Team 2015) using additional packages lattice (Sarkar 2008), mgcv (Wood 2006), and nlme (Pinheiro et al. 2016).

RESULTS

Sediment properties: chemical sediment properties

Oligochaetes significantly increased the solid oxidation depth of the sediment matrix (GLS,

$LR_{df=1} = 23.9$, $P < 0.0001$; Fig. 1). All microcosms had an oxidized layer of 1 mm before the addition of worms. After five days, a burrow network became apparent, which expanded over time. Oxidation depth gradually increased in microcosms with oligochaetes present. At the end of the experiment, mean oxidation depth was 1.2 mm (CL: 1.1, 1.3, $n = 14$) in the control treatment, while solid oxidation was 10.0 mm (CL: 7.7, 12.4) and deep burrow oxidation was 25.4 mm in the oligochaete treatment (CL: 20.5, 30.2; GLS, $LR_{df=1} = 32.2$, $P < 0.0001$; Fig. 1). Variance in solid oxidation depth was approximately 22 (CL: 13, 39) times greater and the variance in deep burrow oxidation was 48 times greater (CL: 28, 84) with bioturbating worms than without (Fig. 1).

There was no difference in ammonium concentrations with and without oligochaetes (ANOVA, $LR_{df=1} = 1.09$, $P = 0.30$; Fig. 2). Mean pore-water ammonium concentrations were 0.52 mmol/L (CL: 0.41, 0.62) and 0.44 mmol/L (CL: 0.35, 0.54) for, respectively, control and oligochaetes. In either treatment, ammonium concentration increased non-linearly with depth (Fig. 2). Ammonium concentrations were lower closer to the sediment surface and gradually increased below 2 cm depth. Statistical models needed a correlation structure because of the dependence among samples. Generally, samples from two adjacent slices have a correlation of 0.86 (CL: 0.64, 0.95), which decreases exponentially with the distance between slices.

Sediment properties: bulk sediment properties

Oligochaetes significantly lowered the mass water content in the upper 2 cm of the sediment matrix (ANOVA, $LR_{df=1} = 42.4$, $P < 0.0001$; Fig. 3). A correlation structure to model dependence between observation was necessary ($\phi = 0.77$, CL: 0.55, 0.90). Exploratory analyses gave no indication for non-linearity in dry bulk density, porosity, and organic matter content but a spatial correlation structure with depth was incorporated in the statistical model. There was no impact of oligochaetes on dry bulk density, porosity, or organic matter content, neither was there an effect of sediment depth (Table 2).

Sediment properties: algal biomass

The presence of oligochaetes significantly decreased algal biomass at the surface (Fig. 4).

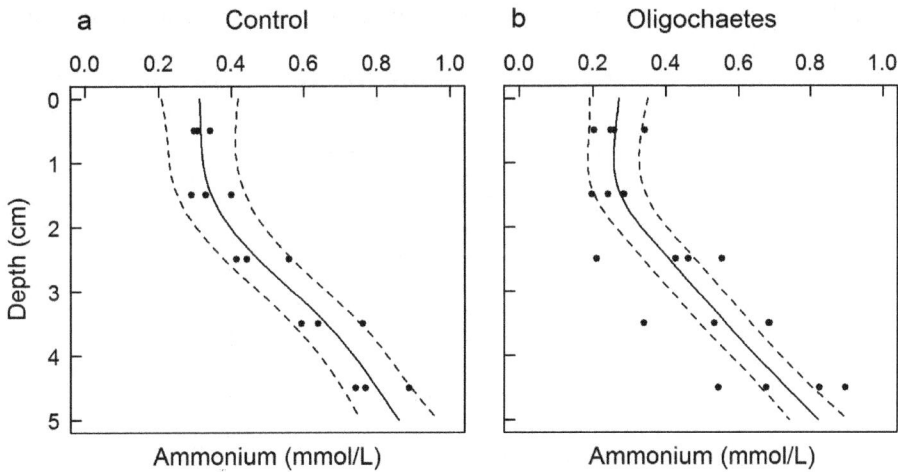

Fig. 2. Ammonium concentrations over depth in (a) control treatment $n = 3$ microcosms and (b) oligochaete treatment $n = 4$ microcosms. The dotted lines indicate the 95% confidence limits, and black dots depict individual measurements.

At the end of the experiment (36 d), algal biomass was 6.1 µg/cm^2 (CL: 5.5,6.8) without worms and 2.6 µg/cm^2 (CL: 2.4, 2.7) with worms (GLS, $LR_{df=1} = 33.9$, $P < 0.0001$). Algal biomass consisted mainly of diatoms, approximately 95%, while the remaining 5% were green algae and cyanobacteria. The presence of oligochaetes led to a decrease in the variation in algal biomass, that is, 0.30 (CL: 0.18, 0.52) times less variation than without worms.

Salicornia germination: algal biomass

The second experiment, including seeds, also showed a clear and similar effect of the worms (Fig. 5), indicating a strong decrease in algal biomass in the presence of oligochaetes (GLS, $LR_{df=3} = 44.6$, $P < 0.0001$). The variation in algal biomass was approximately two times smaller in the treatments with oligochaetes (variance reduction SO: 0.45 [0.20, 1.0]; O: 0.44 [0.20, 0.97]) opposed to treatments without oligochaetes

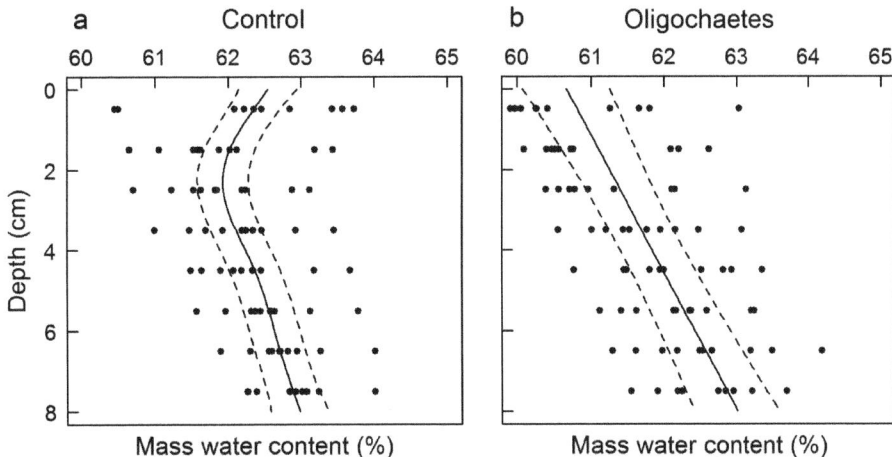

Fig. 3. Mass water content over depth in (a) control treatment and (b) oligochaete treatment, $n = 10$ per treatment. The dotted lines indicate the 95% confidence limits, and black dots depict individual measurements corrected for the random effect.

Table 2. Results of linear regression analyses on sediment properties with factors depth and treatment ($n = 10$ microcosms per treatment).

Sediment property	Mean	95% CL	Depth		Oligochaetes	
			$LR_{df=1}$	P	$LR_{df=1}$	P
Dry bulk density (g/cm^3)	0.49	0.48–0.50	0.8	0.37	0.99	0.32
Organic matter content (%)	16.4	16.1–16.6	2.5	0.11	0.10	0.75
Porosity (%)	81.4	81.0–81.8	0.8	0.37	0.99	0.32

Note: CL, confidence limits; LR, likelihood ratio.

(Fig. 5). In both experiments, the pattern of lower algal biomass in the presence of worms was consistent, though the sediment properties experiment had higher algal biomass in general. This was probably caused by the different methods for defaunating the sediment, that is, sieving and homogenization for sediment properties and freezing for *Salicornia* germination.

Salicornia germination: oligochaete and Salicornia seed distribution

The distribution of worms in the two oligochaete treatments (SO and O) was very similar (Fig. 6), indicating that addition of *Salicornia* seeds did not influence the distribution of worms. We mainly observed burrows in the upper 4 cm of the microcosms, although oligochaetes sporadically occurred as deep as 10 cm. The number of worms differed between the depths; peak

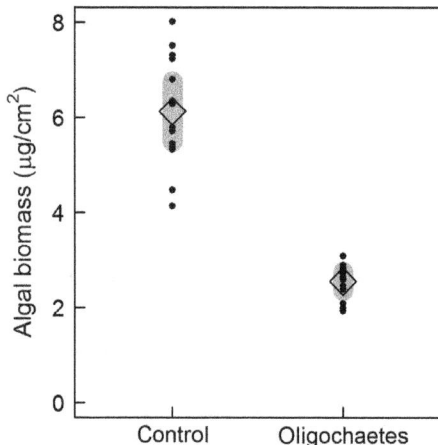

Fig. 4. Algal biomass in μg/cm^2, $n = 14$ per treatment. Diamonds indicate the mean, gray bars indicate the 95% confidence limits, and black dots depict individual measurements.

densities of oligochaetes occurred at a depth of 1–2 cm and then gradually decreased down to the lowest level at 6–12 cm depth (Fig. 6). Approximately 75% of the worms concentrated between 1 and 4 cm depth, and only 3% was recovered below a depth of 6 cm. At the end of the experiment, about 95% of all introduced worms were recollected (Table 1). Of the *Salicornia* seeds, 77% were retrieved (Table 1). Three percent of oligochaetes and 3.9% of seeds were found in the tidal basin, as they drifted out of the microcosms with the simulated tide (Table 1).

After 33 d, there were clear differences in the distribution of intact seeds, seedlings, and germinated seeds in microcosms with and without oligochaetes (Fig. 7). Seeds and seedlings were not found below the upper 0.5 cm without worms and not below 2 cm if worms were present. When oligochaetes were absent, 98.7% of the seeds were retrieved in the top layer (S treatment), opposed to 9.0% when worms were present (SO treatment). Without oligochaetes, all germinated seeds were found in the uppermost sediment layer (0–0.5 cm), while with oligochaetes, most germinated seeds were present below a depth of 0.5 cm (Fig. 7; S and SO treatment). Absence of worms meant that only 1.3% of the seeds were retrieved from the 0.5- to 1-cm sediment layer, opposed to 45.2% with worms. Without worms, no seeds were found below 1 cm, while with worms, 45.8% of seeds were found below 1 cm. The worms transported most of the germinated seeds from the surface into the sediment matrix (Fig. 7). With seedlings, this effect was less pronounced, though some were found at 1–2 cm depth. Bioturbation led to toppling of established seedlings and washing away or subsequent burial through upward conveyor belt deposition. Visual observation of the microcosms indicated that seeds germinated first and were subsequently buried.

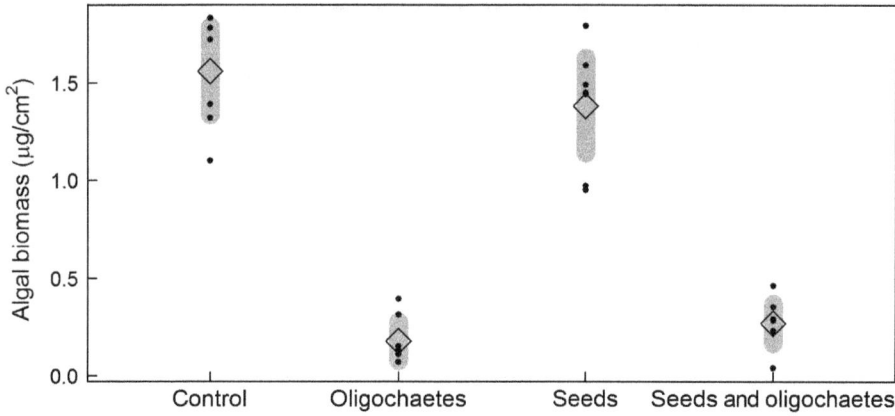

Fig. 5. Algal biomass in μg/cm², $n = 7$ per treatment. Diamonds indicate the mean, gray bars indicate the 95% confidence limits, and black dots depict individual measurements.

DISCUSSION

Algal biomass

Less algal biomass was present at the sediment surface in sediment with oligochaetes (Figs. 4, 5). By reducing diatom biomass, oligochaete bioturbation has an indirect effect on sediment erodibility. Diatoms excrete extra-polymeric substances (Widdows and Brinsley 2002, Weerman et al. 2012), creating a biofilm that stabilizes the sediment. Disturbance of algal biofilms increases erosion (Montserrat et al. 2008, Weerman et al. 2010, 2012, De Lucas Pardo et al. 2013). Observations

during our experiment revealed that a thin layer of sediment eroded during rising tide, when oligochaetes were present and not when worms were absent. Active grazing of the algae by oligochaetes does not seem likely as they feed on detritus and bacteria and showed upward conveyor belt feeding, that is, ingestion at depth and deposition of sediment at the surface (Giere 2006, Dafoe et al. 2011). Diatoms are able to migrate vertically and thus are capable of returning to the sediment surface upon burial (Van Colen et al. 2014). In the current study, the rate of sediment deposition at the sediment surface appeared to

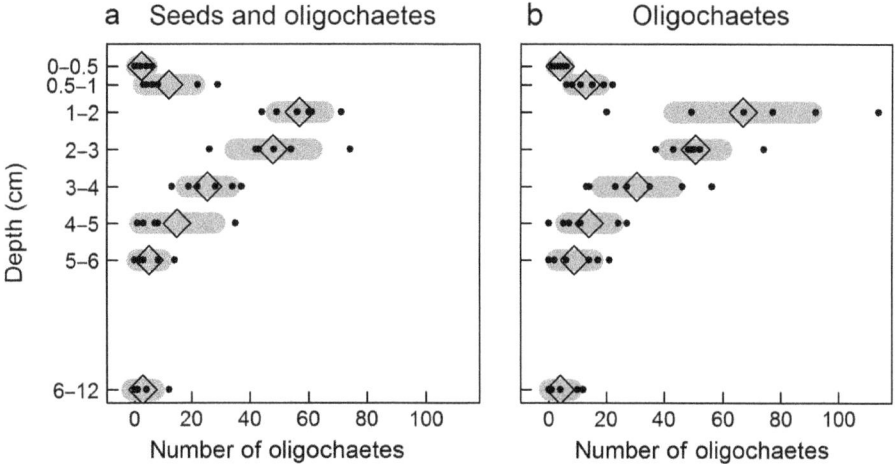

Fig. 6. Vertical distribution of oligochaetes from the (a) seeds and oligochaetes treatment and (b) oligochaetes treatment, $n = 7$ per treatment. Diamonds indicate the mean, gray bars indicate the 95% confidence interval, and black dots depict individual measurements.

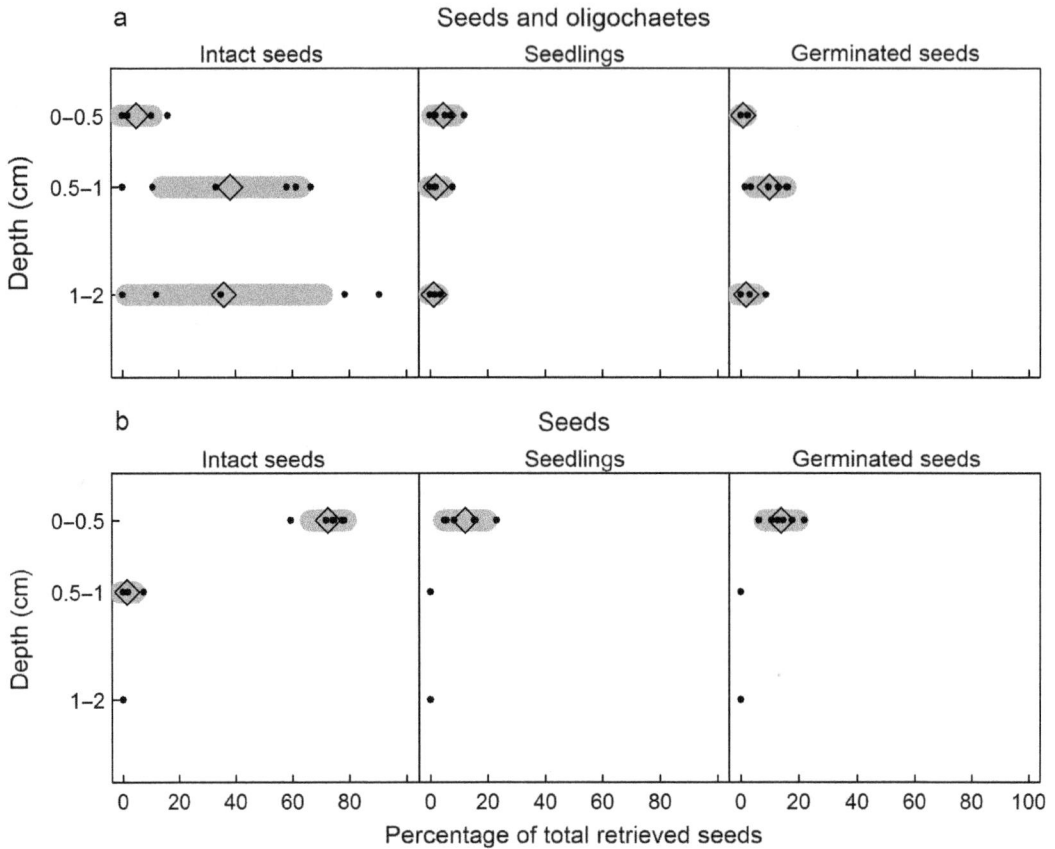

Fig. 7. Distribution of intact seeds, seedlings, and germinated seeds over depth (a) with and (b) without oligochaetes. *Salicornia* seeds were not retrieved below 2 cm depth; therefore, these data points are excluded for clarity. Values are expressed as mean percentage of total retrieved seeds, $n = 7$ per treatment. Diamonds indicate the mean, gray bars indicate the 95% confidence limits, and black dots depict individual measurements; when only one black dot is displayed, all measurements were zero.

exceed diatom production and migration rate so that a biofilm could not fully establish on the surface. In addition, erosion can be temporally increased by the production of fecal pellets by oligochaetes (De Lucas Pardo et al. 2013). Both palletization and bioturbation lead to a rougher microtopography, increasing hydrodynamic roughness and thus potentially increasing erosion (Meysman et al. 2006). The current set-up was not suited to quantify the amount of erosion, but the observations suggest that the presence of oligochaetes in typical field densities may increase sediment erodibility. This in turn affects sedimentation processes and salt-marsh vegetation establishment (Dijkema et al. 1990, Houwing 1999, Widdows and Brinsley 2002).

Sediment properties

To assess potential effects of bioturbation, we examined the depth distribution of worms (Fig. 7). Distribution of the marine oligochaetes roughly corresponded to those of *Tubifex tubifex* and a few other freshwater oligochaete species (Appleby and Brinkhurst 1970, Fisher et al. 1980, Seys et al. 1999, Mermillod-Blondin and Lemoine 2010). The distribution pattern we found, simulating spring conditions, resulted in peak density of worms coinciding with the depth of deep burrow oxidation (Figs. 1, 7). Oligochaete bioturbation increased the oxygen diffusion into marine sediment (Fig. 1), similar to the findings of Davis (1974) in freshwater lake sediments. Because oxidation depth steadily increased, prolongation of the experiment would have likely led to a deeper solid oxidation depth. Bio-irrigation of the burrow network leads to oxygenated burrows (Kristensen et al. 2012) as well as to the upward conveyor belt feeding, which led to even more oxygenated sediment, because anoxic sediment was reworked from deeper layers to the surface. There it was locally oxidized by the overlying air when exposed during low tide. Upward conveyor belt feeding indirectly leads to a draw-down of the organic matter-rich top layer. The former top layer becomes suitable for bacteria and oligochaetes, thereby ensuring their own main food resource (Seys et al. 1999). In the field, oxidation of the sediment will also be affected by erosion and sedimentation processes. Yet, addition of oligochaetes resulted in a tenfold increase in oxidation depth and therefore seems to be an influential factor in intertidal flat oxygen dynamics.

In fine-grained sediments, resource flows are diffusion-dominated and both bioturbation and bio-irrigation increase these flows (Reible et al. 1996, Mermillod-Blondin and Rosenberg 2006, Kristensen et al. 2012, Foshtomi et al. 2015). Organic matter mineralization by the microbial community leads to ammonium production. The ammonium measurements showed no significant difference between bioturbated and non-bioturbated microcosms (Fig. 3). Ammonium concentration was lowest in the upper 2 cm and then increased with depth (Fig. 3; Mermillod-Blondin et al. 2005, Anschutz et al. 2012). There are two mechanisms for ammonium to disappear from the pore water: Either it diffuses upward from pore water into the water column (when subjected to tide) or it is used and transferred from nitrite to nitrate (Foshtomi et al. 2015). Unfortunately, nitrate could not be taken into account in this study because too little pore water was left after ammonium measurements to perform nitrite/nitrate analyses. In the top layer, oligochaetes as well as microbial activity are responsible for the ammonium release (Mermillod-Blondin et al. 2005). It seems that in the current study, the effects of the microbial community on ammonium are more influential than those of the worms. Reports on the impact of oligochaetes on biogeochemical cycles are inconclusive and not unidirectional (Reible et al. 1996, Mermillod-Blondin et al. 2005, 2008, Mermillod-Blondin and Rosenberg 2006, Anschutz et al.

2012, Pigneret et al. 2016). Especially, the cascading effects of organic matter recirculation and nutrient availability on vegetation establishment are of biogeomorphological interest.

Against our expectations, oligochaete bioturbation did not affect bulk sediment properties such as dry bulk density, porosity, and organic matter content. Organic matter content is generally high in silty fine-grained sediments and was probably not a limiting factor for biogeochemical processes. Theoretically, burrows should increase sediment porosity (Giere 2006, De Lucas Pardo et al. 2013). The sediment used in our experiments was already very porous, on average 81% (Table 2). High porosity creates room for consolidation and rapid dewatering of the sediment (Bray and Karig 1985). Bioturbation may form a burrow while adjacently consolidating the sediment, resulting in the same net amount of sediment being present in a slice. In the upper 2 cm, sediment with oligochaetes had lower mass water content (Fig. 5), probably because water evaporated from the burrows. Without worms, a diatom biofilm developed that sealed the sediment surface from evaporation. Differences in water content caused by oligochaetes are small, and the ecological relevance of a slightly higher water content is debatable in an intertidal area.

Oligochaete sediment reworking appears substantial, comparable to those of macrofauna, when up-scaled to field values. Transforming bioturbation measurements into a sediment reworking rate is complex because studies contain various methods and calculations to up-scale reworking rates to field situations (Appleby and Brinkhurst 1970, Cadée 1976, Mermillod-Blondin et al. 2004, Solan et al. 2004, Maire et al. 2006, Teal et al. 2008). We estimated, using solid and deep burrow oxidation depth as a proxy, a reworking rate of $0.0213–0.0562$ $cm^{-1} \cdot d^{-1} (100,000 \text{ individuals})^{-1} \cdot m^{-2}$. This extrapolates to a sediment turnover roughly between 7.8 and 20.5 cm/yr (for 100,000 individuals/m^2). This falls within the estimations for freshwater oligochaete reworking of Appleby and Brinkhurst (1970) of 1.9–25 cm of sediment every year (Cadée 1976). Estimation of oxidation depth is based on the color change from black anoxic to brown oxic sediment, and this results in an underestimation of the actual particle movement (Teal et al. 2008). Our estimation method is therefore precautious and likely an underestimation of the actual impact. The oligochaete reworking rate is then comparable to that of macrofaunal polychaetes such as *Hediste diversicolor*, *Heteromastus filiformis*, and *Arenicola marina* (Cadée 1976, Mermillod-Blondin et al. 2004). These macrofaunal species were hardly present at the study location. Especially, *A. marina* does not reside in soft sediments with high mud content (Cadée 1976). Possibly oligochaetes fulfill the role of bioturbator in highly muddy, anoxic environments where macrofauna cannot thrive. Though, oligochaetes occur in all sediment types (Giere 1993) and their relative contribution is likely to be underestimated, also in other areas. Additionally, the indirect effects of oligochaetes, such as increased erodibility, decreased stability, and the draw-down of organic matter, cascade in biogeomorphological sediment processes. We recommend to include meiofauna in biogeomorphic studies because small, though numerous, organisms can have profound effects on the sediment matrix (Seys et al. 1999, Meysman et al. 2006).

Salicornia *distribution, germination, and establishment*

To our knowledge, this study is the first to suggest that oligochaetes can have a negative impact on the establishment of salt-marsh pioneer vegetation, specifically *Salicornia*. Oligochaete bioturbation reduced the number of seedlings at the surface and buried intact and germinated seeds, up to 2 cm deep (Fig. 7). The critical germination depth for *Salicornia* is estimated to be 1 cm, as deeper burial halts the germination process (Gerdol and Hughes 1993, Wirtz 1994). This indicates that oligochaetes might also hinder the establishment of *Salicornia* seedlings in the field. Seedlings that established with worms present were later often toppled and subsequently buried. Toppling and burial of seedlings is a major disturbance and can inhibit further development and succession (Jensen and Jefferies 1984, Ellison 1987, Van Wesenbeeck et al. 2007, Volkenborn et al. 2007). Additionally, sediment disturbance, that is, bioturbation, can cause loss of root anchorage, which increases the seedling's chance of washing away with the tide (Gerdol and Hughes 1993). We observed this in our experiments as well. Oligochaetes are bacterial and deposit feeders; therefore, direct seed or seedling predation is not a mechanism for reduced seedling establishment unlike the findings of Zhu et al. (2016) for

macrofauna. Oligochaete bioturbation is hereby suggested to shape the biogeomorphic environment by reducing pioneer vegetation establishment and survival.

There is high variation in seedling establishment success in the field due to a limited window of opportunity (Balke et al. 2011, Zhu et al. 2014, Hu et al. 2015). Balke et al. (2011) suggest that this window for salt-marsh vegetation is mainly determined by physical disturbances. Others, however, have shown that biological disturbance of benthic fauna can also play a definite role in seed bank dynamics, both positive and negative (Delefosse and Kristensen 2012, Zhu et al. 2016). Burial of seeds can also function as a mechanism to retain seeds in erosion-prone periods (Zhu et al. 2016). There is a complex interplay of direct and indirect effects that determine vegetation establishment. This can vary per species and developmental stage. Pioneer vegetation may, once sufficiently anchored, benefit from the oligochaete-induced increased oxic conditions (Keiffer et al. 1994, Mermillod-Blondin and Lemoine 2010, Schrama et al. 2015). However, this was not assessed in this study and remains to be further investigated. Nonetheless, first, the negative effects of bioturbation on seedling establishment have to be withstood. We suspect that bioturbation in the field can limit *Salicornia*'s window of opportunity. Especially, *Salicornia* is a pioneer species and the first to colonize the oxygen-poor intertidal flats. Density of worms used in these experiments corresponds to 131.493 individuals/m^2, and as field densities of oligochaetes can range up to a million per square meter, it seems likely that they could influence *Salicornia* establishment in the field. As oligochaetes are not uniformly distributed (Seys et al. 1999), we suspect a negative interaction to probably occur in the field as a patchy effect. Field studies will prove valuable in assessing these results on salt-marsh scale and will reveal the possible positive effects of increased oxygen in the sediment.

In conclusion, the current study shows that oligochaete bioturbation and upward conveyor belt feeding is a significant biogeomorphological process. Depth of oxygen availability in the sediment increased substantially by oligochaete bioturbation. Diatom biofilm production was limited by oligochaete bioturbation, which probably promoted erodibility of the sediment surface. The worms burrowed seeds and reduced *Salicornia* seedling establishment at the sediment surface. Meiofauna needs to be included in biogeomorphic studies because these small, though numerous, organisms can have profound effects on the sediment. Specifically in fine-grained habitats, oligochaetes are an overlooked biotic component that can influence erodibility of the sediment as well as vegetation establishment, and thereby eventually lateral salt-marsh expansion.

ACKNOWLEDGMENTS

This work is part of the research program "Sediment for salt marshes: physical and ecological aspects of a Mud Motor" with Project Number 13888, which is partly financed by The Netherlands Organisation for Scientific Research (NWO). This research received co-funding from Boskalis en Van Oord as part of their contribution to EcoShape: Building with Nature, It Fryske Gea and Waddenfonds. We thank Susanne Kühn and two anonymous reviewers for providing useful comments to improve this manuscript.

LITERATURE CITED

Anschutz, P., A. Ciutat, P. Lecroart, M. Gérino, and A. Boudou. 2012. Effects of tubificid worm bioturbation on freshwater sediment biogeochemistry. Aquatic Geochemistry 18:475–497.

Appleby, A., and R. Brinkhurst. 1970. Defecation rate of three tubificid oligochaetes found in the sediment of Toronto Harbour, Ontario. Journal of the Fisheries Board of Canada 27:1971–1982.

Bagheri, E., and D. McLusky. 1982. Population dynamics of oligochaetes and small polychaetes in the polluted forth estury ecosystem. Netherlands Journal of Sea Research 16:55–66.

Balke, T., T. J. Bouma, E. M. Horstman, E. L. Webb, P. L. Erftemeijer, and P. M. Herman. 2011. Windows of opportunity: thresholds to mangrove seedling establishment on tidal flats. Marine Ecology Progress Series 440:1–9.

Balke, T., M. Stock, K. Jensen, T. J. Bouma, and M. Kleyer. 2016. A global analysis of the seaward salt marsh extent: the importance of tidal range. Water Resources Research 52:1–12.

Boorman, L. A., J. Hazelden, and M. Boorman. 2001. The effect of rates of sedimentation and tidal submersion regimes on the growth of salt marsh plants. Continental Shelf Research 21:2155–2165.

Bouma, T., M. De Vries, E. Low, G. Peralta, I. V. Tánczos, J. van de Koppel, and P. J. Herman. 2005. Trade-offs related to ecosystem engineering: a case study on stiffness of emerging macrophytes. Ecology 86:2187–2199.

Bouma, T., S. Temmerman, L. van Duren, E. Martini, W. Vandenbruwaene, D. Callaghan, T. Balke, G. Biermans, P. Klaassen, and P. van Steeg. 2013. Organism traits determine the strength of scale-dependent bio-geomorphic feedbacks: a flume study on three intertidal plant species. Geomorphology 180:57–65.

Bouma, T., J. van Belzen, T. Balke, J. van Dalen, P. Klaassen, A. Hartog, D. Callaghan, Z. Hu, M. Stive, and S. Temmerman. 2016. Short-term mudflat dynamics drive long-term cyclic salt marsh dynamics. Limnology and Oceanography 61:2261–2275.

Bouma, T. J., J. van Belzen, T. Balke, Z. Zhu, L. Airoldi, A. J. Blight, A. J. Davies, C. Galvan, S. J. Hawkins, and S. P. Hoggart. 2014. Identifying knowledge gaps hampering application of intertidal habitats in coastal protection: opportunities & steps to take. Coastal Engineering 87:147–157.

Bray, C., and D. Karig. 1985. Porosity of sediments in accretionary prisms and some implications for dewatering processes. Journal of Geophysical Research 90:768–778.

Cadée, G. 1976. Sediment reworking by *Arenicola marina* on tidal flats in the Dutch Wadden Sea. Netherlands Journal of Sea Research 10:440–460.

Carter, C. T., and I. A. Ungar. 2003. Germination response of dimorphic seeds of two halophyte species to environmentally controlled and natural conditions. Canadian Journal of Botany 81:918–926.

Chen, H., S. Hagerty, S. M. Crotty, and M. D. Bertness. 2016. Direct and indirect trophic effects of predator depletion on basal trophic levels. Ecology 97:338–346.

Compton, T. J., S. Holthuijsen, A. Koolhaas, A. Dekinga, J. ten Horn, J. Smith, Y. Galama, M. Brugge, D. van der Wal, and J. van der Meer. 2013. Distinctly variable mudscapes: distribution gradients of intertidal macrofauna across the Dutch Wadden Sea. Journal of Sea Research 82:103–116.

Dafoe, L. T., A. L. Rygh, B. Yang, M. K. Gingras, and S. G. Pemberton. 2011. A new technique for assessing tubificid burrowing activities, and recognition of biogenic grading formed by these oligochaetes. Palaios 26:66–80.

Davis, R. B. 1974. Tubificids alter profiles of redox potential and pH in profundal lake sediment. Limnology and Oceanography 2:342–346.

Davy, A. J., G. F. Bishop, and C. S. B. Costa. 2001. *Salicornia* L.(*Salicornia pusilla* J. Woods, *S. ramosissima* J. Woods, *S. europaea* L., *S. obscura* PW Ball & Tutin, *S. nitens* PW Ball & Tutin, *S. fragilis* PW Ball & Tutin and *S. dolichostachya* Moss). Journal of Ecology 89:681–707.

De Groot, A., E. Van der Graaf, R. De Meijer, and M. Maučec. 2009. Sensitivity of in-situ γ-ray spectra to soil density and water content. Nuclear Instruments and Methods in Physics Research A 600:519–523.

De Lucas Pardo, M. A., M. Bakker, T. van Kessel, F. Cozzoli, and J. C. Winterwerp. 2013. Erodibility of soft freshwater sediments in Markermeer: the role of bioturbation by meiobenthic fauna. Ocean Dynamics 63:1137–1150.

Delefosse, M., and E. Kristensen. 2012. Burial of Zostera marina seeds in sediment inhabited by three polychaetes: laboratory and field studies. Journal of Sea Research 71:41–49.

Dijkema, K. S., J. H. Bossinade, P. Bouwsema, R. J. De Glopper, and J. J. Beukema. 1990. Salt marshes in the Netherlands Wadden Sea: rising high-tide levels and accretion enhancement. Pages 173–188 in J. Beukema and J. J. W. M. Brouns, editors. Expected effects of climatic change on marine coastal ecosystems. Kluwer Academic, Dordrecht, The Netherlands.

Elias, E., A. Van der Spek, Z. Wang, and J. De Ronde. 2012. Morphodynamic development and sediment budget of the Dutch Wadden Sea over the last century. Netherlands Journal of Geosciences 91: 293–310.

Ellison, A. M. 1987. Density-dependent dynamics of Salicornia europaea monocultures. Ecology 68:737–741.

Evans, P. R., D. Herdson, P. Knights, and M. Pienkowski. 1979. Short-term effects of reclamation of part of Seal Sands, Teesmouth, on wintering waders and Shelduck. Oecologia 41:183–206.

Fagherazzi, S., M. L. Kirwan, S. M. Mudd, G. R. Guntenspergen, S. Temmerman, A. D'Alpaos, J. van de Koppel, J. M. Rybczyk, E. Reyes, and C. Craft. 2012. Numerical models of salt marsh evolution: ecological, geomorphic, and climatic factors. Reviews of Geophysics 50:1–28.

Fisher, J., W. Lick, P. McCall, and J. Robbins. 1980. Vertical mixing of lake sediments by tubificid oligochaetes. Journal of Geophysical Research 85:3997–4006.

Flemming, B., and M. Delafontaine. 2000. Mass physical properties of muddy intertidal sediments: some applications, misapplications and non-applications. Continental Shelf Research 20:1179–1197.

Foshtomi, M. Y., U. Braeckman, S. Derycke, M. Sapp, D. Van Gansbeke, K. Sabbe, A. Willems, M. Vincx, and J. Vanaverbeke. 2015. The link between microbial diversity and nitrogen cycling in marine sediments is modulated by macrofaunal bioturbation. PLoS ONE 10:1–20.

Friedrichs, C. T., and J. E. Perry. 2001. Tidal salt marsh morphodynamics: a synthesis. Journal of Coastal Research 27:7–37.

Gerdol, V., and R. G. Hughes. 1993. Effect of the amphipod Corophium volutator on the colonization of mud by the halophyte Salicornia europaea. Marine Ecology Progress Series 97:61–69.

Giere, O. 1993. Meiobenthology. The microscopic fauna in aquatic sediments. Springer-Verlag, Berlin, Germany.

Giere, O. 2006. Ecology and biology of marine oligochaeta – an inventory rather than another review. Hydrobiologia 564:103–116.

Heiri, O., A. F. Lotter, and G. Lemcke. 2001. Loss on ignition as a method for estimating organic and carbonate content in sediments: reproducibility and comparability of results. Journal of Paleolimnology 25:101–110.

Houwing, E.-J. 1999. Determination of the critical erosion threshold of cohesive sediments on intertidal mudflats along the Dutch Wadden Sea coast. Estuarine, Coastal and Shelf Science 49:545–555.

Hu, Z., J. Belzen, D. Wal, T. Balke, Z. B. Wang, M. Stive, and T. J. Bouma. 2015. Windows of opportunity for salt marsh vegetation establishment on bare tidal flats: the importance of temporal and spatial variability in hydrodynamic forcing. Journal of Geophysical Research 120:1450–1469.

Jensen, A., and R. Jefferies. 1984. Fecundity and mortality in populations of Salicornia europaea agg. at Skallingen, Denmark. Ecography 7:399–412.

Keiffer, C. H., B. C. McCarthy, and I. A. Ungar. 1994. Effect of salinity and waterlogging on growth and survival of Salicornia europaea L., and inland halophyte. Ohio Journal of Science 94:70–73.

Khan, M. A., B. Gul, and D. J. Weber. 2000. Germination responses of Salicornia rubra to temperature and salinity. Journal of Arid Environments 45: 207–214.

Kirwan, M. L., and G. R. Guntenspergen. 2010. Influence of tidal range on the stability of coastal marshland. Journal of Geophysical Research 115:1–11.

Kristensen, E., G. Penha-Lopes, M. Delefosse, T. Valdemarsen, C. O. Quintana, and G. T. Banta. 2012. What is bioturbation? The need for a precise definition for fauna in aquatic sciences. Marine Ecology Progress Series 446:285–302.

Maire, O., J. C. Duchene, R. Rosenberg, J. B. de Mendonca, and A. Grémare. 2006. Effects of food availability on sediment reworking in Abra ovata and A. nitida. Marine Ecology Progress Series 319: 135–153.

Marani, M., C. Da Lio, and A. D'Alpaos. 2013. Vegetation engineers marsh morphology through multiple competing stable states. Proceedings of the National Academy of Sciences USA 110:3259–3263.

Mariotti, G., and S. Fagherazzi. 2010. A numerical model for the coupled long-term evolution of salt marshes and tidal flats. Journal of Geophysical Research 115:1–15.

Matisoff, G., X. Wang, and P. L. McCall. 1999. Biological redistribution of lake sediments by tubificid oligochaetes: *Branchiura sowerbyi* and *Limnodrilus hoffmeisteri/Tubifex tubifex*. Journal of Great Lakes Research 25:205–219.

McCann, L. D. 1989. Oligochaete influence on settlement, growth and reproduction in a surface-deposit-feeding polychaete. Journal of Experimental Marine Biology and Ecology 131:233–253.

Mermillod-Blondin, F., J. P. Gaudet, M. Gerino, G. Desrosiers, J. Jose, and M. C. des Châtelliers. 2004. Relative influence of bioturbation and predation on organic matter processing in river sediments: a microcosm experiment. Freshwater Biology 49: 895–912.

Mermillod-Blondin, F., and D. G. Lemoine. 2010. Ecosystem engineering by tubificid worms stimulates macrophyte growth in poorly oxygenated wetland sediments. Functional Ecology 24:444–453.

Mermillod-Blondin, F., G. Nogaro, T. Datry, F. Malard, and J. Gibert. 2005. Do tubificid worms influence the fate of organic matter and pollutants in stormwater sediments? Environmental Pollution 134:57–69.

Mermillod-Blondin, F., G. Nogaro, F. Vallier, and J. Gibert. 2008. Laboratory study highlights the key influences of stormwater sediment thickness and bioturbation by tubificid worms on dynamics of nutrients and pollutants in stormwater retention systems. Chemosphere 72:213–223.

Mermillod-Blondin, F., and R. Rosenberg. 2006. Ecosystem engineering: the impact of bioturbation on biogeochemical processes in marine and freshwater benthic habitats. Aquatic Sciences 68:434–442.

Meysman, F. J., J. J. Middelburg, and C. H. Heip. 2006. Bioturbation: a fresh look at Darwin's last idea. Trends in Ecology & Evolution 21:688–695.

Möller, I. 2006. Quantifying saltmarsh vegetation and its effect on wave height dissipation: results from a UK East coast saltmarsh. Estuarine, Coastal and Shelf Science 69:337–351.

Montserrat, F., C. Van Colen, S. Degraer, T. Ysebaert, and P. M. Herman. 2008. Benthic community-mediated sediment dynamics. Marine Ecology Progress Series 372:43–59.

Pigneret, M., F. Mermillod-Blondin, L. Volatier, C. Romestaing, E. Maire, J. Adrien, L. Guillard, D. Roussel, and F. Hervant. 2016. Urban pollution of sediments: impact on the physiology and burrowing activity of tubificid worms and consequences on biogeochemical processes. Science of the Total Environment 568:196–207.

Pinheiro, J. C., and D. M. Bates. 2000. Linear mixed-effects models: basic concepts and examples. Pages 3–56 *in* J. Chambers, W. Eddy, W. Hardle, and S. Sheather, editors. Mixed-effects models in S and S-Plus. Springer-Verlag, New York, New York, USA.

Pinheiro, J., D. Bates, S. DebRoy, D. Sarkar, and R Core Team. 2016. nlme: linear and nonlinear mixed effects models. https://CRAN.R-project.org/package=nlme

R Core Team. 2015. R: a language and environment for statistical computing. R Foundation for Statistical Computing, Vienna, Austria.

Ravera, O. 1955. Amount of mud displaced by some freshwater Oligochaeta in relation to depth. Pages 247–264 *in* Symposium on Biological, Physical and Chemical Characteristics of the Profundal Zone of Lakes. Memorie dell'Istituto Italiano di Idrobiologia, Pallanza, Italy.

Reible, D. D., V. Popov, K. T. Valsaraj, L. J. Thibodeaux, F. Lin, M. Dikshit, M. A. Todaro, and J. W. Fleeger. 1996. Contaminant fluxes from sediment due to tubificid oligochaete bioturbation. Water Research 30:704–714.

Reise, K., M. Baptist, P. Burbridge, N. Dankers, L. Fischer, B. Flemming, A. Oost, and C. Smit. 2010. The Wadden Sea: a universally outstanding tidal wetland. Wadden Sea Ecosystem Number 29. 0946-896X. Common Wadden Sea Secretariat, Wilhelmshaven, Germany.

Sarkar, D. 2008. Lattice: multivariate data visualization with R. Springer, New York, New York, USA.

Schrama, M., L. A. Boheemen, H. Olff, and M. P. Berg. 2015. How the litter-feeding bioturbator *Orchestia gammarellus* promotes late-successional saltmarsh vegetation. Journal of Ecology 103:915–924.

Schuerch, M., T. Dolch, K. Reise, and A. T. Vafeidis. 2014. Unravelling interactions between salt marsh evolution and sedimentary processes in the Wadden Sea (southeastern North Sea). Progress in Physical Geography 38:691–715.

Seys, J., M. Vincx, and P. Meire. 1999. Spatial distribution of oligochaetes (Clitellata) in the tidal freshwater and brackish parts of the Schelde estuary (Belgium). Hydrobiologia 406:119–132.

Solan, M., B. D. Wigham, I. R. Hudson, R. Kennedy, C. H. Coulon, K. Norling, H. C. Nilsson, and R. Rosenberg. 2004. In situ quantification of bioturbation using time-lapse fluorescent sediment profile imaging (f-SPI), luminophore tracers and model simulation. Marine Ecology Progress Series 271:1–12.

Teal, L., M. T. Bulling, E. Parker, and M. Solan. 2008. Global patterns of bioturbation intensity and mixed depth of marine soft sediments. Aquatic Biology 2:207–218.

Temmerman, S., P. Meire, T. J. Bouma, P. M. Herman, T. Ysebaert, and H. J. De Vriend. 2013. Ecosystem-based coastal defence in the face of global change. Nature 504:79–83.

Van Colen, C., G. J. Underwood, J. Serôdio, and D. M. Paterson. 2014. Ecology of intertidal microbial biofilms: mechanisms, patterns and future research needs. Journal of Sea Research 92:2–5.

van de Koppel, J., D. van der Wal, J. P. Bakker, and P. M. Herman. 2005. Self-organization and vegetation collapse in salt marsh ecosystems. American Naturalist 165:E1–E12.

van der Wal, D., and P. M. Herman. 2012. Ecosystem engineering effects of *Aster tripolium* and *Salicornia procumbens* salt marsh on macrofaunal community structure. Estuaries and Coasts 35:714–726.

van Haaren, T. 2016. Oligochaeten van Brakke en Zoute Wateren in Nederland. Pages 115–164 *in* R. M. J. C. Kleukers, editor. Nederlandse Faunistische Mededelingen. EIS Kenniscentrum Insecten en andere ongewervelden, Naturalis Biodiversity Center, Leiden, Germany.

Van Wesenbeeck, B., J. Van de Koppel, P. Herman, J. Bakker, and T. Bouma. 2007. Biomechanical warfare in ecology; negative interactions between species by habitat modification. Oikos 116:742–750.

Van Wesenbeeck, B. K., J. Van de Koppel, P. M. J. Herman, M. D. Bertness, D. Van der Wal, J. P. Bakker, and T. J. Bouma. 2008a. Potential for sudden shifts in transient systems: distinguishing between local and landscape-scale processes. Ecosystems 11:1133–1141.

Van Wesenbeeck, B. K., J. Van de Koppel, P. M. J. Herman, and T. J. Bouma. 2008b. Does scale-dependent feedback explain spatial complexity in salt-marsh ecosystems? Oikos 117:152–159.

Volkenborn, N., S. Hedtkamp, J. Van Beusekom, and K. Reise. 2007. Effects of bioturbation and bioirrigation by lugworms (*Arenicola marina*) on physical and chemical sediment properties and implications for intertidal habitat succession. Estuarine, Coastal and Shelf Science 74:331–343.

Weerman, E. J., J. Van Belzen, M. Rietkerk, S. Temmerman, S. Kefi, P. M. J. Herman, and J. Van de Koppel. 2012. Changes in diatom patch-size distribution and degradation in a spatially self-organized intertidal mudflat ecosystem. Ecology 93:608–618.

Weerman, E. J., J. Van de Koppel, M. B. Eppinga, F. Montserrat, Q. X. Liu, and P. M. Herman. 2010. Spatial self-organization on intertidal mudflats through biophysical stress divergence. American Naturalist 176:E15–E32.

Widdows, J., and M. Brinsley. 2002. Impact of biotic and abiotic processes on sediment dynamics and the consequences to the structure and functioning of the intertidal zone. Journal of Sea Research 48:143–156.

Wirtz, J. A. P. 1994. Zaadproductie en verspreiding van *Salicornia dolichostachya*. Thesis. University of Wageningen, Wageningen, The Netherlands.

Wolters, M., A. Garbutt, and J. P. Bakker. 2005. Salt-marsh restoration: evaluating the success of de-embankments in north-west Europe. Biological Conservation 123:249–268.

Wood, S. 2006. Generalized additive models: an introduction with R. CRC Press, Boca Raton, Florida, USA.

Zhu, Z., T. J. Bouma, T. Ysebaert, L. Zhang, and P. M. Herman. 2014. Seed arrival and persistence at the tidal mudflat: identifying key processes for pioneer seedling establishment in salt marshes. Marine Ecology Progress Series 513:97–109.

Zhu, Z., J. van Belzen, T. Hong, T. Kunihiro, T. Ysebaert, P. M. Herman, and T. J. Bouma. 2016. Sprouting as a gardening strategy to obtain superior supplementary food: evidence from a seed-caching marine worm. Ecology 97:3278–3284.

Zuur, A. F., E. N. Ieno, and C. S. Elphick. 2010. A protocol for data exploration to avoid common statistical problems. Methods in Ecology and Evolution 1:3–14.

5

Building biodiversity: drivers of bird and butterfly diversity on tropical urban roof gardens

James Wei Wang,[1,2] Choon Hock Poh,[1] Chloe Yi Ting Tan,[3] Vivien Naomi Lee,[3] Anuj Jain,[3,4] and Edward L. Webb[3,]†

[1]*National Parks Board, 1 Cluny Road, Singapore S259569*
[2]*Department of Geography, National University of Singapore, 1 Arts Link, Kent Ridge, Singapore S117570*
[3]*Department of Biological Sciences, National University of Singapore, Science Drive 4, Singapore S117543*

Abstract. Conservation of faunal diversity in highly urbanized landscapes is facilitated through the integration of anthropogenic and natural elements in urban green spaces. Roof gardens have the potential to provide resources for urban wildlife populations, yet they typically offer marginal ecological conditions relative to ground-level habitat so it is necessary to identify the factors associated with focal taxa diversity and abundance to better inform their planning, design, and management. We conducted 20 monthly surveys of diurnal bird and butterfly visitors to 30 urban roof gardens in the tropical city-state of Singapore. Sixty-five site variables were evaluated for their relative importance to four community diversity metrics (species richness, Simpson's diversity, abundance, and functional dispersion) using a conceptual framework relevant to key stakeholders involved in roof garden planning, design, and management. Surveys from a total of 80 observation hours per site recorded more than 23,000 independent bird and butterfly encounters, comprising 53 bird and 57 butterfly species, which represents 13% and 18% of the avifauna and butterfly fauna in Singapore, respectively. Twenty-four species (12 birds, 12 butterflies) were considered uncommon or rare. Reproductive behaviors (courtship, mating, or nesting) were noted in 35 species (20 birds, 15 butterflies). Understory birds were under-represented, whereas understory butterflies were over-represented compared to the overall Singapore fauna. Early morning noise showed strong negative linear correlations with diversity of both taxa, but this variable was also negatively associated with planted area, perhaps indicating proxy effects. Faunal diversity decreased asymptotically with height. Both managed and spontaneous floral diversity had clear positive effects on bird and butterfly diversity. Our results provide clear indication that tropical urban roof gardens can support a diverse subset of the bird and butterfly species assemblages. Site height, site area, and plant selection are suggested to play dominant roles in determining the attractiveness of roof gardens for wildlife. To a first approximation, roof gardens should be built lower than 50 m in height and contain planted areas larger than 1100 m^2, as well as shrubs for birds and both managed and spontaneous nectar plant species for butterflies.

Key words: conservation; green roofs; reconciliation ecology; urban ecology; urban green space.

† **E-mail:** ted.webb@nus.edu.sg

Introduction

Urbanization is a major driver of global biodiversity loss (McKinney 2006, McDonald et al. 2008, Seto et al. 2012). Contemporary debates addressing this problem have highlighted the critical roles of urban nature reserves and managed green spaces for supporting both local and regional biodiversity (Bonthoux et al. 2014, Beninde et al. 2015, Rupprecht et al. 2015). Yet, in many urbanizing areas, the patchworks of rural and informal green spaces that structure much urban biodiversity are being rapidly converted to more highly managed forms of land use (Pauchard et al. 2006, Nagendra et al. 2014, Xun et al. 2014). It follows that the role of managed green spaces such as parks, gardens, and streetscapes in providing space and resources for the long-term preservation of urban wildlife will become increasingly important to inform strategies for effective urban nature conservation (Alvey 2006, Goddard et al. 2010). These sites also present important opportunities to connect urban dwellers with nature and grow their appreciation for more natural ecosystems (Dearborn and Kark 2010).

Managed green spaces play three major roles important for urban biodiversity conservation. First, by softening the quality of the matrix, they could facilitate the movement of wildlife between otherwise isolated populations, thereby buffering metapopulation viability (Munshi-South 2012, Caryl et al. 2013, Jha and Kremen 2013). Second, they could provide stable supplementary resources for wildlife whose core habitats are located within foraging distance (Davies et al. 2009, Stagoll et al. 2012). Third, they could provide breeding sites for a limited suite of species (Bland et al. 2004, Gaston et al. 2005). Nevertheless, the degree to which these roles can be fulfilled can be expected to depend on the way in which these spaces are planned, designed, and maintained. Indeed, Quigley (2011) argues that because the vegetation in these sites is often chosen and managed to meet a narrow set of aesthetic requirements, such landscapes may not support the long-term self-assembly of a broader set of species and may be of very limited conservation value.

Roof gardens (also known as intensive green roofs or sky gardens) may contribute to urban biodiversity conservation as compact urban green spaces. These gardens are incorporated into the structure of a building at various heights and designed principally for active human use. Intensive roof gardens in which growing mediums exceed 300 mm depth (the focus of this study) can sustain an arboreous typology described by Madre et al. (2014) and may therefore present

complex structure and diverse plant assemblages that could attract wildlife, particularly birds and invertebrates. This is in contrast to extensive green roofs, which have more shallow growing mediums (<300 mm in depth) and exhibit more limited structural plant diversity.

Invertebrate diversity on temperate extensive green roofs has been associated with building height, soil depth, and vegetation structural heterogeneity (Madre et al. 2013, MacIvor 2015, MacIvor and Ksiazek 2015), and one study has corroborated the role for these habitats in providing habitat connectivity (Braaker et al. 2014). Studies conducted on temperate green roofs have documented up to 29 species of birds, although records of breeding success are sparse (Baumann 2006, Fernandez-Canero and Gonzalez-Redondo 2010). Both vegetation structure and composition have been suggested to affect bird diversity on green roofs from the perspective of the provision of foraging resources (Coffman and Waite 2011).

To date, nearly all research has focused on temperate extensive green roofs, and little information is available on the ecology of tropical roof gardens; the lack of research attention belies their rising popularity, particularly in modern high-density Asian cities (Yuen and Hien 2005, Tian and Jim 2012). Understanding how roof gardens and other managed urban green spaces may contribute to urban biodiversity conservation in the tropics is crucial, because the tropics harbor the vast majority of the Earth's biodiversity (Gaston 2000) and yet are projected to incur the highest rate of habitat loss due to future urban expansion (Seto et al. 2012).

The city-state of Singapore provides an ideal venue to assess the socio-ecological factors that may contribute to the biological potential of roof gardens. Located in the Sundaland biodiversity hotspot in the Indo-Malayan archipelago (Sodhi et al. 2004), Singapore harbors a diverse and well-documented natural assemblage of species, including 392 birds and 323 butterflies (NSS Bird Group 2015, Butterfly Circle 2016). Significant proportions of the total bird and butterfly faunal diversity (35% and 33%, respectively) have been observed in ground-level urbanized conditions (Chong et al. 2014). Crucially, at the time of the present study, approximately 60 ha of green roofs (intensive and extensive) were available, encompassing a wide range of socio-ecological conditions. Birds and butterflies were chosen as focal taxa because the habitat requirements of at least a broad subset of these species are known to be consistent with urban environments. Also, these taxa generally hold a cross-cultural appeal such that their presence in public roof gardens would be accepted or encouraged.

In this context, the objectives of the present study were to assess the diversity of birds and butterflies associated with roof gardens and to identify the socio-ecological variables associated with faunal abundance and diversity. We sought to provide evidence-based support for practical conservation measures applicable to the three main stakeholder groups involved in roof garden design and maintenance: planners, architects, and managers. We developed hypotheses and structured our analyses to correspond with this framework of key stakeholders, which we explain below.

The stakeholder framework and hypotheses

The stakeholder framework was developed based on authors' JWW and CHP experience with the local statutory authority for the regulation and promotion of green roof design and construction (the National Parks Board, see also Steed 2015). This framework is explicitly social but consists of environmental variables. The Planning level broadly corresponds to environmental variables described as landscape-scale drivers, whereas the Architecture and Management levels translate to local-scale drivers, in terms common to the broader ecological literature. Whereas the roles of design and management are conflated in some urban green spaces such as private gardens (Goddard et al. 2010), they are administratively separated in public and commercial recreational spaces. For example, the roof gardens we surveyed were generally managed by horticultural contractors on behalf of the building owners. We therefore made a distinction between environmental variables specified during the design of a site (Architecture), and those which could be modified by subsequent anthropogenic influence such as horticultural practices (Management), sensu Beninde et al. (2015). The intent of making this distinction explicit upstream in our analyses was to target our findings more precisely to the stakeholders with practical leverage over roof garden ecosystems.

For each stakeholder group, we developed hypotheses that identified a set of related variables we considered most likely to show statistically significant independent effects on the community structure of both taxa, based on relevant literature. These hypotheses are articulated as follows:

1. Planning: At the landscape scale, habitat area (the proportion of land area surrounding the building that was covered by either trees or grass) would exert a stronger (positive) effect on faunal community structure than habitat isolation or noise. This is because we expected the suite of species colonizing the roof gardens to be primarily disturbance-tolerant, widespread, and highly mobile and thus less sensitive to variables associated with urbanization (e.g., noise) or dispersal limitation (Sattler et al. 2010, Swan et al. 2011).

2. Architecture: At the local scale, the total vegetated area within a roof garden would exert a dominant (positive) influence on the community structure of both taxa compared to site height, vegetation structure, or floristic composition. This hypothesis is based on well-established findings across many urban areas (Nielsen et al. 2014, Beninde et al. 2015).

3. Management: (a) Variables relating to vegetation structure such as tree cover (Fontana et al. 2011), herbaceous plant height (Carbó-Ramírez and Zuria 2011), and understory volume (Threlfall et al. 2016) would show stronger (positive) effects than floristic composition on bird community structure; (b) total species richness of nectar plants (rather than other metrics of floristic diversity that did not directly account for nectar availability as inferred from butterfly visitation, Garbuzov and Ratnieks 2014) would show a strong (positive) effect on butterfly community structure.

Methods

Site description

The study area was conducted in the tropical city-state of Singapore (1°22' N, 103°48' E), which has a total area of 718.3 km^2 (Department of Statistics 2015). Singapore has an equatorial climate, with a diurnal temperature range of 23–30°C and a total annual rainfall of 2300 mm (Fong et al. 2012). Original land cover was a mix of lowland/hill dipterocarp forest (82%), mangrove forest (13%) and freshwater swamp forest (5%; Corlett 1991). However, the landscape has been radically transformed over the past 150 yr, with original forest cover reduced to <2% of its original extent, and replaced with non-vegetated urban surfaces (39%), managed vegetation (parks and streetscapes, 27%), secondary forests (21%), and scrubland (6%), which now constitute ~93% of the total land area (Yee et al. 2011). There is a main bird migratory season for passage migrants and winter visitors between September and April (Lim 2009).

Site selection

Study sites were selected through a three-step process. First, we obtained a list of roof gardens from the National Parks Board, the Housing Development Board, and the Urban Redevelopment Authority. Using an internet search, we also compiled a list of hotels, shopping malls, and hospitals that contained roof gardens. We filtered this master list according to three preliminary criteria: (1) The garden was accessible to building residents (or the public), (2) the garden was accessible to birds and butterflies, that is, not enclosed by skylights or netting, and (3) vegetation in the garden was grown in permanent planting pits. This resulted in an initial list of 259 sites.

Subsequently, through telephone interviews with site managers and online map observation, we retained potential sites if they conformed to the following criteria: (1) The garden was at least on the fourth story above ground level, (2) total vegetated area of the garden was estimated to be at least 300 m² (although 10 selected sites were later determined to have <300 m² planted area, Table 1), (3) not completely enclosed by walls, and (4) the garden was not actively managed to exclude wildlife (e.g., by cutting off flowers). The criterion of a garden height at least four stories was because gardens built by the Housing Development Board were typically built at four stories; this height was also generally above the canopies of ground-level trees, thereby offering a clear distinction from ground-level gardens. After applying these criteria, 76 sites were retained.

Finally, the remaining potential sites were divided into low (<24 m above ground, approximately six stories) and high (>24 m above ground). This height was chosen because we expected that several environmental variables relevant for bird and butterfly visitation would scale with respect to height. Moreover, local Fire Code regulations mandated significantly more stringent fire-fighting provisions above 24 m, such as increased water tank capacity and smoke-stop lobbies. This threshold impacts building form and would influence the size and height of roof gardens accordingly. This division resulted in 60 low sites and 16 high sites. We randomly selected 16 low sites from the list, requiring at least 200 m between sites to maintain landscape-scale variability between sites. For the high sites, eight were deemed unsuitable owing to restrictions from building management or proximity (<200 m) to other previously selected sites. Thus, we undertook extensive field surveys to locate an additional eight high sites. Two sites were eventually eliminated because construction at those sites resulted in less than the full number of surveys being conducted. This resulted in a final total sample size of 30 sites (15 low, 15 high; Fig. 1).

Our study sites encompassed a range of biophysical conditions (Fig. 2), including variations in site area (235–5997 m²), site height (9–189 m), age (0.5–28 yr), and floristic richness (6–126 plant species; Table 1). The plant species composition of each roof garden was established through an exhaustive botanical census. The vegetation planted in these sites was largely pan-tropical and dominated by exotic species, which comprised 89% of the total number of plant taxa identified across sites.

Wildlife surveys

A total of 20 surveys were conducted at each site at intervals of 3–5 weeks, from May 2014 to December 2015, which covered two migratory bird seasons. To ensure that robust, comparable data were collected at each site, we followed the general guideline that surveys were conducted only in non-rainy weather, so bouts were suspended during rain events. If the rain cleared within 15 min, the bout was restarted. If the rain lasted for more than 15 continuous minutes, the survey was discontinued and repeated at the next earliest opportunity. Observations were made using Nikon Aculon A211 8 × 42 binoculars (Nikon Vision Co. Ltd, Tokyo, Japan). Photographs of the overall

Table 1. Characteristics of the 30 tropical urban roof gardens surveyed in this study.

	Site characteristics				Plants	Birds		Butterflies	
No.	Area (m^2)	Height (m)	Age (yr)	Total planted area (m^2)	Richness	Richness	Abundance	Richness	Abundance
1	1169	26	9	291	74	29	604	22	117
2	3204	16	7	1347	42	26	517	19	141
3	5736	12	22	3737	65	25	927	22	378
4	2463	14	4	1530	87	24	759	37	355
5	4840	15	18	2836	34	24	927	29	682
6	2913	15	0.5	1660	40	24	1248	20	138
7	4583	13	13	1701	63	23	572	22	340
8	1334	16	28	677	56	22	400	13	85
9	4813	14	4	1709	66	22	551	18	155
10	1747	30	21	1138	68	20	451	10	35
11	2045	14	19	621	89	20	678	15	52
12	1401	12	3	715	38	19	231	19	186
13	3333	28	1	1482	126	19	334	24	420
14	2405	70	4	1764	26	18	256	14	82
15	4033	14	12	2106	46	16	235	23	324
16	1525	21	6	368	38	16	333	13	39
17	544	40	6	139	6	14	190	6	18
18	1220	25	11	130	46	13	327	11	41
19	2925	15	6	1261	57	12	185	19	227
20	2171	61	5	945	84	10	200	11	39
21	714	51	2	427	19	9	103	8	46
22	4680	143	11	883	33	8	112	8	226
23	4494	9	2	1673	56	7	135	21	436
24	668	48	3	211	25	6	115	2	3
25	5997	189	4	145	54	6	36	5	65
26	455	55	1	228	6	5	78	2	4
27	236	141	1	103	9	4	11	1	2
28	235	51	9	131	32	4	32	6	13
29	540	139	10	257	15	4	16	11	16
30	459	77	6	233	28	2	3	2	2

Note: Site age was measured at intervals of one year, except for Site 6, which was less than a year old.

landscape were taken from fixed points to monitor if the horticultural maintenance regimes remained constant; plant species that were flowering or fruiting were noted.

Bird surveys were conducted in four bouts of 25 min each between 07:00 and 09:00 hours. Within each bout, surveyors actively searched the site and noted all birds seen or heard, covering as much garden area as possible. Only birds that were observed to interact with the garden were recorded; bird sightings above tree level were ignored. Interactions with plant species for fruit and nectar were recorded, and lists of these plants were compiled for each site. Butterfly surveys were conducted in four bouts of 25 min each between 10:00 and 12:00 hours. Within each bout, surveyors actively searched the site and noted all butterflies seen or heard, covering as much garden area as possible. Only butterflies that were observed to interact with the garden were recorded. Larvae or pupae sighted were recorded and photographed, and host plant species were noted. Butterfly nets (BioQuip collapsible net, 18" diameter, 17" handle, Bioquip Products, Rancho Dominguez, California, USA) were used to aid in the identification of species that were difficult to identify with binoculars. All butterflies were released after identification; we detected no injuries to any butterflies from our capture technique. Interactions with plant species for nectar were recorded, and lists of these plants were compiled for each site. For both birds and butterflies, taxa that were not identified at least to genus level or that were sighted outside of the designated survey times were excluded from the analyses.

Fig. 1. Distribution of site locations in Singapore.

Fig. 2. Selected site photographs: (clockwise from top left): Sengkang 437 (semi-open); Shaw House (arboreous); The Metropolis (covered); Pinnacles at Duxton (high); Orchard Central (water gardens); Pan Pacific Hotel (manicured); Secom Centre (mixed vegetation); Grange Infinite (mid-height). (Photographs taken by the authors of this article.)

Our data collection method sought to record the number of unique individuals at each site. Thus, the count from a bout contributed toward the total sum for that day only if it increased the number of sightings of that species than the bout immediately preceding it. Thus, for example, if the first bout recorded two individuals of Species A and the second bout recorded one individual of Species A, the second bout contributed zero counts to the total. However, if the third bout recorded two individuals of Species A, that bout contributed one count to the total because it was one more individual than the preceding bout. This approach avoided double-counting of sightings reported within 60 min of each other, although an individual that left the site and returned after more than 60 min could be double-counted. We also made notes of behaviors, summarized in Appendix S1: Tables S1, S2.

We standardized sampling effort by time rather than area, because all sites were <0.6 ha in size, and differences in accessibility to locations within sites precluded the establishment of standardized point or transect-based count schemes. Detectability was assumed to be constant among sites, which was considered reasonable due to their small sizes and generally open landscape character; average tree height did not exceed 9 m at any site.

Bird and butterfly attributes

The species attributes we designated are summarized in Fig. 3, with sources provided in Appendix S1: Table S1. All these attributes were

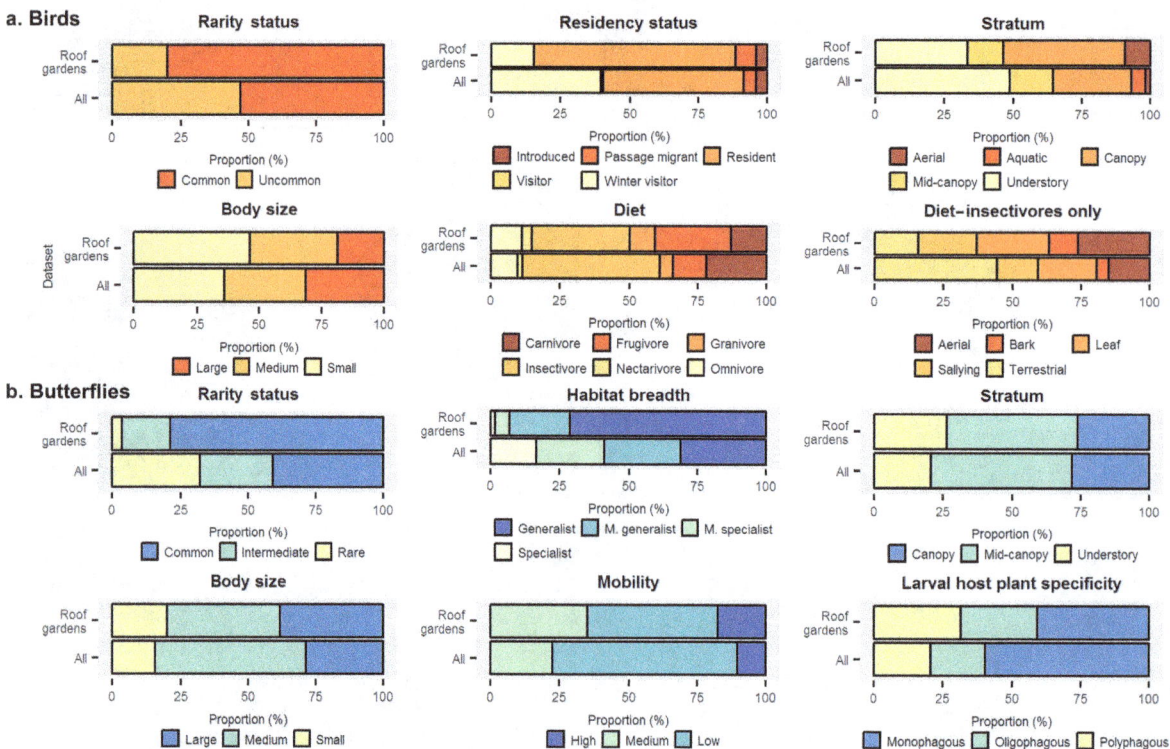

Fig. 3. Comparison of species attributes of fauna observed in roof gardens (Roof gardens) and those of the wider Singapore fauna (All) for birds (a) and butterflies (b). For birds, rarity status and residency assignments excluded rare species and followed Wang and Hails (2007), foraging stratum and diet followed Strange (2000) and Yong et al. (2013); body size was summarized as large, ≥30 cm; medium, 20–29.9 cm; small, <20 cm. For butterflies, attribute assignments followed Jain et al. (2017); larval host plant specificity was designated as monophagous = one genus in one family, oligophagous = more than one genus in one family, polyphagous = two or more genera in more than one family, for habitat breadth, M. generalist = moderate generalist, M. specialist = moderate specialist.

used for quantitative comparisons of species attribute differences between species observed in the roof garden surveys and those of the overall Singapore fauna. A subset of these attributes were designated as traits after Violle et al. (2007), that is, trait defined as any morphological, physiological, or phenological feature measurable at the individual level. For birds, these were body size, foraging stratum, and diet, whereas for butterflies, these were body size, flight height, mobility, and number of larval host plant genera (see Appendix S1: Table S3 for complete definitions).

Site attributes

We collected and collated a total of 65 attribute variables for each garden (Table 2). To facilitate application of this research by various stakeholders, we non-exclusively categorized these variables into three analytical groups—Planning, Architecture, and Management (See *Introduction*). Planning variables were defined as those under the control of urban planners who set and implement broadscale urban development conditions, including general levels of human activity at a site as proxied by noise levels, as well as overall levels of vegetation cover and habitat isolation. Architecture variables were defined as those related to the detailed physical form of roof gardens in terms of both building materials and initial plant selection, as specified by architects. Variables exclusively assigned to the Architecture group included height, exposure, planted area, temperature, human presence, and water type. Finally, Management variables were defined as those under the purview of on-site supervisors administering the long-term landscape regimes. Variables exclusively assigned to the Management group included maintenance levels, frequency of pesticide application, site age, and spontaneous vegetation. We did not include the number of host plant species at each site as a variable because its importance is already well established in the literature (Koh and Sodhi 2004), and the scope of such an analysis would need to be more focused on species-specific foraging range and breeding success to yield useful guidelines on host plant selection. Statistical analyses were run using this Planning–Architecture–Management framework. More details on the

methods used for environmental data collection are available in Appendix S1.

Analyses

Species accumulation curves were used to gauge the completeness of the species presence/absence records at each site using abundance/sample-based estimator Chao1 computed in EstimateS (Colwell 2006). For each taxon, four response variables were considered: species richness (total number of species), abundance (total number of individuals observed), Simpson's diversity ($1/D$) of individual species as calculated using the R package vegan (Oksanen et al. 2015), and Functional Dispersion (FDis) computed using the R package FD (Laliberté and Shipley 2011). Functional dispersion was defined as the mean distance of individual species to the centroid of all species in the community in terms of the species attributes designated as traits (Laliberté and Legendre 2010); thus, a site with higher FDis would contain a species assemblage with a greater diversity of the selected traits. The influence of predictor variables on these biodiversity metrics was examined separately following the Planning–Architecture–Management framework described above, as further detailed below. Unless otherwise stated, all statistical analyses were conducted in R.3.2.2 (R Core Team 2015), and graphs were plotted using the R package ggplot2 (Wickham 2009).

For faunal species response, Simpson's index was preferred because we considered it desirable from a conservation perspective to prioritize an even relative abundance across species, to mitigate against urban adapters sequestering all available resources and crowding out rarer species that might otherwise benefit from the gardens. Functional dispersion (Laliberté and Legendre 2010) was chosen to indicate guild diversity in order to compare communities with less than three functionally similar species, which would not be possible with functional richness (Mason et al. 2005). Shannon's Diversity Index (H) was used to summarize tree species community data as a predictor, since it was considered relevant to emphasize species richness over relative abundance, since a single mature tree of a suitable species was expected to have a greater effect on faunal diversity compared to a higher number of immature trees of a different species.

Table 2. Environmental variables used to analyze bird and butterfly communities on roof gardens and method-
ological details of environmental data collection are provided in Appendix S1.

P	A	M	Variable code	Explanation	Unit
*			Px_Tree	Proportion of tree-covered land within buffers of $x = 125$, 250, 500, and 1000 m of the site polygon boundaries	%
*			Px_Non-tree_Veg	Proportion of non-tree vegetated land within buffers of $x = 125, 250, 500$, and 1000 m of the site polygon boundaries	%
*			Px_Non-veg_Land	Proportion of non-vegetated land within buffers of $x = 125$, 250, 500, and 1000 m of the site polygon boundaries	%
*			GrassY	Distance to nearest non-tree vegetated patch of an area $\geq Y = 1, 5, 10$, and 20 ha	m
*			TreesY	Distance to nearest tree-covered patch of an area $\geq Y = 1, 5$, 10, and 20 ha	m
*			GCY	Distance to nearest vegetated patch of an area $\geq Y = 1, 5, 10$, and 20 ha	m
*	*		Noise_8am	Average decibel levels over a five-min period (08:00–08:05 hours)	dB
*	*		Noise_11am	Average decibel levels over a five-min period (11:00–11:05 hours)	dB
*			Site_area	Total area of garden surveyed (includes footpaths and hardscape features, but not mechanical and electrical service structures)	m^2
*			Site_height	Relative height of garden above ground level	m
*			Exposure	Whether the garden was partially or completely covered by a ceiling, 0 = no, 1 = yes	factor
*			Water	Presence of water, 0 = no water, 1 = chlorinated water, 2 = pond	factor
*			Water_area	Total area of waterbodies on site	m^2
*			Human_presence	Presence of humans, 0 = negligible, 1 = sporadic, 2 = regular	ordinal
*			Lawn_cover	Total area of vegetation maintained as lawn	m^2
*			Shrub_cover	Total area of shrubs, climbers, ferns, and other non-tree vegetation	m^2
*			Total_planted_area	Total area of lawn, shrubs, climbers, ferns, and other non-tree vegetation	m^2
*			Average_temp	Average ambient temperature recorded at each site	°C
*			Max_temp	Max ambient temperature recorded at each site	°C
*			Min_temp	Min ambient temperature recorded at each site	°C
*	*		Tree_height	Average height of all trees and palms in the garden	m
*	*		Tree_height diversity	Standard deviation of the heights of all trees and palms in the garden	m
*	*		Tree_crown area	Total crown area of trees and palms in the garden, averaged from five individuals and multiplied by species-specific abundance	m^2
*	*		Tree_abundance	Total number of trees and palms in the garden	individuals
*	*		Plant_species_richness_IS	Total number of plant species, including spontaneous plants	species
*	*		Butt_plants_IS	Number of confirmed butterfly nectar plants including spontaneous plants	species
*	*		Butt_plants_NS	Number of confirmed butterfly nectar plants excluding spontaneous plants	species
*	*		Bird_plants	Number of confirmed bird fruit/nectar plants	species
*	*		Plant_species_richness_NS	Total number of plant species, excluding spontaneous plants	species
*	*		Native_species_richness	Total number of native plant species	species
*	*		Habits_H	Shannon's D of six plant habits	. . .
*	*		Habits_N	Total number of plant habits represented at each site	habits
*	*		Climbers	Number of climber species recorded at each site	species
*	*		Epiphytes	Number of epiphyte species recorded at each site	species
*	*		Herbs	Number of herbaceous species recorded at each site	species
*	*		Shrubs	Number of shrub species recorded at each site	species

(Table 2. *Continued*)

Set P	Set A	Set M	Variable code	Explanation	Unit
*	*		Trees	Number of tree species recorded at each site	species
*	*		Trees_H	Shannon's *D* of tree species at each site	...
		*	Butt_plants_S	Number of spontaneous species confirmed to be butterfly nectar plants	species
	*		Plant_species_richness_S	Number of spontaneous plant species	species
	*		Site_age	Number of years since the garden was constructed, since 2014	yr
		*	Maintenance	Frequency/intensity of vegetation pruning, 0 = low, 1 = medium, 2 = high	ordinal
		*	Pesticide_use	Pesticide application frequency, all-natural = 0; as and when needed = 1; monthly = 2; fortnightly = 3	ordinal

Note: The Set column indicates which analysis set each predictor was grouped with for analyses; some variables were defined to belong to more than one set.

Hierarchical partitioning was used as the main analytical strategy, which required the pre-selection of up to nine candidate variables using the R package hier.part (Olea et al. 2010, Walsh and Mac Nally 2013). Since hierarchical partitioning only recognizes monotonic relationships, relationships between all candidate predictors and the respective response variables were first visually inspected for indications of quadratic relationships; other polynomial forms were ruled out as ecologically implausible. For each combination of two taxonomic scales, four response variables and three predictor variable sets, Spearman's rank correlations were first used to exclude predictor variables which could be correlated with other predictors but not with the dependent variable (Murray and Conner 2009). Predictor variables with $\rho > 0.3$ were selected as candidate predictors of influence; in the event that only one predictor variable was selected on this basis, it was retained if that correlation (ρ) was statistically significant ($P < 0.05$).

A stepwise variance inflation factor procedure was then applied, which removed all predictors with variance inflation factors exceeding five (Zuur et al. 2010); the remaining variables were retained in order of their correlation coefficients up to the maximum of nine predictors. Hierarchical partitioning was then applied to assess the total independent contributions of the predictor variables selected for each of the 12 model sets per taxon, and the statistical significance of each predictor was determined with a randomization routine comprising 1000 runs (Mac Nally 2002).

Hierarchical partitioning could yield spurious results associated with the existence of suppressor variables, that is, variables not directly correlated with the response, but correlated with other predictors such that their addition consistently improved overall model fit (Chevan and Sutherland 1991). To mitigate against this, we calculated another metric quantifying relative variable importance, the sum of AIC_c weights, using the R package MuMIn (Barton 2016). We considered the independent effect of a predictor to be statistically important if there was agreement with the rank order of its importance assigned by both hierarchical partitioning and the sum of AIC_c weights. The sum of AIC_c weights for variables was considered to change the rank order of the variables identified as statistically significant by hierarchical partitioning only if they differed by ≤ 0.05.

Relationships between all predictors identified to be statistically important for each model were fitted to regression equations. Quasipoisson errors were specified where abundance was the response variable, and non-continuous variables were coded as ordered factor levels. For relationships with linear fits, regression coefficients (βs), P values, and r^2 were reported, whereas for relationships with exponential asymptotic fits, the value of X prior to the marginal rate of change decreasing below the average rate of change for the data series was reported based on the marginal value theorem (Charnov 1976). Estimates for all these parameters were established with 1000 bootstrap runs using the R packages car (Fox and Weisberg 2011) and boot (Davison and

Hinkley 1997, Canty and Ripley 2016), except for exposure for which jack-knife (leave-one-out) resampling was applied owing to the limited number of non-exposed sites compared to exposed sites.

Although the above approach effectively identified predictor effects, which were the most statistically independent, it would be erroneous to interpret them as ecologically independent of one another, since many predictors were mutually correlated. Thus, correlations between the predictors determined to be statistically important based on the above approach were examined using a principal components analysis bi-plot. Spatial autocorrelation in the residuals of the original (non-bootstrapped) fits was assessed using 1000 random permutations of the local Moran's I statistic at two neighborhood levels (3 and 6) representing 10% and 20% of all sites, respectively, with package spdep (Bivand and Piras 2015).

RESULTS

A total of 53 bird and 57 butterfly species were recorded over 40 observation hours per taxa per site (Appendix S1: Tables S1, S2). Total independent encounters exceeded 15,000 for birds and 8000 for butterflies. Site-level species richness ranged from 2 to 29 for birds, and 1 to 37 for butterflies. Except for a single site with exceptionally low bird species richness (species observed = 1, predicted Chao1 = 2), the minimum percentage of observed species as a fraction of the lower 95% confidence interval for the Chao1 estimator after 20 surveys was 79% for birds, and 86% for butterflies, so we consider that our data provide a reasonably complete record of the taxa present at the sites.

For birds, the total number of species encountered in the roof garden sites represented 13% of the total avifauna of Singapore. Courtship, mating, or nesting behaviors were observed in 20 species, and 12 out of a total of 53 records were of species regarded as locally rare or uncommon, of which nine were native species (Appendix S1: Table S1). Three uncommon native residents (*Zosterops palpebrosus*, *Loriculus galgulus*, *Copsychus saularis*) and one migrant (*Ficedula mugimaki*) were found in at least five sites with a total abundance exceeding 30 individuals. The Blue Rock

Thrush (*Monticola solitarius*), a rare passage migrant, was sighted at two sites.

Birds on roof gardens were more likely to be residents and canopy dwellers, with fewer understory species relative to the overall avifauna (Fig. 3a). Large-bodied species were relatively less represented on roof gardens than in Singapore overall (Fig. 3a). Finally, roof gardens were relatively enriched with frugivores and aerial insectivores, whereas the overall Singapore avifauna contained relatively more carnivorous and terrestrial insectivores (Fig. 3a).

The total number of butterfly species observed in the roof garden sites (57) represented 18% of the total butterfly fauna of Singapore. Courtship, mating, or ovipositing behaviors were observed in 15 species (Appendix S1: Table S2). Twelve species were either intermediately abundant or rare (Appendix S1: Table S2), of which 11 were native, and two were present at abundances of more than 10 individuals across three or more sites (*Pelopidas assamensis* and *Pachliopta aristolochiae asteris*). The rare *Tajuria dominus* was present at one site. Roof gardens attracted a proportionately higher richness of understory butterflies compared to the overall fauna (Fig. 3b). Species with a larger wingspan and which were more mobile were better represented in roof gardens (Fig. 3b). Finally, of the species observed in roof gardens, proportionally fewer butterfly species were monophagous (40% vs. 60%), while more were polyphagous (31% vs. 20%).

Principal components analysis indicated that the first two axes explained 49.9% of the variance among the 22 statistically important predictor variables (Fig. 4). Two sets of correlations were relevant to constrain interpretations of the bivariate relationships highlighted in Table 3: (1) Noise at 08:00 hours (early morning), site height, low exposure, and distance to the nearest non-tree vegetation patch ≥5 ha were positively interrelated, and (2) the number of shrub species was positively associated with the number of managed butterfly nectar plants, while both were anti-correlated with average tree height. The strength of some bivariate relationships depended on geographic location; these were mainly associated with vegetation variables including the number of bird fruit and nectar plants, the number of butterfly nectar plants, plant species richness, and tree species diversity,

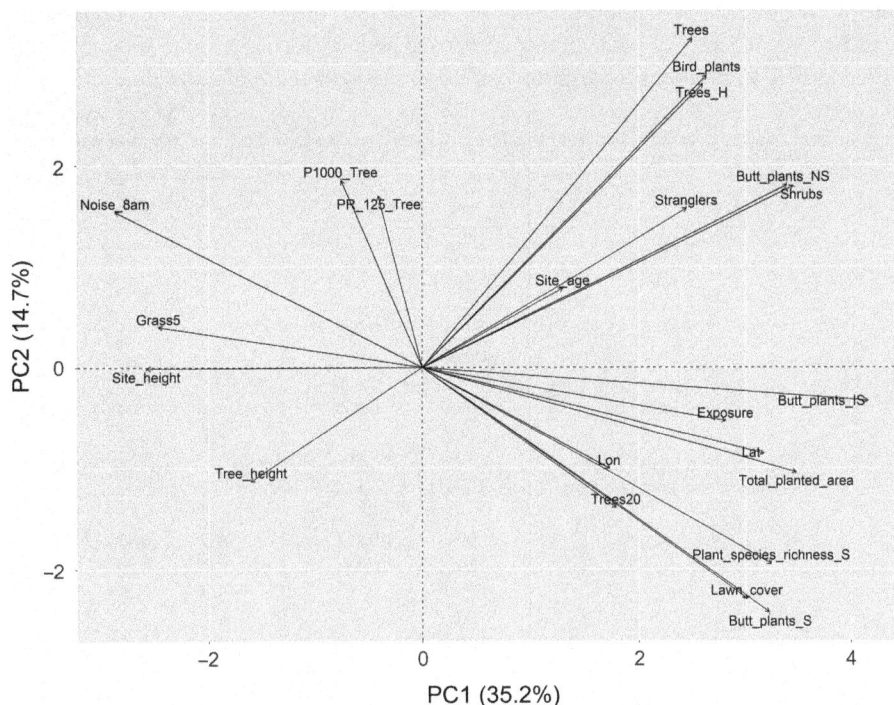

Fig. 4. Principal components analysis bi-plot visualizing correlation structure between statistically significant predictor variables listed in Table 3.

as well as site age (Table 3; Appendix S1: Fig. S1).

Among the Planning set, the proportion of locally vegetated land did not have a statistically important influence on either bird or butterfly communities (Table 3), leading us to reject hypothesis 1. In comparison, distance to the nearest non-tree vegetation patch ≥5 ha was the most important predictor for all butterfly community metrics except Simpson's diversity. The importance of proximity to these patches for roof garden butterfly community composition exhibited a marginal increase up to approximately 1 km (Fig. 5b). Distance to the nearest patch of tree-covered vegetation ≥20 ha was positively and non-linearly associated with Simpson's diversity of butterfly species, with this marginal effect declining beyond 1.3 km (Table 3). Early morning noise was the only statistically important variable for both birds and butterflies in terms of affecting species richness and Simpson's diversity (Table 3, Fig. 5a). The effect of noise as signified by the Spearman's correlation coefficient (r) between noise and species abundance

was not significantly different between bird species observed to engage in reproductive behaviors, compared to those that did not (Kruskal–Wallis test, $P = 0.49$), nor was it different between predatory (carnivores, insectivores, omnivores) and non-predatory bird species (Kruskal–Wallis test, $P = 0.12$). There was also no evidence that the relationship between the abundance of predatory bird species and total butterfly abundance depended on noise levels ($P = 0.345$).

For the Architecture set, total planted area did not have a statistically important influence on any community metric for either taxa at the scales examined (Tables 2 and 3), leading us to reject hypothesis 2. Nevertheless, bird species richness increased asymptotically with total planted area, with above-average marginal increases up to 1100 m^2; similar results were seen for butterflies, with an equivalent threshold at 1400 m^2 (Fig. 5d). Site height was important for both taxa, negatively affecting species richness, Simpson's diversity, and abundance for birds, and species richness and FDis for butterflies

Table 3. Summarized relative variable importance measures for four bird and butterfly community metrics: species richness (s), Simpson's diversity (1/D), abundance (total abundance of all individuals counted per site) and functional dispersion (FDis, mean distance in multidimensional trait space of individual species to the centroid of all species).

Taxon/Metric	Rank	Variable	P_H	SW/NSW	R	P_β	R^2	O	M3	M6
Birds										
Planning										
Richness	1	Noise_8am	<0.001	0.900	NL	0.030	0.269		0.440	0.425
Simpson's diversity	1	Noise_8am	$P_\rho = 0.004$	...	NL	0.046	0.244		0.495	0.574
Architecture										
Richness	1	Site_height	<0.001	0.926	NA			66	0.582	0.599
	2	Exposure	<0.001	0.974	PL	<0.001			0.958	0.837
Simpson's diversity	1	Site_height	0.005	0.830	NA			69	0.130	0.209
	2	Bird_plants	0.023	0.764	PL	0.01	0.347		0.167	0.045
	3	Exposure	0.049	0.482	PL	<0.001	0.202		0.903	0.71
Abundance	1	Site_height	<0.001	0.998	NA			59	0.761	0.814
FDis*	1	Average tree height	0.055	0.613	PA			2.1	0.397	0.230
Management										
Richness	1	Shrubs	0.004	0.714	PL	0.001	0.410		0.937	0.512
	2	Butterfly plants_NS	0.003	0.431	PL	0.012	0.304		0.032	0.014
Simpson's diversity	1	Bird_plants	<0.001	0.797	PL	0.002	0.261		0.008	0.001
	2	Site_age	0.002	0.842	PA			6	0.228	0.044
Abundance	1	Butterfly_plants_IS	<0.001	0.890	PL	0.016	0.287		0.863	0.279
FDis*	1	Average_tree_height	0.087	0.611	PA			2.1		
Butterflies										
Planning										
Richness	1	Grass5	0.017	0.673	NA			1100	0.511	0.304
	2	Noise_8am	0.019	0.712	NL	0.046	0.281			
Simpson's diversity	1	Noise_8am	<0.001	0.892	NL	0.043	0.274		0.440	0.425
	2	Trees20	0.016	0.815	PA			1300	0.414	0.093
Abundance	1	Grass5	<0.001	0.697	NA			900	0.330	0.253
FDis	1	Grass5	<0.001	0.827	NA			1200	0.750	0.498
Architecture										
Richness	1	Butterfly_plants_NS	<0.001	0.989	PL	<0.001	0.457		0.117	0.013
	2	Site_height	<0.001	0.919	NA			49	0.933	0.926
Simpson's diversity	1	Butterfly_plants_NS	0.018	0.848	PL	0.004	0.429		<0.001	<0.001
Abundance	1	Lawn_cover	<0.001	0.993	PL	0.000	0.694		0.305	0.104
FDis	1	Site_height	<0.001	0.993	NL	0.006	0.454		0.586	0.212
Management										
Richness	1	Butterfly_plants_S	<0.001	0.998	PA			4	0.026	0.011
	2	Butterfly_plants_NS	<0.001	0.978	PL	0.248	0.121		0.743	0.626
Simpson's diversity	1	Butterfly_plants_S	<0.001	0.856	PL	0.004	0.429		<0.001	<0.001
	2	Butterfly_plants_NS	0.01	0.742	PA			6	0.645	
Abundance	1	Butterfly_plants_S	<0.001	0.999	PL	0.002	0.427		0.95	0.918
FDis	1	Plant_species richness_S	0.131	0.516	PA			5	0.104	0.033
	2	Trees_H	0.143	0.420	PA			0.5	0.033	0.021
	3	Butterfly plants_NS	0.177	0.348	PA			8	0.54	0.101

Notes: Under a hierarchical partitioning framework, the probability values (P_H) indicate the probability of the independent effect of that predictor being higher than the mean independent effects of all other variables considered based on the one-tailed upper 95% confidence limit. The sum of weights (SW) indicates the probability of that predictor having an effect on the response relative to all other predictors normalized to one. Predictors shown were those for which $P < 0.05$, and the rank orders of both P and SW were in agreement. Response variables for which predictors returned $P < 0.1$ are indicated with asterisks in the Predictor column. Four types of bivariate relationships (R) were denoted as positive linear (PL), negative linear (NL), positive asymptotic (PA), and negative asymptotic (NA). For linear fits, two-tailed P-values indicating $\beta \neq 0$ are reported (P_β) and adjusted R^2 values are shown. For non-linear fits, the x value approximating the optimization rule based on marginal value theorem is reported (O). All selected predictors were continuous, and fitted parameters were estimated from bootstrapped or jack-knifed runs ($n = 1000$, jack-knife applied for Exposure only). P values indicating non-stationarity in the residuals of each model fit at neighborhoods of three (M3) and six (M6) based on local Moran's I, respectively, based on 1000 permutations of the spatial weighting scheme.

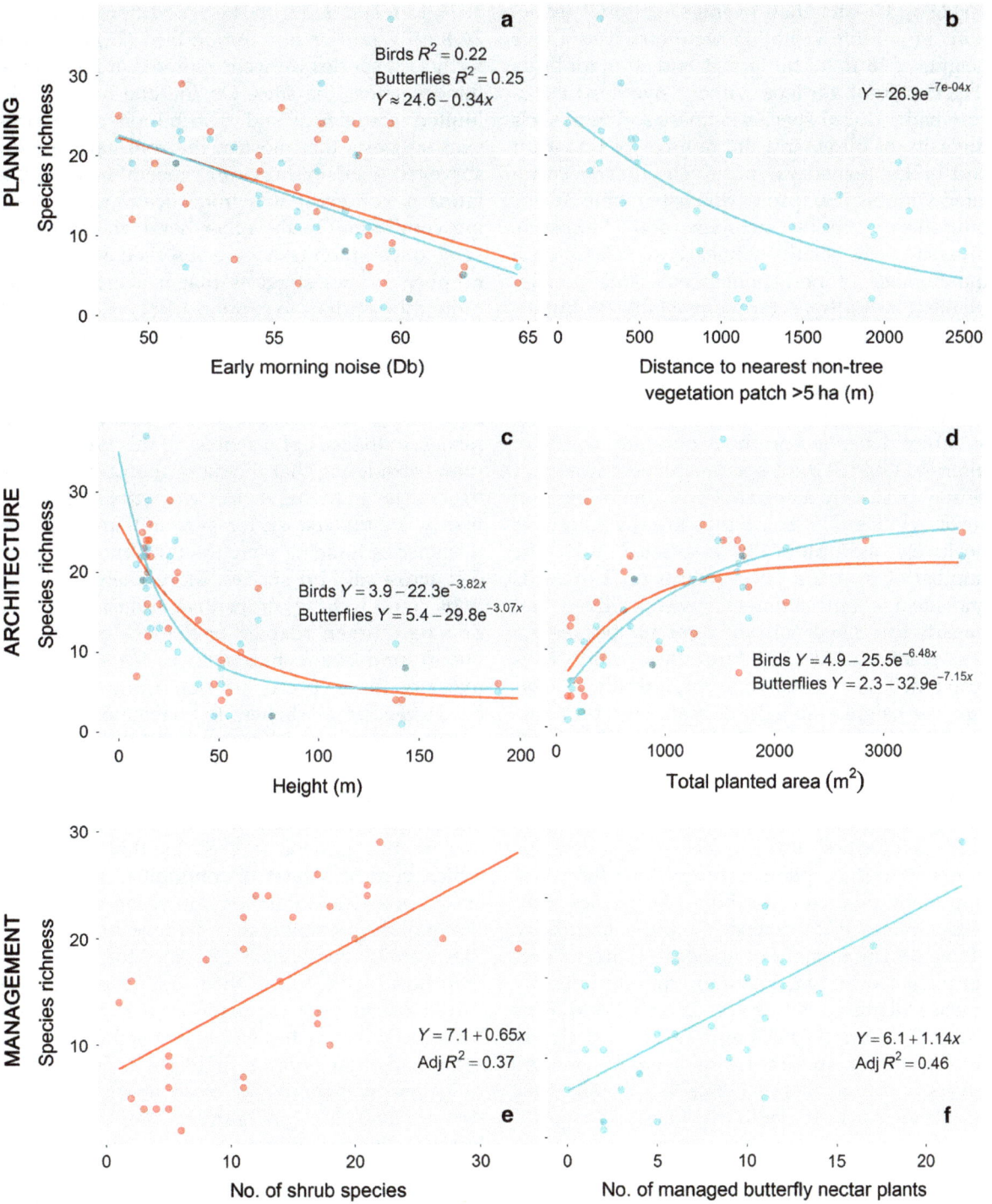

Fig. 5. Relationships between core predictors and bird (red) and butterfly (blue) species richness, organized by variable set: early morning noise at 08:00 hours (a), distance to nearest non-tree vegetation patch ≥5 ha (b), roof garden height (c), total planted area (d), no. of shrub species (e), and no. of managed butterfly nectar plants (f).

(Table 3). Greater than average marginal reductions in wildlife visitation were observed above heights of 48 m for butterflies and 60 m for birds (Fig. 5c). Roof gardens without overhead exposure had reduced species richness and Simpson's diversity of birds, and the number of bird fruit and nectar plants was positively linearly correlated with bird Simpson's diversity (Table 3). For butterflies, species richness and Simpson's diversity were positively linearly correlated with the number of non-spontaneous (intentionally planted) butterfly nectar plants (Table 3). Butterfly abundance was positively related to the amount of lawn cover (Table 3).

For the Management set, the most important environmental predictor of bird community structure depended on the community metric of interest (Table 3). Bird species richness was most closely positively associated with the number of shrub species (Fig. 5e), Simpson's diversity was positively asymptotically associated with the number of fruit and nectar plants used by birds and site age, bird abundance was positively and linearly correlated with the total number of nectar plants used by butterflies, and FDis responded positively and asymptotically to average tree height (Table 3). Overall, both bird species richness and abundance responded more strongly to floristic composition than to habitat structural variables such as tree height, leading us to reject hypothesis 3a.

For butterflies, the number of spontaneous butterfly nectar plant species was the most important predictor for butterfly species richness, Simpson's diversity, and abundance (Table 3). The number of managed butterfly nectar plant species was also important for butterfly species richness, Simpson's diversity, and FDis (Table 3). Butterfly FDis responded most clearly to total plant species richness and Shannon's diversity of tree species (Table 3). Overall, these results are consistent with hypothesis 3(b).

DISCUSSION

Species attributes of roof garden fauna

Bird and butterfly species that visited roof gardens in Singapore tended to be common, mobile generalists with typical urban adapter profiles, which is consistent with previous local studies on these taxa from urban parks (Koh and Sodhi 2004, Lim and Sodhi 2004). In addition, however, 24 locally rare or uncommon bird (12) and butterfly (12) species were encountered at low abundances across the sites. On the one hand, their limited abundance and distribution across the sites suggests that most of the roof gardens we surveyed were not providing general benefits for fauna of conservation significance in their existing condition. On the other hand, the fact that some uncommon taxa were observed at a limited number of sites suggests that it is possible for more roof gardens to support key taxa of interest —if they are planned, designed, and managed appropriately.

The finding that understory birds were underrepresented in roof gardens is unlikely to be related to the lack of resources at this height stratum, considering that the most abundant species overall (Javan myna, *Acridotheres javanicus*) is primarily an understory forager, and instances of this species foraging were the third most numerous across all bird species. More likely explanations could include competitive exclusion by this dominant urban adapter, or the lack of contiguous ground-level connectivity to the sites, since many of these species are not strong fliers and need vegetated shelter to navigate between resources (Hails and Kavanagh 2013). For butterflies, the prevalence of understory butterfly species particularly grassland specialists could be attributed to the availability of sunny, open habitats on the gardens favored by this guild, for which vertical habitat discontinuities are apparently not a strict limitation. However, we cannot discount the possibility that larvae of some butterflies were unintentionally imported together with their host plants, which then continued to establish self-sustaining populations in roof gardens high above their natural foraging heights.

Overall, these findings highlight both opportunities and constraints for biodiversity conservation in intensive roof gardens. Populations of butterfly species should be readily supported in roof gardens if their nectar plants are planted in abundance, and not treated regularly to remove butterfly larvae. In contrast, frugivorous birds may be well adapted to take advantage of suitable resources provided by roof gardens, but it may not be practical to target understory bird diversity given the lack of vertical connectivity required by most members of this group.

Conservation goals for roof gardens should therefore take into account that these novel ecosystems may be environmentally predisposed to host certain suites of species, rather than aiming to directly replicate ground-level ecological communities (Williams et al. 2014).

Urban planning for roof garden diversity

We expected to find that vegetation cover within the more local buffers would show a stronger influence on faunal diversity in the roof gardens, because these would provide source habitats for a homogeneous suite of urban-adapted species to disperse to the sites. Generally, we found that metrics relating to isolation up to 3.5 km had stronger (negative) effects on faunal diversity than did landscape composition at buffer sizes up to 1 km. This implies that the assemblage of species we observed at the roof gardens was not merely a subset of urban-adapted species already present in the local vicinity at ground level, but included species from other less-urbanized habitats. We identified the threshold size of these habitats to be five hectares for non-tree vegetation for butterflies, implying that the retention of ground-level patches at this spatial scale would be important to maintain the diversity of butterfly species observed to visit roof gardens.

We found that early morning noise was more strongly associated with the faunal communities observed in the roof garden sites, than was either the composition of the surrounding habitat or isolation of the roof garden. Mechanistic explanations for negative effects of urban noise on birds include the masking of vocalization signals used for reproduction such as territory defense, mate location, and feeding calls (Warren et al. 2006, Lowry et al. 2013), as well as auditory signals used for prey detection (Francis 2015). However, neither of these hypotheses were supported by our data, in that the effect of noise on species abundance did not differ between bird species observed to reproduce at the study sites and those that did not, or between predatory and non-predatory bird species (see *Results*). Urban noise could affect the predator-evasion response of at least some butterfly species by masking the sounds of bird flight (Fournier et al. 2013); however, at these sites, the relationship between butterfly abundance and the abundance of predatory bird species was not significantly moderated by noise (see *Results*). Also, we found that noise affected birds and butterflies equally (i.e., exhibited very similar slopes in Fig. 5), indicating a need to consider other interpretations.

We suggest that noise exerted proxy effects on species richness in the roof gardens through its relationship with total planted area, in that noise and total planted area were negatively correlated, whereas total planted area was positively correlated with species richness of birds and butterflies (Figs. 4, 5d and discussion in next section). The reasoning for this is that in noisier, more heavily developed urban precincts, the opportunity cost for architects to incorporate planted areas that do not offer direct commercial returns is greater; hence, there is greater pressure to design roof gardens with more limited planting space. If this interpretation is correct, the implication for planners is that policies that incentivize developers to create and maintain more sizeable vegetated areas (as opposed to total roof garden area) are important for roof garden biodiversity and should be encouraged. Nevertheless, given that noise was also correlated with several other environmental variables (Fig. 4), this hypothesis needs to be further investigated with an experimental approach.

Designing faunally diverse roof gardens

Contrary to expectations, the amount of vegetation on the roof gardens was not the most important predictor of community structure for either taxa. However, we did find that total planted area exhibited generally well-defined positive asymptotic relationships with both bird and butterfly species richness (Fig. 5d). The size of a roof garden is typically dictated by human use constraints rather than wildlife value. Nevertheless, the existence of size thresholds for enhancing biodiversity on roof gardens could provide empirical incentives for roof garden designers to expand the size of planted areas through the design of new buildings or renovating existing gardens. Application of marginal value theory suggests increasing returns on wildlife enhancement up to 1100 m^2 for birds and 1400 m^2 for butterflies. We emphasize that these thresholds apply for the absolute vegetated areas allocated at the planting bed level, and not the total area of each garden. Indeed, several sites with total areas exceeding 4000 m^2 had among the lowest species richness of birds and butterflies due to the low proportion of planted area at these sites. In the context of space constraints typical of high-density urban development, these thresholds offer first approximations as to the optimal sizing of compact managed green spaces for biodiversity conservation and enhancement.

Site height had the strongest (negative) effects on the community structure of both birds and butterflies on our roof garden sites, including species richness, Simpson's diversity, and abundance for birds, and richness and FDis for butterflies (Table 3). These findings contribute to growing evidence that the height of a green roof affects its value to a range of wildlife taxa including bees (MacIvor 2015), bats (Pearce and Walters 2012), spiders, and wasps (Madre et al. 2013). Lee and Lin (2015) did not find an effect of roof garden height on butterfly species richness in Taipei City, Taiwan, but their study included only 11 sites with a height range of 7–34 m. Although site height was closely correlated with other environmental variables including noise, total planted area, and exposure, relationships between those predictors and species richness suggest that the effect of height was the strongest and most direct (Table 3, Fig. 5). For example, whereas sites with moderately high noise levels still recorded moderate species richness of birds and butterflies (Fig. 5a), the drop in species numbers with height was much more pronounced (Fig. 5c).

The species whose abundances were most significantly affected by site height were in fact the most common for both taxa, which suggests that a general exclusion mechanism, as opposed to a guild-specific filter, could be responsible for this pattern. Specifically, we posit that the energetic cost to volant fauna (including birds and butterflies) of visiting roof gardens should increase in tandem with the rise in the urban wind velocity profile with height (Pelliccioni et al. 2016). In this context, optimal foraging theory predicts that they should preferentially search horizontally in urban areas, and only search vertically in the absence of adequate ground-level resources. This hypothesis predicts that the creation of (sheltered) gardens at mid-levels (vertical stepping-stones) would be expected to have a positive influence on wildlife access to higher roof gardens. Indeed, Site 14, which had a continuous vegetated corridor wrapping around the building from the ground level to the roof garden, exhibited the highest (positive) deviance from the predicted values (Table 1, Site 14). In contrast, greater than average marginal reductions in wildlife visitation were observed in stand-alone roof gardens above heights of 48 m for butterflies and 60 m for birds (Table 3). Generally, these findings suggest that roof gardens designed to be below 50 m in height offer the best opportunities for supporting bird and butterfly diversity.

Roof garden management

We found that floristic diversity—as represented by the number of shrub species and the number of fruit and nectar plants (Appendix S1: Table S4)—generally had a greater influence on bird community structure compared to variables related to vegetation structure, such as tree crown cover, tree height diversity, and growth form diversity. It is worth noting that bird richness and abundance showed stronger associations with butterfly nectar plant richness than with bird fruit or nectar plant richness, which suggests that nectar plants commonly associated with butterflies could play an indirect role in avifaunal biodiversity support, for example, by enhancing the general entomofauna for insectivores. Paker et al. (2014) also found that the bird community in an urban green space responded more strongly to shrub species richness than to shrub cover. Together, these results imply that to encourage bird diversity in roof gardens, it would be more effective to focus on planting a diverse selection of shrub species, compared to providing proportionally large areas of homogeneous shrub cover.

The significance of spontaneous butterfly nectar plant species for butterfly community structure on roof gardens is not surprising, considering that many of those plant species were continuously flowering ruderal herbs with short corollas, which would be accessible to a wide range of butterfly species. The slopes of the relationships indicate relatively low thresholds in terms of the number of spontaneous plant species required to significantly enhance butterfly species richness and FDis (4–6 spp., Table 3). Thus, roof gardens could be selectively weeded to retain a handful of key ruderal nectar plant species (Appendix S1: Table S5), which may attract a relatively diverse and abundant urban butterfly assemblage. Nevertheless, these results also support the role of managed nectar plants in supporting butterfly diversity on roof gardens. The observation that the effects of managed nectar plant species were positive and linear compared to the asymptotically saturating effects of spontaneous nectar plant species (Table 3) suggests that the judicious planting of butterfly nectar plants should have positive effects on butterfly diversity, beyond what would be supported by spontaneous nectar plants alone.

Conclusion

Tropical roof gardens in Singapore hosted a diverse assemblage of bird and butterfly species. Most of the 53 bird and 57 butterfly species observed were common urban generalists, but 24 uncommon or rare species (12 bird species, 12 butterfly species) were also recorded. Many of the factors identified to be important for wildlife such as height, total planted area, and plant structural and floristic diversity could be considered complementary to the environmental preferences of human users of roof gardens, suggesting that if properly conceived, roof gardens do have some potential to reconcile urban development and biodiversity conservation. Nevertheless, in their existing form, their role vis-à-vis large ground-level habitats was generally supplementary rather than substitutionary. The results of this study provide practical first approximations for planners, architects, and managers seeking to improve the biodiversity potential of managed urban green spaces.

Acknowledgments

This research was funded by the National Parks Board, Singapore. The authors wish to thank Jolene Lim and Anisha Rajbhandari for fieldwork assistance; Benjamin Lee for advice on noise measurements; Lok Yan Ling, Tay Bee Choo, and Rachel Teo for information on roof gardens in Singapore; and the building owners and management of all our sites for access to their roof gardens, including Ang Mo Kio Town Council, Chua Chu Kang Town Council, Marine Parade Town Council, Punggol-Pasir Ris Town Council, Tanjong Pajar Town Council, Grange Infinite, Hotel Jen, NTUC Centre, Orchard Central, Khoo Teck Phuat Hospital, Secom (Singapore) Pte Ltd, Solaris@One-North, The Metropolis, Ho Bee (One-North) Pte Ltd, and others.

Literature Cited

Alvey, A. A. 2006. Promoting and preserving biodiversity in the urban forest. Urban Forestry & Urban Greening 5:195–201.

Barton, K. 2016. MuMIn: Multi-Model Inference. https://cran.r-project.org/package=MuMIn

Baumann, N. 2006. Ground-nesting birds on green roofs in Switzerland: preliminary observations. Urban Habitats 4:37–50.

Beninde, J., M. Veith, and A. Hochkirch. 2015. Biodiversity in cities needs space: a meta-analysis of factors determining intra-urban biodiversity variation. Ecology Letters 18:581–592.

Bivand, R., and G. Piras. 2015. Comparing implementations of estimation methods for spatial econometrics. Journal of Statistical Software 63:1–36. http://www.jstatsoft.org/v63/i18/

Bland, R. L., J. Tully, and J. J. Greenwood. 2004. Birds breeding in British gardens: An underestimated population? Bird Study 51:97–106.

Bonthoux, S., M. Brun, F. Di Pietro, S. Greulich, and S. Bouché-Pillon. 2014. How can wastelands promote biodiversity in cities? A review. Landscape and Urban Planning 132:79–88.

Braaker, S., J. Ghazoul, M. Obrist, and M. Moretti. 2014. Habitat connectivity shapes urban arthropod communities: the key role of green roofs. Ecology 95:1010–1021.

Butterfly Circle. 2016. Butterfly Circle Checklist. http://www.checklist.butterflycircle.com/

Canty, A. and B. Ripley. 2016. boot: Bootstrap R (S-Plus). https://cran.r-project.org/web/packages/boot/index.html

Carbó-Ramírez, P., and I. Zuria. 2011. The value of small urban greenspaces for birds in a Mexican city. Landscape and Urban Planning 100:213–222.

Caryl, F. M., K. Thomson, and R. Ree. 2013. Permeability of the urban matrix to arboreal gliding mammals: sugar gliders in Melbourne, Australia. Austral Ecology 38:609–616.

Charnov, E. L. 1976. Optimal foraging, the marginal value theorem. Theoretical Population Biology 9:129–136.

Chevan, A., and M. Sutherland. 1991. Hierarchical partitioning. American Statistician 45:90–96.

Chong, K. Y., S. Teo, B. Kurukulasuriya, Y. F. Chung, S. Rajathurai, and H. T. W. Tan. 2014. Not all green is as good: different effects of the natural and cultivated components of urban vegetation on bird and butterfly diversity. Biological Conservation 171:299–309.

Coffman, R. R., and T. Waite. 2011. Vegetated roofs as reconciled habitats: rapid assays beyond mere species counts. Urban Habitats 6. http://urbanhabitats.org/v06n01/vegetatedroofs_full.html

Colwell, R. K. 2006. EstimateS: statistical estimation of species richness and shared species from simples, version 8.0. http://purl.oclc.org/estimates

Corlett, R. T. 1991. Vegetation. Pages 134–154 in L. S. Chia, A. Rahman, and B. H. Tay, editors. The biophysical environment of Singapore. Singapore University Press, Singapore, Singapore.

Davies, Z. G., R. A. Fuller, A. Loram, K. N. Irvine, V. Sims, and K. J. Gaston. 2009. A national scale inventory of resource provision for biodiversity within domestic gardens. Biological Conservation 142:761–771.

Davison, A. C., and D. V. Hinkley. 1997. Bootstrap methods and their application. Cambridge University Press, Cambridge, UK.

Dearborn, D. C., and S. Kark. 2010. Motivations for conserving urban biodiversity. Conservation Biology 24:432–440.

Department of Statistics. 2015. Yearbook of statistics, Singapore, 2015. Department of Statistics, Ministry of Trade & Industry, Singapore, Singapore.

Fernandez-Canero, R., and P. Gonzalez-Redondo. 2010. Green roofs as a habitat for birds: a review. Journal of Animal and Veterinary Advances 9:2041–2052.

Fong, M., L. K. Ng, and Meteorological Service Singapore. 2012. The weather and climate of Singapore. Meteorological Service Singapore, Singapore, Singapore.

Fontana, S., T. Sattler, F. Bontadina, and M. Moretti. 2011. How to manage the urban green to improve bird diversity and community structure. Landscape and Urban Planning 101:278–285.

Fournier, J. P., J. W. Dawson, A. Mikhail, and J. E. Yack. 2013. If a bird flies in the forest, does an insect hear it? Biology Letters 9:20130319.

Fox, J., and S. Weisberg. 2011. An R companion to applied regression. Second edition. Sage, Thousand Oaks, California, USA.

Francis, C. D. 2015. Vocal traits and diet explain avian sensitivities to anthropogenic noise. Global Change Biology 21:1809–1820.

Garbuzov, M., and F. L. Ratnieks. 2014. Quantifying variation among garden plants in attractiveness to bees and other flower-visiting insects. Functional Ecology 28:364–374.

Gaston, K. J. 2000. Global patterns in biodiversity. Nature 405:220–227.

Gaston, K. J., R. M. Smith, K. Thompson, and P. H. Warren. 2005. Urban domestic gardens (II): experimental tests of methods for increasing biodiversity. Biodiversity & Conservation 14:395–413.

Goddard, M. A., A. J. Dougill, and T. G. Benton. 2010. Scaling up from gardens: biodiversity conservation in urban environments. Trends in Ecology & Evolution 25:90–98.

Hails, C. J., and M. Kavanagh. 2013. Bring back the birds! Planning for trees and other plants to support Southeast Asian wildlife in urban areas. The Raffles Bulletin of Zoology 29:243–258.

Jain, A., F. K. Lim, and E. L. Webb. 2017. Species-habitat relationships and ecological correlates of butterfly abundance in a transformed tropical landscape. Biotropica 49:355–364.

Jha, S., and C. Kremen. 2013. Resource diversity and landscape-level homogeneity drive native bee foraging. Proceedings of the National Academy of Sciences USA 110:555–558.

Koh, L. P., and N. S. Sodhi. 2004. Importance of reserves, fragments, and parks for butterfly conservation in a tropical urban landscape. Ecological Applications 14:1695–1708.

Laliberté, E., and P. Legendre. 2010. A distance-based framework for measuring functional diversity from multiple traits. Ecology 91:299–305.

Laliberté, E., and B. Shipley. 2011. FD: measuring functional diversity from multiple traits, and other tools for functional ecology. R package version 1.0-11. http://CRAN.R-project.org/package=FD

Lee, L.-H., and J.-C. Lin. 2015. Green roof performance towards good habitat for butterflies in the compact city. International Journal of Biology 7:103.

Lim, K. S. 2009. The avifauna of Singapore. Nature Society (Singapore), Bird Group Records Committee, Singapore, Singapore.

Lim, H. C., and N. S. Sodhi. 2004. Responses of avian guilds to urbanisation in a tropical city. Landscape and Urban Planning 66:199–215.

Lowry, H., A. Lill, and B. Wong. 2013. Behavioural responses of wildlife to urban environments. Biological Reviews 88:537–549.

Mac Nally, R. 2002. Multiple regression and inference in ecology and conservation biology: further comments on identifying important predictor variables. Biodiversity & Conservation 11:1397–1401.

MacIvor, J. S. 2015. Building height matters: nesting activity of bees and wasps on vegetated roofs. Israel Journal of Ecology & Evolution 62:88–96.

MacIvor, J. S., and K. Ksiazek. 2015. Invertebrates on green roofs. Pages 333–355 in R. K. Sutton, editor. Green roof ecosystems. Springer, New York, New York, USA.

Madre, F., A. Vergnes, N. Machon, and P. Clergeau. 2013. A comparison of 3 types of green roof as habitats for arthropods. Ecological Engineering 57:109–117.

Madre, F., A. Vergnes, N. Machon, and P. Clergeau. 2014. Green roofs as habitats for wild plant species in urban landscapes: first insights from a large-scale sampling. Landscape and Urban Planning 122:100–107.

Mason, N. W., D. Mouillot, W. G. Lee, and J. B. Wilson. 2005. Functional richness, functional evenness and functional divergence: the primary components of functional diversity. Oikos 111:112–118.

McDonald, R. I., P. Kareiva, and R. T. Forman. 2008. The implications of current and future urbanization for global protected areas and biodiversity conservation. Biological Conservation 141:1695–1703.

McKinney, M. L. 2006. Urbanization as a major cause of biotic homogenization. Biological Conservation 127:247–260.

Munshi-South, J. 2012. Urban landscape genetics: Canopy cover predicts gene flow between white-footed mouse (Peromyscus leucopus) populations in New York City. Molecular Ecology 21:1360–1378.

Murray, K., and M. M. Conner. 2009. Methods to quantify variable importance: implications for the analysis of noisy ecological data. Ecology 90: 348–355.

Nagendra, H., H. S. Sudhira, M. Katti, M. Tengö, and M. Schewenius. 2014. Urbanization and its impacts on land use, biodiversity and ecosystems in India. INTERdisciplina 2:305–313.

Nielsen, A. B., M. van den Bosch, S. Maruthaveeran, and C. K. van den Bosch. 2014. Species richness in urban parks and its drivers: a review of empirical evidence. Urban Ecosystems 17:305–327.

NSS Bird Group. 2015. Checklist of the birds of Singapore. Nature Society, Singapore, Singapore.

Oksanen, J., F. G. Blanchet, R. Kindt, P. Legendre, P. R. Minchin, R. B. O'Hara, G. L. Simpson, P. Solymos, M. H. H. Stevens, and H. Wagner. 2015. vegan: community ecology package. https://cran.r-project.org/package=vegan

Olea, P. P., P. Mateo-Tomás, and Á. De Frutos. 2010. Estimating and modelling bias of the hierarchical partitioning public-domain software: implications in environmental management and conservation. PLoS ONE 5:e11698.

Paker, Y., Y. Yom-Tov, T. Alon-Mozes, and A. Barnea. 2014. The effect of plant richness and urban garden structure on bird species richness, diversity and community structure. Landscape and Urban Planning 122:186–195.

Pauchard, A., M. Aguayo, E. Peña, and R. Urrutia. 2006. Multiple effects of urbanization on the biodiversity of developing countries: the case of a fast-growing metropolitan area (Concepción, Chile). Biological Conservation 127:272–281.

Pearce, H., and C. L. Walters. 2012. Do green roofs provide habitat for bats in urban areas? Acta Chiropterologica 14:469–478.

Pelliccioni, A., P. Monti, and G. Leuzzi. 2016. Wind-speed profile and roughness sublayer depth modelling in urban boundary layers. Boundary-Layer Meteorology 160:225–248.

Quigley, M. F. 2011. Potemkin gardens: biodiversity in small designed landscapes. Pages 85–92 in J. Niemelä, editor. Urban Ecology: patterns, processes and applications. Oxford University Press, Oxford, UK.

R Core Team. 2015. R: a language and environment for statistical computing. R Foundation for Statistical Computing, Vienna, Austria. http://www.R-project.org/

Rupprecht, C. D., J. A. Byrne, J. G. Garden, and J.-M. Hero. 2015. Informal urban green space: a trilingual systematic review of its role for biodiversity and trends in the literature. Urban Forestry & Urban Greening 14:883–908.

Sattler, T., D. Borcard, R. Arlettaz, F. Bontadina, P. Legendre, M. K. Obrist, and M. Moretti. 2010. Spider, bee, and bird communities in cities are shaped by environmental control and high stochasticity. Ecology 91:3343–3353.

Seto, K. C., B. Güneralp, and L. R. Hutyra. 2012. Global forecasts of urban expansion to 2030 and direct impacts on biodiversity and carbon pools. Proceedings of the National Academy of Sciences USA 109:16083–16088.

Sodhi, N. S., L. P. Koh, B. W. Brook, and P. K. Ng. 2004. Southeast Asian biodiversity: an impending disaster. Trends in Ecology & Evolution 19: 654–660.

Stagoll, K., D. B. Lindenmayer, E. Knight, J. Fischer, and A. D. Manning. 2012. Large trees are keystone structures in urban parks: urban keystone structures. Conservation Letters 5:115–122.

Steed, H. 2015. Greening the vertical garden city: the planning, design, and management of planting in high density tropical cities. Straits Times Press Pte. Ltd, Singapore, Singapore.

Strange, M. 2000. A photographic guide to the birds of Southeast Asia: including the Philippines and Borneo. Princeton University Press, Princeton, New Jersey, USA.

Swan, C. M., S. T. A. Pickett, K. Szlavecz, P. Warren, and K. T. Willey. 2011. Biodiversity and community composition in urban ecosystems: coupled human, spatial, and metacommunity processes. Pages 179–186 in J. H. Breuste, T. Elmqvist, G. Guntenspergen, P. James, and N. E. McIntyre, editors. Urban ecology. Oxford University Press, Oxford, UK.

Threlfall, C. G., N. S. Williams, A. K. Hahs, and S. J. Livesley. 2016. Approaches to urban vegetation management and the impacts on urban bird and bat assemblages. Landscape and Urban Planning 153:28–39.

Tian, Y., and C. Jim. 2012. Development potential of sky gardens in the compact city of Hong Kong. Urban Forestry & Urban Greening 11:223–233.

Violle, C., M. L. Navas, D. Vile, E. Kazakou, C. Fortunel, I. Hummel, and E. Garnier. 2007. Let the concept of trait be functional! Oikos 116:882–892.

Walsh, C., and R. M. Mac Nally. 2013. hier.part: hierarchical partitioning. https://cran.r-project.org/package=hier.part

Wang, L. K., and C. J. Hails. 2007. An annotated checklist of the birds of Singapore. Raffles Bulletin of Zoology 15(Supplement):1–179.

Warren, P. S., M. Katti, M. Ermann, and A. Brazel. 2006. Urban bioacoustics: It's not just noise. Animal Behaviour 71:491–502.

Wickham, H. 2009. ggplot2: elegant graphics for data analysis. Springer, New York, New York, USA.

Williams, N. S. G., J. Lundholm, and J. S. MacIvor. 2014. Do green roofs help urban biodiversity conservation? Journal of Applied Ecology 51:1643–1649.

Xun, B., D. Yu, Y. Liu, R. Hao, and Y. Sun. 2014. Quantifying isolation effect of urban growth on key ecological areas. Ecological Engineering 69:46–54.

Yee, A. T. K., et al. 2011. The vegetation of Singapore: an updated map. Gardens' Bulletin Singapore 63:205–212.

Yong, D. L., K. C. Lim, and T. K. Lee. 2013. A naturalist's guide to the birds of Singapore. John Beaufoy Pub., Oxford, Great Britain.

Yuen, B., and W. N. Hien. 2005. Resident perceptions and expectations of rooftop gardens in Singapore. Landscape and Urban Planning 73:263–276.

Zuur, A. F., E. N. Ieno, and C. S. Elphick. 2010. A protocol for data exploration to avoid common statistical problems. Methods in Ecology and Evolution 1:3–14.

6

Present-day and future contribution of climate and fires to vegetation composition in the boreal forest of China

Chao Wu [ID],[1,2] Sergey Venevsky,[1,†] Stephen Sitch,[2] Yang Yang,[1] Menghui Wang,[1] Lei Wang,[1] and Yu Gao[1]

[1]*Ministry of Education Key Laboratory for Earth System Modeling, Department of Earth System Science, Tsinghua University, Beijing 100084 China*
[2]*College of Life and Environmental Sciences, University of Exeter, Exeter EX4 4QF UK*

Abstract. Climate is well known as an important determinant of biogeography. Although climate is directly important for vegetation composition in the boreal forests, these ecosystems are strongly sensitive to an indirect effect of climate via fire disturbance. However, the driving balance of fire disturbance and climate on composition is poorly understood. In this study, we quantitatively analyzed their individual contributions for the boreal forests of the Heilongjiang Province, China, and their response to climate change using four warming scenarios (+1.5°, 2°, 3°, and 4°C). This study employs the statistical methods of Redundancy Analysis (RDA) and variation partitioning combined with simulation results from a SErgey VERsion Dynamic Global Vegetation Model (SEVER-DGVM), and remote sensing datasets of global land cover (GLC2000) and the third version of Global Fire Emissions Database (GFED3). Results show that the vegetation distribution for the present day is mainly determined directly by climate (35%) rather than fire (1–10.9%). However, with a future global warming of 1.5°C, local vegetation composition will be determined by fires rather than climate (36.3% >29.3%). Above 1.5°C warming, temperature will be more important than fires in regulating vegetation distribution although other factors such as precipitation can also contribute. The spatial pattern in vegetation composition over the region, as evaluated by Moran's Eigenvector Map (MEM), has a significant impact on local vegetation coverage; for example, composition at any individual location is highly related to that in its neighborhood. It represents the largest contribution to vegetation distribution in all scenarios, but will not change the driving balance between climate and fires. Our results are highly relevant for forest and wildfires' management.

Key words: boreal forests; China; climate change; dynamic global vegetation models (DGVMs); fires; individual contribution.

† E-mail: venevsky@tsinghua.edu.cn

Introduction

The boreal forest, as one of the important flammable ecosystems around the world, occupies 30% of the global forest areas (Gauthier et al. 2015). The vegetation structure and distribution are influenced by many factors. It is widely considered that climate, especially temperature and precipitation, directly controls the vegetation composition and distribution (Scheiter and Higgins 2009); hence, vegetation classifications are mainly based on such climate variables (e.g., Holdridge life zones [Holdridge 1947]). Temperature impacts the vegetation growth and distribution by changing the rates of photosynthesis, respiration, regulating phenology, tissue growth, regeneration, and mortality processes (e.g., frost damage). Plant Function Types (PFTs), which are assigned different bioclimatic limits (e.g., minimum coldest month temperature and maximum coldest month temperature), will determine whether they are able to survive and regenerate based on the climatic conditions. In addition to temperature, precipitation controls vegetation distribution by changing the water balance of the ecosystem (Stephenson 1990). Recent research indicates that annual rainfall is the dominant factor in regulating the relative distribution of global tropical forests and savannas (Hirota et al. 2011). Although climate is undoubtedly an important driver in regulating vegetation structure, within a single climate zone, different combinations of species can exist together, suggesting a decoupling of climate and vegetation, which means that other controls are also important in determining the local vegetation composition within any biome (Murphy and Bowman 2012, Scheffer et al. 2012).

Furthermore, vegetation distribution is indirectly determined by changes in the local disturbance regimes (Weber and Flannigan 1997, Dale et al. 2001, Gauthier et al. 2015). Fire is a particularly important natural disturbance and has a significant impact on the extent of forest cover in the flammable boreal forest ecosystems, helping to shape the vegetation structure and accelerate the natural carbon cycle (Bowman et al. 2009). Moreover, fires instantaneously link the atmosphere and biosphere by carbon emissions and have strong feedbacks to climate. In addition, fires will impact the climate by changing biophysical characters of the land surface, for example, albedo. Currently, all sources of fires (landscape and biomass combustion) cause CO_2 emissions (2–4 PgC/yr) equivalent to around one-third of emissions from fossil-fuel combustion (~10 PgC/yr; Van der Werf et al. 2006, Bowman et al. 2009, Le Quéré et al. 2015).

Dynamic global vegetation models (DGVMs) are important tools to simulate potential vegetation and carbon cycles in the terrestrial ecosystems. Meanwhile, DGVMs integrate biophysical, physiological, and ecological processes on a large scale, including vegetation physiology, phenology, vegetation dynamics, and competition. Vegetation distribution, carbon pools, and carbon fluxes are typically simulated at $0.5° \times 0.5°$ spatial resolution. The area unit of the model is a grid cell, and

vegetation distribution in each grid cell is described as the fraction of different PFTs, or Foliage Projective Cover (FPC). Competition, as an important part of vegetation dynamics, is the most widely documented biotic factor affecting vegetation range by changing range limits and thus may impact range shifts (Ettinger and HilleRisLambers 2017). Zielinski et al. (2017) suggested that competition for resource was a significant control on "warm-edge" limits based on large-scale non-invasive surveys and home range data (Zielinski et al. 2017). Resource competition among PFTs in DGVMs, including water, light, nutrients, and individual response to disturbance (e.g., fire), impacts their relative FPC in each grid cell yearly (Sitch et al. 2003). Competition of woody PFT individuals depends on carbon gain, which depends on water, nutrients and light. Carbon gain is allocated to crowns and competition takes place as a self-thinning when potential FPC summed over all PFTs exceeds 1. We assume there are no differences for competition from leaves, roots, and wood in DGVMs, although the main use of carbon gain in roots is for fecundity and growth (Dybzinski et al. 2011). Carbon pools subdivided by PFTs exist in each grid cell, including leaves, sapwood, heartwood, fine roots, a fast and a slow decomposing above-ground litter pool, and a below-ground litter pool. Soil carbon pools in each grid cell collect the inputs from the litter pools of PFTs residing in the grid cell, and carbon fluxes connect terrestrial ecosystems and atmosphere, including net primary production (NPP) of PFTs, soil heterotrophic respiration, and combustion emissions.

In particular, when fire models have been incorporated into DGVMs, fire regime and vegetation–fire interactions can be represented (Scheiter et al. 2013, 2015, Bachelet et al. 2015, Wu et al. 2015). According to the DGVM simulations, forest cover (around 80–100% of tree cover) would more than double from 26.9% to 56.4% in a world without fire (Bond and Keeley 2005). Existing research revealed that some flammable ecosystems (including boreal forests, eucalypt woodlands, shrublands, grasslands, and savannas) are actually determined by fires (Bond et al. 2005). However, there is a difficulty in isolating the controls on vegetation distribution (Mills et al. 2006) and a limited number of studies have focused on analyzing the driving balance between fire disturbance and climate on boreal forests. For example, Bond-Lamberty et al. (2007) explored the impact of environmental factors in driving the carbon balance of central Canadian boreal forests based on factorial experiments; they proved that the carbon balance of this area was determined by changes of fire disturbance between 1948 and 2005. Similarly, Weber and Flannigan (1997) illustrated that compared with the direct impact of climatic change, the change of fire regime might be more important in driving or facilitating vegetation distribution changes, migration, shift, and extinction. In addition, Bergeron and Dansereau (1993) ascribed the difference in composition of the Canadian boreal forest to varying fire cycles. Besides, boreal biosphere interactions with climate, fire disturbance, insect disturbance, and permafrost were assessed by Scheffer et al. (2012) and Soja et al. (2007) based on historical predictions. Furthermore, Scheffer et al. (2012) described thresholds for boreal biome transitions based on satellite data and multi-models, suggesting the change of tree cover was strongly dependent on temperature. Factorial experiments have been widely used to quantify the individual contributions of environmental factors. Generally, the indicators or objects of factorial experiments are usually one type of independent variable, for example, leaf area index (Mao et al. 2013, Zhu et al. 2016), terrestrial evapotranspiration (ET; Jiafu et al. 2015), net biome production, NPP, and vegetation dominance (Bond-Lamberty et al. 2007). However, in this study, we are devoted to analyzing whether the vegetation distribution in boreal forest ecosystems, described by fractional cover/FPC of different plant functional types, is mainly determined by climate or fires and quantifying their individual contributions. Multiple vegetation types exist in the boreal forest of China, for example, needle-leaved evergreen and deciduous conifers and deciduous broadleaf species. Under this circumstance of multiple vegetation types and their properties, we adopt instead the statistical methods of Redundancy Analysis (RDA) and variation partitioning to explore the above-mentioned questions.

Global warming is likely to significantly impact the stability and health of boreal forests (Gauthier et al. 2015). The Paris Agreement aims to control the global warming below 2°C and to pursue efforts to achieve a limit of 1.5°C (Hulme 2016).

Therefore, in this study, we aimed to explore the questions of the potential change of vegetation distribution and fire regime in the boreal forest ecosystems of China, the driving balance between climate and fire disturbance, and quantify their distinctive contributions to biome composition in six scenarios, including two present-day scenarios and four different global warming targets (1.5°, 2.0°, 3.0°, and 4.0°C, relative to pre-industrial climate).

MATERIALS AND METHODS

Study area

The study area is located in Heilongjiang Province between 42°30′–51°20′ N and 121°40′–128°30′ E in northeast China, covering an area of around 4.54×10^5 km^2. The summers are usually hot and humid, while the winters are cold and dry. The annual average temperature is between -4 and $+5$°C from north to south, and the annual precipitation ranges from 400 to 700 mm from west to east (Zhang et al. 2015). The main vegetation types include evergreen needleleaf forest (ENF), deciduous needleleaf forest (DNF), deciduous broadleaf forest (DBF), and cultivated and managed areas/grassland (see Fig. 1b and Appendix S1: Table S2) based on GLC2000 (Bartholomé and Belward 2005). Existing research shows that historically, the most common fire type was frequent, low-intensity surface fires mixed with infrequent stand-replacing fires in this area, and burnt area was usually large with fire return interval ranging from 30 to 120 yr and the average number of fires was 317 per year during 1980–1987 (Xu et al. 1997, Liu et al. 2012). However, after a catastrophic fire which was occurred in 1987 in this area, burning a total area of 1.3 Mha, forest harvesting and fire suppression have changed the fire regime of this area. Currently, fire regime is characterized by infrequent but more intense fires and larger burnt area, with a fire return interval of more than 500 yr (Chang et al. 2008, Liu et al. 2012). The total number of grid cells in Heilongjiang Province is 216 at the 0.5° × 0.5° spatial resolution.

Data and tools

Present-day PFT coverages and burnt area simulation.—SEVER-DGVM, which is an intermediate-complexity DGVM and is developed from the Lund–Potsdam–Jena Dynamic Global

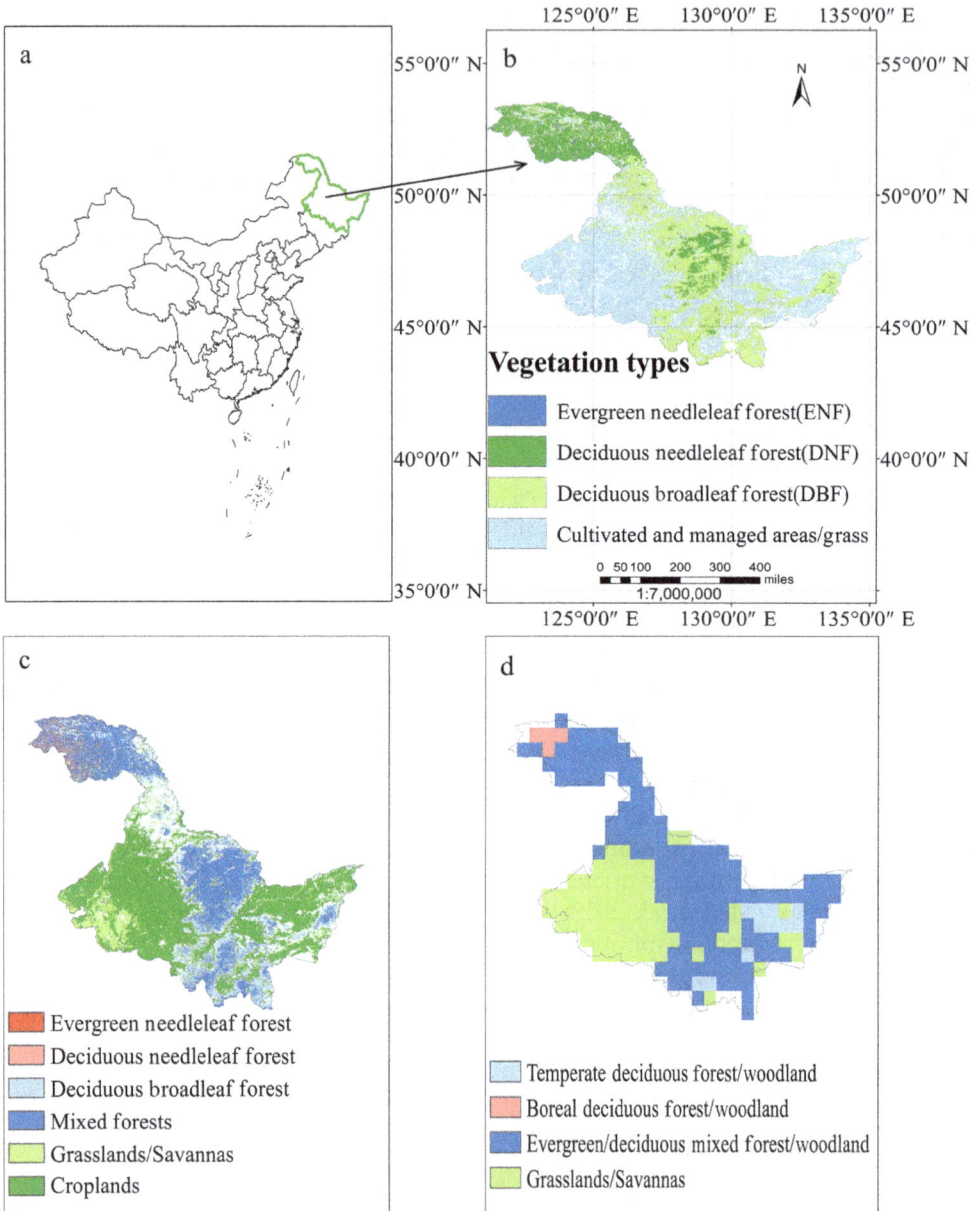

Fig. 1. Location and main vegetation types in Heilongjiang Province. The sources of datasets are (a) Administrative divisions of China, (b) GLC2000, (c) 0.5-km MODIS-based global land cover climatology, and (d) global potential vegetation dataset.

Vegetation Model (LPJ-DGVM; Sitch et al. 2003) with much improvement for high latitudes (Venevsky and Maksyutov 2007), for example, including a daily time step description for dynamics of soil temperature, potential ET, and fire disturbance, is used to simulate PFT coverages. Meanwhile, burnt area is simulated by Glob-FIRM (Global FIRe Model; Thonicke et al. 2001), which

is incorporated into SEVER-DGVM. Here, some simplifying hypotheses are made. First, fire occurrence is only dependent upon fuel load and litter moisture (i.e., the amount of dry material available), which combines both the influence of climate and vegetation. Ignition is assumed to take place sooner or later, without specific consideration. Secondly, fire effects are mainly driven by the length of the fire season and the PFT-dependent fire resistances. Thirdly, we assumed that the smallest burnt area in each grid cell is 250 ha and fire intensity is not considered in this study.

All input datasets were provided at a $0.5° \times 0.5°$ spatial resolution. We used NCEP/NCAR (National Centers for Environmental Prediction/National Center for Atmospheric Research) Reanalysis data (http://www.esrl.noaa.gov/psd/) as the input climate data in SEVER-DGVM, including daily temperature, precipitation, and shortwave radiation during 1957–2002, which were downscaled to a 0.5° grid based on Kalnay et al. (1996). The soil physical and thermo-dynamic characteristics were determined by simplified FAO soil dataset (FAO 1991). We used historical observed CO_2 concentration from 1957 to 2002 (Meinshausen et al. 2011). The global DGVM applications often misrepresent vegetation dynamics on a regional scale (Seiler et al. 2014). Therefore, a PFT parameterization, suitable for Eurasian boreal forests, was used here, based on Khvostikov et al. (2015). A typical simulation with SEVER-DGVM started from "bare ground" (no plant biomass present) and "spined up" 1012 yr until approximate equilibrium was reached with respect to carbon pools and vegetation cover. We used climate data during 1957–2002 repeated 22 times, and a prescribed constant atmospheric CO_2 concentration of the year 1957 was used. The present-day simulation by SEVER-DGVM is run in the transient phase 1957–2002 with historical changes in atmospheric CO_2 and climate.

Future PFT coverages and burnt area projection induced by climate change.—Four different global warming targets (1.5°, 2.0°, 3.0°, and 4.0°C, relative to pre-industrial climate) were used in modeling the response of future PFT coverages and burnt area to climate change. We selected 22 general circulation models (GCMs) of Coupled Model Intercomparison Project Phase 5 (CMIP5) (see Appendix S1: Table S1), which have been bias-corrected, to project the future climate data in this study though the number of GCMs actually used changed depending on different global warming targets (see Table 1). The year, when a specific global warming target was reached, from the multi-model ensemble was recorded (see Table 1 and Appendix S1: Table S1). Daily precipitation and temperature used in SEVER-DGVM for each global warming target were the integrated climate

Table 1. Future SEVER-DGVM simulation in different global warming targets.

Global warming target	No. GCMs used	Year when warming target was reached	Running years
Temperature + 1.5°C	22	2026	24
Temperature + 2.0°C	22	2040	38
Temperature + 3.0°C	18	2063	61
Temperature + 4.0°C	13	2085	83

Note: DGVM, dynamic global vegetation model; GCMs, general circulation models.

data from the GCMs when the specific global warming target was reached. Here, we ignored the future climatic inter-annual variation and used a simple method to recycle the daily precipitation and temperature of the target year in each warming scenario for the future simulations (starting from 2002) (see Table 1). Daily shortwave radiation values used in SEVER-DGVM were kept the same values with the year 2002. CO_2 concentration data in different global warming targets during running years were from the RCP8.5 emission scenario (Riahi et al. 2007). Soil data and parameters needed in SEVER-DGVM stayed the same as present day.

Validation of present-day PFT coverages and burnt area

The accuracy of simulated PFT coverages against current remote sensing products is an important component to reduce the uncertainty of terrestrial biogeochemistry to climate change (Poulter et al. 2011). Three independent datasets were used for validation of present-day PFT coverages in Heilongjiang Province. Firstly, we selected an observed potential vegetation dataset by Ramankutty and Foley (1999), which was based on satellite data at $0.5° \times 0.5°$ spatial resolution and includes 15 categories of vegetation. Four main categories were extracted for Heilongjiang Province (see Fig. 1d). Secondly, in order to obtain the latest land cover which also considers human disturbance on forests in study area, we used a 0.5-km MODIS-based global land cover climatology (Broxton et al. 2014), which was based on 10 yr (2001–2010) of Collection 5.1 MCD12Q1 land cover type data as compared with potential vegetation datasets (see Fig. 1c). We found that the large grasslands and savannas areas were actually replaced by cropland. However, our study is mainly focused on forest ecosystems. What is more, cropland and grassland are usually assessed as the greatest uncertainty in PFT classification (Poulter et al. 2011). Therefore, we extracted the forest areas based on the grasslands/savannas category in Fig. 1d. Finally, we used GLC2000, which was based on SPOT 4 satellite and provides the year 2000 global land cover, to validate the present-day distribution of different PFTs from the simulation of SEVER-DGVM. Although GLC2000 contains 17 different global categories of vegetation, only four main categories were used for validation in Heilongjiang Province (see Fig. 1b). Using the PFT mapping methods in DGVMs by Poulter et al. (2011), GLC2000 datasets were first reclassified into phenology-based categories consistent with the PFTs used in SEVER-DGVM (see Appendix S1: Table S2). And then, three main forests (DNF, DBF, and ENF) were translated to a spatial resolution of 0.5° by summing the area of each PFT class within corresponding 0.5° grid cell and dividing by the grid cell area (Poulter et al. 2011).

In recognition of fires as a large-scale and important agent of change in earth system, it has led to the development of long-term, spatially and temporally explicit global burnt area datasets based on satellite (Justice et al. 2002, Roy et al. 2008, Giglio et al. 2009, 2013, Randerson et al. 2012, Boschetti et al. 2015). The third version of Global Fire Emissions Database (GFED3; Giglio et al. 2010) was used to validate the burnt area simulated by SEVER-DGVM in this study. Global Fire Emissions Database provides global monthly burnt area estimates in 0.5° spatial resolution from July 1996 to mid-2009. We first used Glob-FIRM to compare the similarity of the annual

burnt area between GFED3 and SEVER-DGVM during 1996–2002 using a Student's t test, and then over the same period, we conducted a monthly burnt area validation (see Appendix S1).

Quantifying individual contributions

Contributions of climate and fire to explaining present-day vegetation distribution were detected by two independent experiments. One was based on the observed remote sensing dataset, and the other was dependent upon the simulation by SEVER-DGVM and Glob-FIRM which successfully reproduce contemporary vegetation distribution and fire regimes. The latter was also applied in the global warming scenario simulations (see Table 2).

We selected Plant Function Types Coverages (PFTC) of the last year of the simulation period as the response variable and mean annual burnt area (BA, ha), simulated by Glob-FIRM, as the explanatory variable, representing the impact of fires. Mean annual temperature (MAT, °C), mean annual shortwave radiation (MAR, W/m^2), and mean annual precipitation (MAP, mm) were used to be the climatic explanatory variables. Mean annual burnt area (BA, ha) in Glob-FIRM is determined by PFT specific soil moisture and flammability threshold and, thus, depends on MAT and MAP in a non-linear way.

First, we exclude explanatory variables with strong covariation. Redundancy analysis, whose aim is to explore a series of linear combinations of the explanatory variables that can best explain the variation in the response variables (Borcard et al. 2011a), and variation partitioning were used to quantify the individual contributions of

Table 2. Data used to produce RDA in different experimental designs.

Scenarios	PFTC	BA	MAT	MAP	MAR	Periods
Present-day 1	GLC2000†	GFED3‡	NCEP/NCAR	NCEP/NCAR	NCEP/NCAR	1957–2002
Present-day 2	SEVER-DGVM	Glob-FIRM‡	NCEP/NCAR	NCEP/NCAR	NCEP/NCAR	1957–2002
Temperature + 1.5°C	SEVER-DGVM	Glob-FIRM	NCEP/NCAR	NCEP/NCAR	NCEP/NCAR	2002–2026
Temperature + 2.0°C	SEVER-DGVM	Glob-FIRM	NCEP/NCAR	NCEP/NCAR	NCEP/NCAR	2002–2040
Temperature + 3.0°C	SEVER-DGVM	Glob-FIRM	NCEP/NCAR	NCEP/NCAR	NCEP/NCAR	2002–2063
Temperature + 4.0°C	SEVER-DGVM	Glob-FIRM	NCEP/NCAR	NCEP/NCAR	NCEP/NCAR	2002–2085

Notes: BA, mean annual burnt area; DGVM, dynamic global vegetation model; GFED3, The third version of Global Fire Emissions Database; MAR, mean annual shortwave radiation; MAP, mean annual precipitation; MAT, Mean annual temperature; PFTC, Plant Function Types Coverages; RDA, redundancy analysis. All the data are provided at a 0.5° × 0.5° spatial resolution. The number of samples is 216.

† The year of PFTC in the "present-day 1" scenario is the projected year of land cover in GLC2000 product.

‡ The periods for GFED3 and Glob-FIRM are 1996–2002, which are the overlapping years between GFED3 and present-day simulation.

climate and fires to the vegetation distribution in the boreal forest ecosystem of China in different scenarios. The data used to produce RDA are shown in Table 2. Second, we used the R package "vegan" version 2.3-2 (Oksanen et al. 2015) to build RDA sequence: rda (formula = PFTC ~ BA + MAT + MAP + MAR). Next, we produced forward selection and collinearity test to determine critical factors. Collinearity test on explanatory variables was based on Variance Inflation Factors (VIF), and it is considered that there is little/no collinearity when VIF < 4. Finally, we tested the RDA results using "Permutable test" with 999 runs. R package "vegan" version 2.3-2 was also applied in variation partitioning, which was shown as Venn diagrams by R package "VennDiagram" version 1.6.17 (Chen and Boutros 2011). Here, we only focused on the impacts of climate and fires on the vegetation distribution; the contributions of other controls (e.g., soil and human activities) were described as residuals.

Spatial structures play a crucial role in the analysis of ecological data. If external forcings (e.g., climatic, physical, and chemical) are spatially structured, biomes that are actually controlled by these factors will be spatially structured at many scales (Borcard et al. 2011b). This can be confirmed by autocorrelation of quantitative biome characteristics in space, which is referred to as spatial pattern (Borcard et al. 2011b). In order to analyze individual contributions of vegetation spatial pattern, climate, and fire in regulating vegetation distribution, we used variation partitioning and Moran's Eigenvector Map (MEM) spatial analysis method (Dray et al. 2006), which was based on sets of variables describing spatial structures in the way of deriving from the coordinates of the samples or from the neighborhood relationships among samples and could model structures at multiple scales and allowed the modeling of any type of spatial structures (Borcard and Legendre 2002), by R package "vegan" version 2.3-2 (Oksanen et al. 2015). However, a linear trend in vegetation distribution should be considered as a source of variation if the test for response data was significant (Borcard et al. 2011b). The whole explanatory variables were standardized firstly. All calculations were based on RStudio version 0.99.489 software environment (RStudio 2015).

RESULTS

Present-day PFT coverages and burnt area distribution

We simulated PFT coverages in forest areas of Heilongjiang Province in 2002 and compared them with the main forest types from GLC2000 (see Fig. 2a). We found that simulated PFT coverages were relatively consistent with GLC2000 for

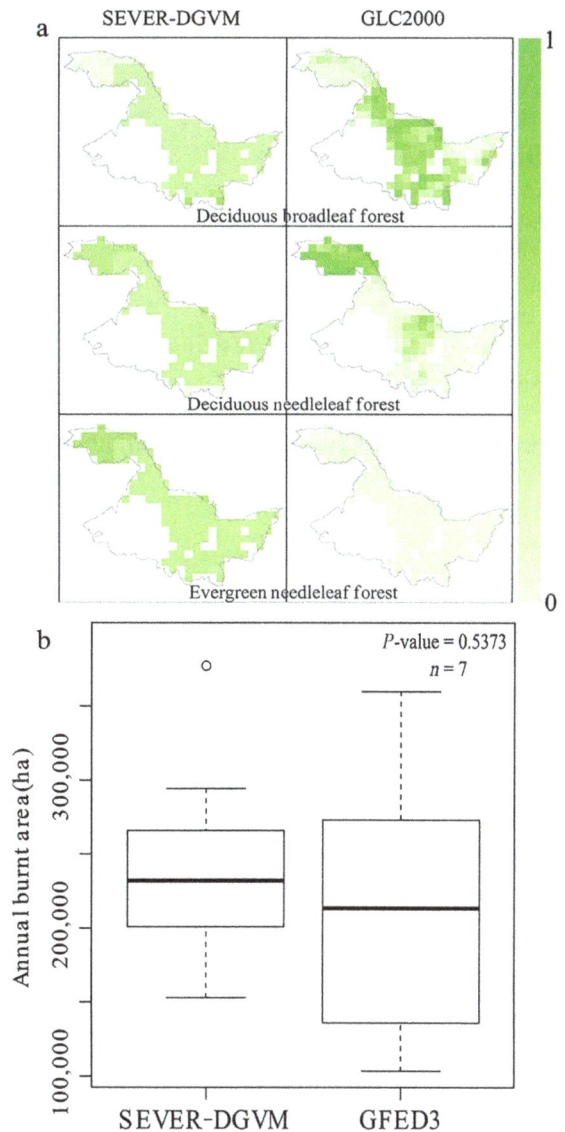

Fig. 2. (a) Present-day Plant Function Type coverages and (b) mean annual burnt area during 1996–2002 in forest areas of Heilongjiang Province. DGVM, Dynamic global vegetation model; GFED3, The third version of Global Fire Emissions Database.

categories DBF and DNF, especially in the north-west parts of the Heilongjiang Province. How-ever, a large difference existed for ENF and we only captured the ENF distribution in the north-west; these might be the results of the misclassifi-cation in GLC2000 between ENF and mixed forests in other parts of the study areas. Based on remote sensing products (see Fig. 1c, d), we find that large areas of mixed forests are actually distributed in Heilongjiang Province, which has been proved by the vegetation atlas of China (Tan et al. 2007). Also, mixed forests were twice as large in extent as DNF in this area (Xiao et al. 2002).

The results of the validation in the BA between GFED3 and simulated by SEVER-DGVM are shown in Fig. 2b. We suggest that the simulated total burnt area reproduced GFED3, and Student's t test demonstrated that there was not a significant difference between SEVER-DGVM and GFED3 at the 90% confidence level ($t = 0.63512$, $n = 7$, df = 12, $P = 0.5373$). Monthly burnt area com-parison is described in Appendix S1: Fig. S1.

Present-day individual contributions of climate and fire in regulating PFTC

First, we use the "present-day 1" experiment to quantify present-day individual contributions of

climate and fire in regulating PFTC. The correla-tion analysis between climate and fire factors revealed that MAR and MAT had a strong relationship (adjusted $R^2 = 0.91$) (see Appendix S1: Fig. S2). Considering the ecological meanings, we excluded MAR from the explanatory variables. Thus, MAP, MAT, and BA are the explanatory variables in regulating PFTC in the boreal forest ecosystems of Heilongjiang Province, China.

Results from a Principal Component Analysis (PCA) showed that the first two components (PCA1 and PCA2) could together explain 77.8% of the total variation. The RDA for PFTC and explanatory variables demonstrated that RDA1 and RDA2 were able to explain 35.5% of the total variation. Then, forward selection revealed that MAP, MAT, and BA were the significant factors in determining PFTC ($P = 0.015$); meanwhile, none had obvious collinearity among explanatory vari-ables (VIF < 4). Next, the "Permutable test" of the RDA results was significant ($F = 40.436$, df = 3, $P < 0.001$) and all the canonical axes were signifi-cant as well. The adjusted bimultivariate redun-dancy statistic (R^2) was 0.3550. The RDA results are shown in Fig. 3a. Similarly, experiment results from "present-day 2" are shown in Fig. 3b. The RDA results are described as "triplot," including

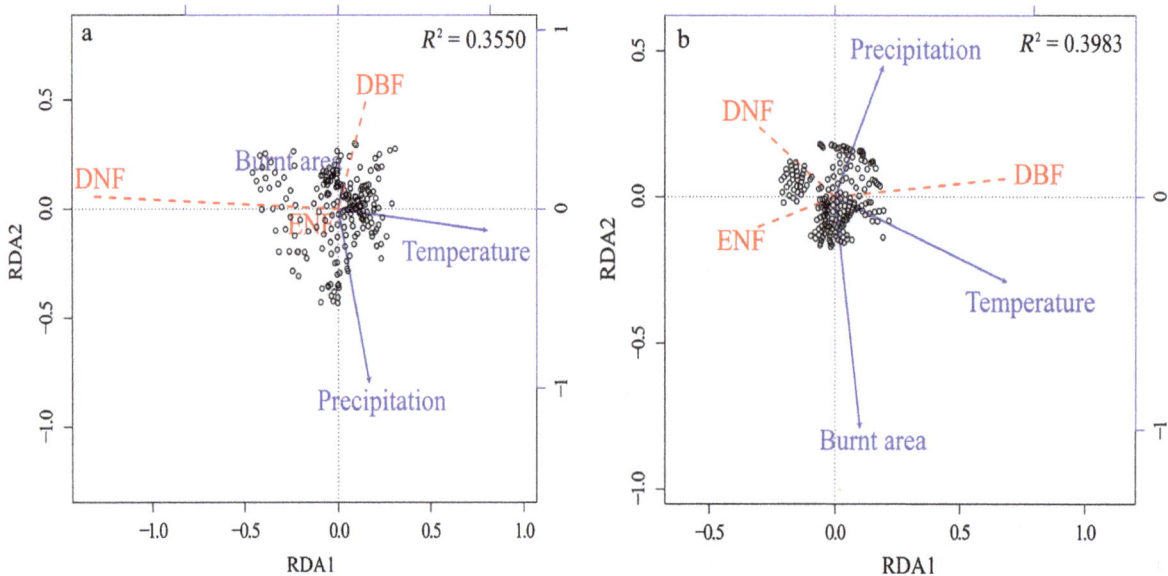

Fig. 3. Correlation triplot based on a redundancy analysis (RDA) depicting the relationship between the selected climate and fire variables and the variation of coverages among different Plant Function Types. (a) Pre-sent-day 1; and (b) present-day 2. DNF, deciduous needleleaf forest; ENF, evergreen needleleaf forest; DBF, deciduous broadleaf forest.

three different entities: sites, response variables (without arrowheads, red), and explanatory variables (with arrowheads, blue). The "triplot" was interpreted as "scaling 2—correlation biplot," in which the angles between variables (explanatory and/or response variables) reflect their correlations. We found that DNF distribution was negatively related to temperature, while temperature would contribute to the growth of DBF as well. Meanwhile, fire would decrease the distribution of flammable DNF according to both present-day experiments. However, the influence of precipitation on biome composition was uncertain in these two scenarios.

Variation partitioning illustrated that fire and climate factors could explain 35.5% of the total variation in PFTC from "present-day 1." Furthermore, fires alone could only explain around 1%, while climatic individual contributions were around 35.8% of the total variation (see Fig. 4a). Here, we ignored the minus values and it could be considered as zero, for the joint contributions of explanatory factors, which indicated that the explanatory variables did worse than random normal variables (Borcard et al. 2011a). Moreover,

temperature was much more important than precipitation in regulating the PFTC (30.6% >4.5%). All the results were significant (P < 0.001) based on permutation tests.

Next, based on the experiment "present-day 2," we found that 39.8% of the total variation could be explained by climate and fires in determining vegetation distribution (see Fig. 4b). Different from "present-day 1," fire could describe 10.9% of the total variation based on modeling. However, individual contributions of climate factors did not change a great deal. Therefore, based on two independent present-day experiments, the results illustrate that the distribution of boreal forests in China is more determined by climate rather than fires; meanwhile, the response of vegetation is more sensitive to temperature than precipitation at the present day. Even though the contribution of fire in regulating PFTC is strongly dependent on the data source and accuracy of burnt area, changing from 1% to 10.9% in our study, climate contributes around 30%, largely driven by temperature, to the distribution of vegetation in boreal forests of China.

Spatial pattern in regulating PFTC

Spatial pattern over the region is important in the analysis of individual contributions of fires and climate factors in regulating local vegetation distribution; that is, one can interpret this as the importance of the vegetation in the surrounding area for the composition at a specific location. Essentially it gives an indication of the level of spatial homogeneity in vegetation across the region, and its importance for determining local vegetation cover. Based on the "present-day 1" scenario, there is a strong linear trend in the spatial distribution of vegetation (F = 57.121, df = 2, P < 0.001). After detrending the data (Borcard et al. 2011b), the MEM spatial analysis showed that spatial pattern could explain 82.0% of the distribution of vegetation. Spatial scale and regression analysis help to analyze whether the spatial pattern of vegetation distribution is related to environmental factors (climate and fire disturbance in this study) at different scales. The different scales here are defined as the significant MEM variables map according to the scales of the patterns they represent. And the results illustrated that the vegetation distribution was significant at the broad scale (F = 19.066, df = 11, P < 0.001) and produced two significant canonical axes to

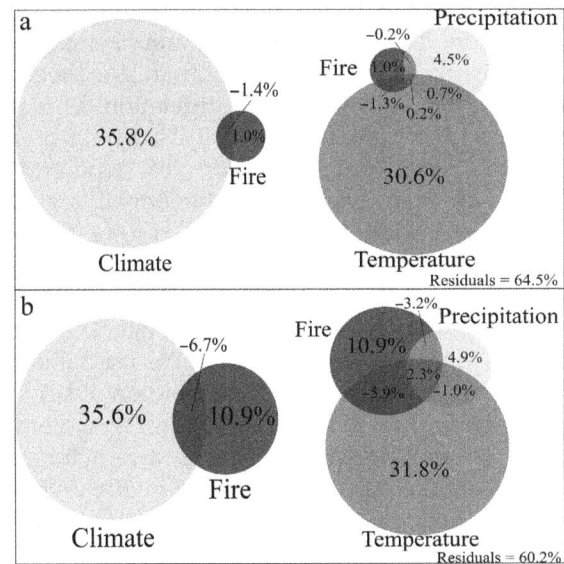

Fig. 4. Individual contributions of fires and climate in regulating Plant Function Types Coverages in (a) "present-day 1" and (b) "present-day 2" (left: Climate consists of comprehensive effects of temperature and precipitation; right: individual effects of temperature and precipitation).

explain the spatial pattern. The first canonical axis was significant to MAP ($R^2 = 0.3033$, $P < 0.001$), while the second canonical axis was significant to BA, MAT, and MAP ($R^2 = 0.2229$, $P < 0.001$). However, there were no obvious spatial differences at medium and fine scales ($P > 0.05$). The reason might be that at finer spatial scale, vegetation often displayed properties of inertia, contingency, and hysteresis, most frequently because of climatic variability across multiple timescales and the episodic nature of disturbance and establishment (Jackson 2013).

A similar analysis was produced for the "present-day 2" scenario. Variation partitioning results are shown in Table 3. Spatial pattern has a strong impact on the quantitative analysis of the individual contributions of fire and climate factors to PFTC. Fires, climate, and spatial information could explain around 90% of the total variation. Compared to the small influence of fire disturbance (0–4.2%), climate could explain around 30% of the total variation. However, as an important source of variation, the linear trend in vegetation spatial distribution itself cannot be ignored. Therefore, we should consider spatial information, including spatial pattern and trend in vegetation distribution when we quantify the contributions of explanatory variables even if there is no significant influence on the driving balance between climate and fires. Here, the fact that contributions of fires, climate, spatial pattern, and trend might add up to more than 100% is the reason to consider the overlap among individual factors.

Table 3. Individual contributions (%) of fires, climate, and spatial pattern on PFTC in present-day and warming scenarios.

Scenarios	Fire	Climate	Spatial	Trend	Residuals
Present-day 1	0	34.5	88.4	34.3	11.2
Present-day 2	4.2	28.9	83.6	42.0	13.6
Temperature + 1.5°C	36.3	29.3	89.7	32.7	7.4
Temperature + 2.0°C	7.2	42.1	90.1	45.1	8.5
Temperature + 3.0°C	3.4	62.3	90.4	59.5	8.6
Temperature + 4.0°C	5.6	59.4	93.7	55.9	5.3

Note: PFTC, Plant Function Types Coverages. Individual contributions had included the joint contributions (Fire: fire contributions; Climate: climate contributions; Spatial: spatial pattern contributions; Trend: the linear trend of PFTC contributions; Residuals: the contributions that could not be explained).

Future PFT coverages and burnt area projection induced by climate change

Climate change, especially global warming, is simulated to significantly impact vegetation distribution, possibly leading to important biome-level changes (Gauthier et al. 2015). PFT coverages projections for four different global warming targets (1.5°, 2°, 3°, and 4°C) are shown in Fig. 5. We find that the dominant forests in Heilongjiang Province are ENF in the present day, which is consistent with actual vegetation distribution, such as *Pinus koraiensis*, *Pinus sylvestris* var. *mongolica*, *Picea koraiensis*, and *Abies nephrolepis*. Besides, DBF occupies large areas as well, such as *Juglans mandshurica* and *Quercus mongolica Ledeb*. In response to climate change, the biome composition would change in different scenarios. When temperature increased, DNF would decrease from the south to the north until "temperature + 4°C"; almost no DNF existed in this area. ENF would increase at lower temperature increases; however, above 2°C, large areas of ENF would be replaced by temperate forests. Meanwhile, as temperature increased, DBF would start to shift from the south to the north until the time when temperature increased by 3°C; DBF would be the most dominant PFT in this region. There are also some minor areas of grass distributed in the transition areas between boreal forests and temperate vegetation. This is in line with the evidence that there would be a gradual northward migration of temperate deciduous tree species into the boreal region (Gauthier et al. 2015). There are also some minor areas of grass distributed in the transition areas between boreal forests and temperate vegetation.

Climate change would also change the fire regime, in particular burnt area. We used Glob-FIRM to simulate the spatial distribution of BA in Heilongjiang Province. As shown in Fig. 6, compared to the present-day burnt area, climate change would lead to a decrease in BA. When temperature increased by 1.5°C, the areas most disturbed by wildfires are centered in the areas with large flammable boreal forests, such as DNF and ENF (see Fig. 5). However, when temperature increased by more than 2°C, BA decreases rapidly and hotspots of wildfires move from the south to the north, which is consistent with the shift from boreal forests to temperate forests. That is, the wildfires induced by global warming are strongly

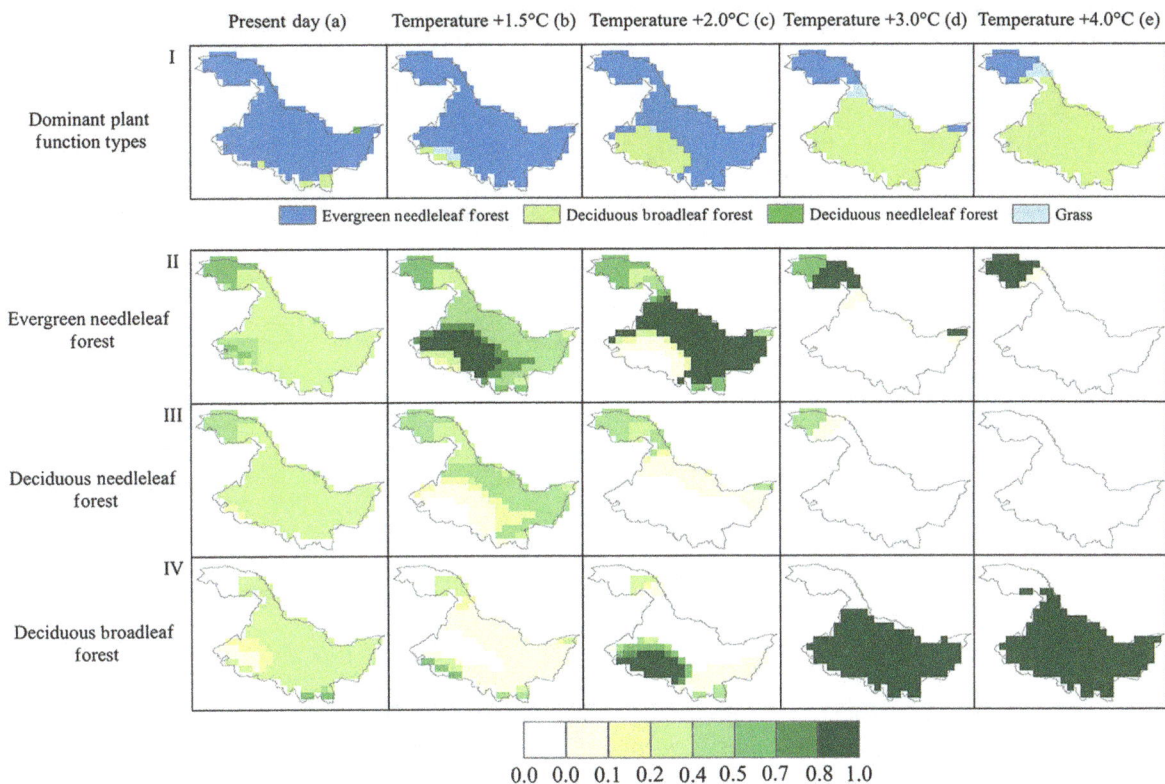

Fig. 5. Plant Function Type (PFT) distribution simulated by SEVER-DGVM in Heilongjiang Province, China. (a) Present day; (b) temperature increase by 1.5°C; (c) temperature increase by 2.0°C; (d) temperature increase by 3.0°C; (e) temperature increase by 4.0°C; and (I) dominant PFTs; tree cover in each grid cell: II, III, IV.

dependent on the PFTs and flammability of the vegetation. In addition, although global warming will decrease precipitation compared to the present day, there is a slight increase in precipitation when temperature increases from +1.5° to +4°C (see Appendix S1: Fig. S3), which might lead to the decrease in burnt area in different scenarios.

Individual contributions of climate and fire in regulating PFTC induced by climate change

Results of the correlation analysis among explanatory variables (MAP, MAT, MAR, and BA) in regulating PFTC for the four different scenarios are shown in Appendix S1: Fig. S4. Similar to the present-day scenarios, we excluded MAR from the explanatory variables. Redundancy analysis results from the four global warming experiments are shown in Fig. 7. Deciduous needleleaf forest, as one type of main flammable forest, was strongly and negatively related to burnt area and temperature in all scenarios. With temperature increasing, precipitation would firstly contribute to the growth

of DNF, but when temperature increased by 2°C, precipitation would limit the distribution of DNF. The growth of DBF is strongly dependent on temperature and more DBF will exist in a warmer world, which is consistent with Fig. 5. In addition, when the temperature increased by higher than 2°C, temperature and precipitation would be the dominant limitation factors in regulating ENF. Redundancy analysis results are significant by permutation test ($P < 0.001$).

Variation partitioning can be applied to analyze the individual contributions of climate factors and fires in regulating PFTC in different warming scenarios quantitatively (see Fig. 8). Compared to fire influence, generally climate factors are more important in regulating PFTC, whose individual contributions are larger than fires' except the scenario of "temperature +1.5°C" (Fire 36.3% >Climate 29.3%). When temperature increased by 1.5°C, the vegetation distribution of Heilongjiang Province would be mainly determined by fires rather than climate. When

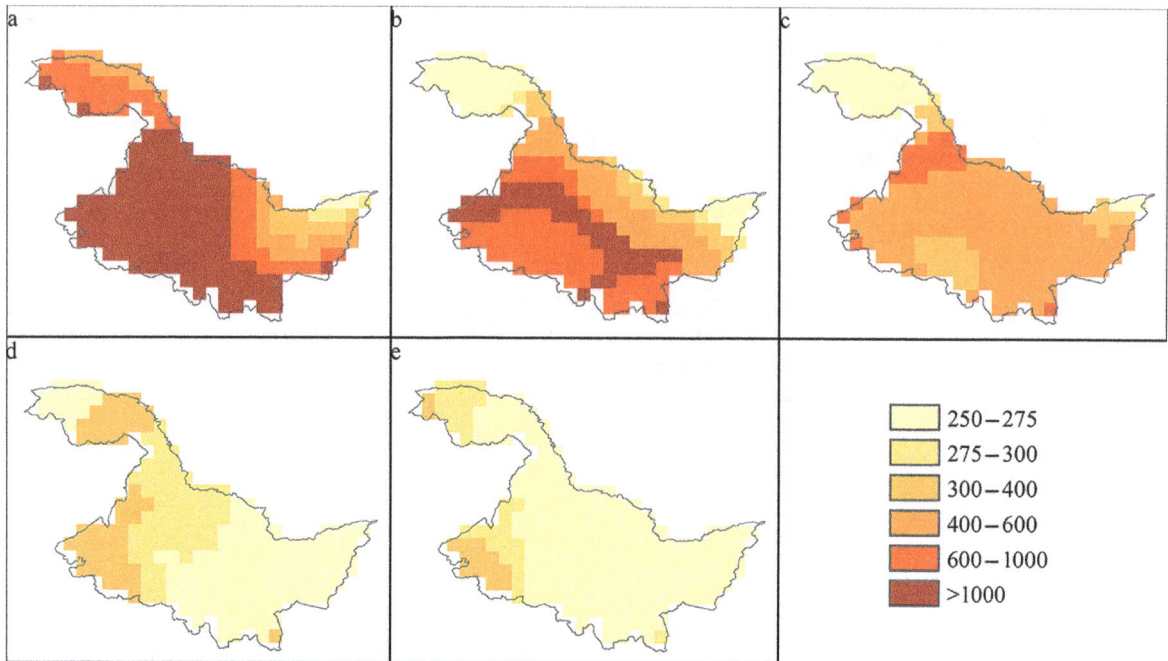

Fig. 6. Mean annual burnt area (ha) spatial distribution simulated by Glob-FIRM in Heilongjiang Province, China. (a) Present day; (b) temperature increase by 1.5°C; (c) temperature increase by 2.0°C; (d) temperature increase by 3.0°C; and (e) temperature increase by 4.0°C.

temperature increased by higher than 1.5°C, the individual contributions of climate would be much larger than the contributions of fires. What is more, we found temperature to be more important than precipitation in regulating PFTC. All the results from permutation tests were significant ($P < 0.001$).

Spatial pattern in regulating PFTC induced by climate change

We cannot ignore the impact of the spatial pattern when analyzing the individual contributions of climate and fire disturbance factors to the PFTC under a changing climate. The results of MEM spatial analysis are also shown in Table 3. Results show how spatial pattern represents the largest contribution to regulating PFTC for the whole four scenarios; that is, the vegetation distribution is strongly dependent on the spatial difference. Except the scenario of "temperature +1.5°C," the other three scenarios showed that climate factors were more important than fires on PFTC. When temperature increased by 1.5°C, the vegetation distribution was strongly dependent on wildfires (Fire 36.29% >Climate

29.31%), which is consistent with the results in Fig. 8. All the permutation test results were significant ($P < 0.001$).

DISCUSSION

Climatic, weather, and fire impacts on forest ecosystems

Climatic change is likely to exert a strong influence on plant physiology and vegetation coverage (Betts et al. 2000). Temperature will affect biome composition via impacts on plant physiological processes, especially photosynthesis (Collatz et al. 1991) and respiration (Tjoelker et al. 2001). Besides, temperature, combined with growing degree days, has a significant influence on plant phenology and growing season length (Chen and Pan 2002, Tao et al. 2006). Precipitation controls vegetation distribution by impacting water supply, ET, and runoff, leading to the water balance change of the ecosystem (Stephenson 1990). Generally, savanna would be replaced by the forest when annual precipitation exceeds around 1500 mm/yr (Lewis 2006). Moreover, Hirota et al. (2011) implied that actually the

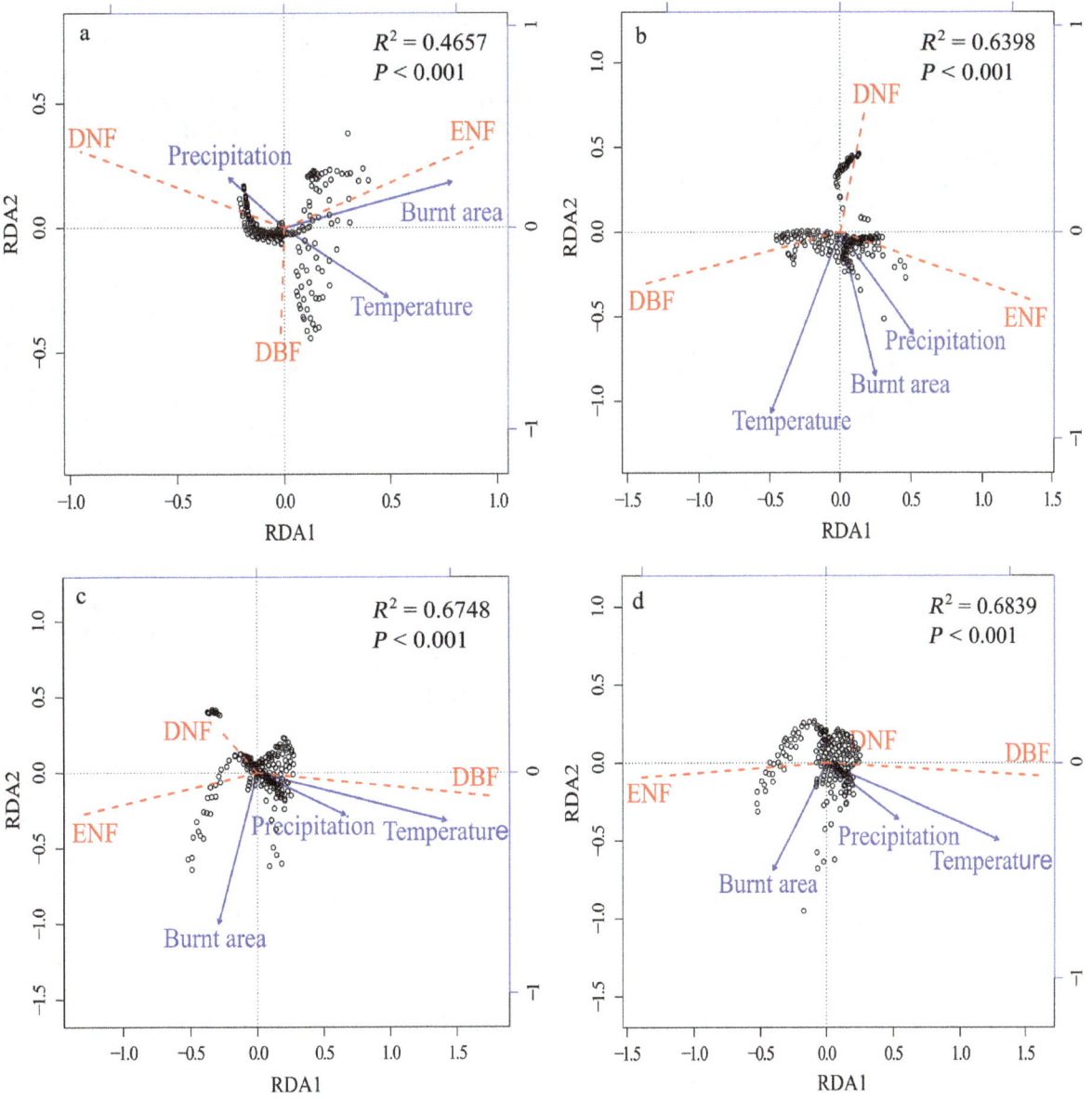

Fig. 7. Correlation triplot based on a redundancy analysis (RDA) depicting the relationship between the selected climate and fire variables and the variation of coverages among different Plant Function Types. (a) Temperature increase by 1.5°C; (b) temperature increase by 2.0°C; (c) temperature increase by 3.0°C; and (d) temperature increase by 4.0°C. For the expansions of the abbreviations used in the Figure 7, refer the caption of Figure 2.

global tropical forests and savannas are controlled by annual precipitation. In addition, the boundaries of boreal forests are modified by precipitation as influenced by oceans and mountains (Volney and Fleming 2000). Drought stress, which is strongly related to precipitation and temperature, will also impact biome composition

(Allen and Breshears 1998). Different from the long-term impact from climate, generally the impact of weather happens in a short period of time. For example, climate determines what will probably grow well in a given area, but plants can still be damaged or killed by extreme weather. The impact of weather is always related

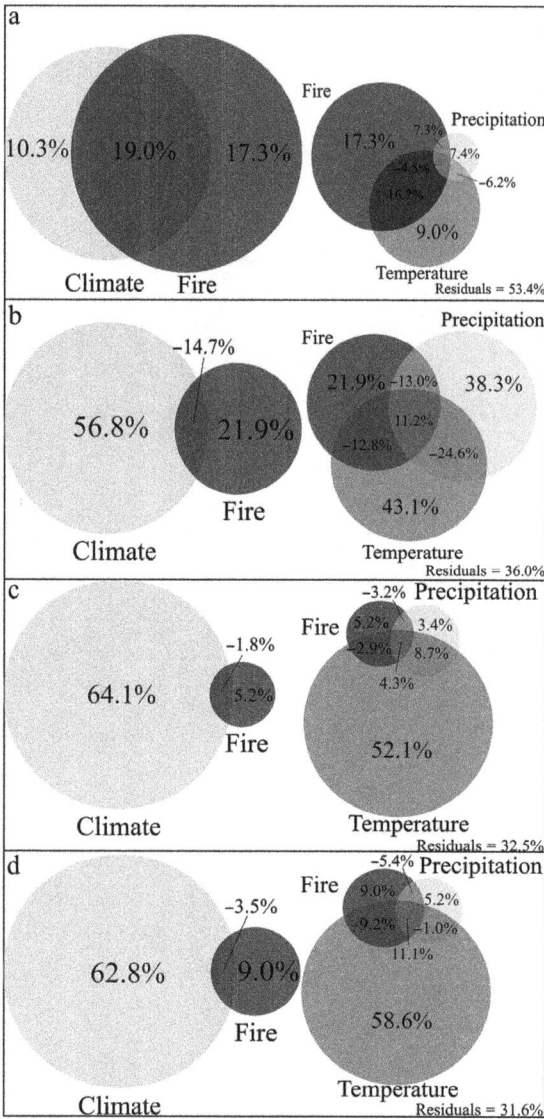

Fig. 8. Individual contributions of fires and climate in regulating Plant Function Types Coverages (climate consists of comprehensive effects of temperature and precipitation). (a) Temperature increase by 1.5°C; (b) temperature increase by 2.0°C; (c) temperature increase by 3.0°C; and (d) temperature increase by 4.0°C.

to extreme weather evens, for example, drought, flood, and storm, which would strongly impact the vegetation composition (Parmesan et al. 2000). Besides, weather-related stress can also make plants more susceptible to disease and insect problems. However, vegetation will also modify weather by changing surface albedo, transpiration and evaporation of water vapor,

aerodynamic effects, and emission of hydrocarbons whose oxidation can form aerosol particles (Brown et al. 2005). Extreme, large-scale weather events are likely to trigger ecosystem-level disturbance, for example, wildfires, which may affect the species composition and diversity (Parmesan et al. 2000).

Fire is an important and necessary natural disturbance in forest ecosystems, especially for flammable communities (e.g., boreal forests, grasslands, savannas, and Mediterranean shrublands). All of these fire-prone ecosystems cover around 40% of the world's land surface (Bond et al. 2005). Fires help to shape global biome distribution and maintain the structure and function of fire-prone communities. Meanwhile, climate plays a decisive impact on vegetation growth and distribution, which is also an important factor for vegetation classification around the world (Raich and Schlesinger 1992, Cramer et al. 2001, Nemani et al. 2003, Parmesan and Yohe 2003, Zhengqiu et al. 2015). Bond et al. (2005) suggested that some flammable ecosystems were actually determined by fires. However, few studies focus on the analysis of the actual contributions of fire to ecosystems later.

Our quantitative analysis further supplements Bond's ideas on a regional scale. Regardless of spatial pattern, similar to climate impact, fires also have a substantial impact on vegetation distribution. Contributions of climate to the vegetation distribution are larger than contributions of fires for the present-day boreal forests (around 35% vs. 1–10.9%; see Fig. 4). The reasons might be that the climate influence is a long-term and permanent process, which has been widely considered as the dominant factor for vegetation growth and distribution by the ecologists. Thus, the local vegetation has adapted to the local climatic environment and vegetation coverages and will remain stable under normal growth unless encountering sudden disturbance, such as wildfires. Fire will not only change directly the forest distribution, but will also affect subsequent (postfire) nutrient availability, soil moisture, soil temperature, rates of mineralization, and light availability, and all of these potential influences will lead to competition among vegetation (Mills et al. 2006). What is more, vegetation coverages will not change a great deal once vegetation is established and occupies the area in a certain climatic environment. But fires will kill plants and decrease PFTC quickly, especially in the flammable ecosystems, and heat-stimulated germination is globally widespread in numerous fire-prone ecosystems (Bond and Keeley 2005), making fire the necessary condition for growth and distribution of vegetation in these ecosystems. This might be another reason why fires will have a decisive impact on vegetation components. However, fire occurrence is also dependent on the regional climatic conditions to some degree, such as temperature and precipitation (Liu et al. 2012), and this might be an important reason to explain the larger contribution of climate than fires. Although Mills et al. (2006) suggested that there might be a danger when we use fire to account for the existence of the vegetation state because fire is not an ultimate cause, fire has been treated as a separate effect to explore quantitatively contributions of controls to carbon balance and biome composition (Bond-Lamberty et al. 2007, Soja et al. 2007, Murphy and Bowman 2012). In addition, the contribution of fire is uncertain in the two independent present-day experiments (1% vs. 10.9%). This might be the reason for the accuracy of the projections of burnt area and ENF, and the latter is quite different between DGVM simulation and GLC2000 land cover dataset (see Fig. 2a). Although the evaluation of the annual burnt area is acceptable in Fig. 2b, there are differences in spatial distribution in burnt area.

Other controls in regulating vegetation distribution

Many controls, including soil, topography, insect outbreak, permafrost, and human activity, also play important roles in vegetation distribution and dynamics in the boreal region, in addition to climate and fire disturbance (Murphy and Bowman 2012, Scheffer et al. 2012). We find large residuals (around 60%) exist in this study, which means that there are other controls that are beyond the scope of our study. However, it does not mean these factors are not important. Bond (2008) used the terms "bottom-up" and "top-down" to classify the controls into resource-based (e.g., water, nutrient) and disturbance-based (e.g., fire, insects). As mentioned earlier, water availability is related to climate. Nutrient availability is strongly dependent on soils, and it provides a large nutrient pool (such as soil organic matter) to facilitate the growth and

productivity of the forests, which in turn impact rates of succession (Bond 2008, Murphy and Bowman 2012). However, in this study, the soil may not be the dominant control, and Chen et al. (2015) also proved that the vegetation–atmosphere carbon fluxes (Gross Primary Productivity [GPP], Ecosystem Respiration [ER], Net Ecosystem Production [NEP]) in the Northern Hemisphere were not significantly related to soil factors ($R = -0.11$ to 0.14, $P > 0.05$). In addition, large areas of permafrost exist in Heilongjiang Province, which is an important control on vegetation and soil carbon dynamics as well by influencing hydrology and soil thermal conditions in boreal forests (Jiang et al. 2016); meanwhile, permafrost is quite sensitive to climate change (Ran et al. 2012). Recent trends of continuous and island permafrost degradation in northeast China are pushing boreal ecosystems into a disequilibrium state. This may influence the relative role of climate factors and fires in determining vegetation distribution. Recent thaw of permafrost in northeast China can be relatively fast. So, winter baseflow at two watersheds in permafrost area of northeast China had a distinct annual increasing trend, 1–2%, and lagged MAT increase by only two years (Duan et al. 2017). However, the area where these processes are happening is relatively small and does not change our "present-day" results significantly. Similar to fire, insect outbreaks alter the accumulation and distribution of vegetation and strongly disrupt and redirect succession in forest ecosystem (McCullough et al. 1998). Moreover, human activities in forest areas impact not only vegetation composition and forest structure (e.g., deforestation and forest management), but also ecological processes, nutrient availability, and biodiversity (Josefsson et al. 2009). However, the controls as noted earlier are actually interacting with each other in a complicated manner in boreal forests (Soja et al. 2007, Murphy and Bowman 2012); for example, the influence of topography affects vegetation state by impacting water, nutrient availability, and fire activity.

PFTs vs. Species

The vegetation distribution is described as the fraction of different PFTs from the simulation of SEVER-DGVM or the classification of GLC2000 remote sensing product. Although using PFTs, which are widely used in DGVMs and earth system models, as vegetation classification can help us to easily broaden the study area and provide larger scale vegetation composition, the limitations of PFTs are also obvious. Modelers define PFTs to account, in a very general way, for the variation of structure and function among plants (Sitch et al. 2003). Especially, fixed values of leaf-level traits such as carboxylation capacity and nitrogen content per unit leaf area to PFTs do not adequately describe the plasticity of such traits and the variation within PFTs (Prentice and Cowling 2013). Therefore, the new trend in local- and global-scale simulations of vegetation dynamics is to replace the "static" representation of functional diversity with trait continua, that is, incorporating functional trait variation into DGVMs (Fyllas et al. 2017). Meanwhile, species-level analysis has been widely used in the studies of vegetation compositions and their impact factors (Bond-Lamberty et al. 2007, Soja et al. 2007, Mitchell et al. 2017). Rogers et al. (2015) used remote sensing imagery, climate reanalysis data, and field inventories to evaluate the differences in boreal fire dynamics between North America and Eurasia and their main drivers based on the species level and found the difference of fire regime between two regions and then suggested that species-level traits should be considered in the evaluation of fire impacts and response to climate change. However, the difference of fire regime is inferred from the large different species between two regions. Although the analysis based on species level provides the insights into a smaller spatial scale and closer to the fire itself, many other impact factors relevant to species level, such as the impact of soil organic matter, peatland fires, permafrost, the runoff in the forest, and the interactions between different species, should be also considered.

Climate change impacts on PFTC

Climate change impacts the boreal forest health and plant physiology (Piao et al. 2008). Conceivably, boreal forests are more sensitive to climate change than other biome communities (Magnani et al. 2007, Pan et al. 2011, Bradshaw and Warkentin 2015, Steffen et al. 2015). Under a globally averaged prediction of a warming of 4°C, boreal regions would experience temperature increased from 4° to 11°C, which will lead to boreal functional groups being replaced by other more temperate PFTs, such as woodland/shrubland biomes (Scheffer et al. 2012, Gauthier et al. 2015). In our study, we further analyzed the future PFT distribution in four different global warming scenarios (see Fig. 5). Our results suggest that boreal forests in Heilongjiang Province will face a severe challenge to be replaced by other biomes due to climate change. The fastest increasing biomes will be thermophilic PFT, such as DBF, which quickly reacts to temperature.

Climate change will also impact the PFTC by altering global fire regime (Jolly et al. 2015). However, there is great uncertainty on fire regimes in boreal forests of northeast China (Liu et al. 2012). Fire frequency, burnt area, and severity are projected to increase considerably induced by warming (Héon et al. 2014, Gauthier et al. 2015). However, based on a statistic model between fire activity and different environmental controls, Krawchuk et al. (2009) predicted that fire would decrease in this region. In our simulation, burnt area will be the greatest when temperature increased by 1.5°C, because vegetation combustion is dependent not only on temperature but on fuel characters as well. Compared to temperate forests, boreal biomes, such as DNF, are more flammable and can provide more fuels to potential fires (see Fig. 5). Meanwhile, the precipitation is lowest in this scenario (see Appendix S1: Fig. S3). The relative importance of fire and climate change acts in a non-linear way between +1.5° and +4°C with a general decrease in fire influence with a small increase thereafter (see Table 3). This can be explained by the appearance of a small patch of ENF forest substituting grasslands in the most northern mountainous part of Heilongjiang Province (see Fig. 5). Furthermore, human disturbance and simulation uncertainty should be considered as well (Syphard et al. 2007, Knorr et al. 2014).

This is the first time research which has focused on whether the future boreal forests in China will be determined by fires or climate and their individual contributions to vegetation distribution. Our study suggests that the present-day boreal forest ecosystem of China is mainly determined by climate rather than fire disturbance. However, climate change may change the driving balance between climate and fires. When temperature increases by 1.5°C relative to pre-industrial climate, the boreal forest will be mainly determined by fires. It is likely due to a peak in coverage by flammable PFTs (DNF), which accelerate fire spread. What is more, climate change will impact the temperature sensitivity of Soil Organic Carbon (SOC) decomposition (Karhu et al. 2010) and further influence the fire regime by changing fuel loads. However, when the temperature increases by higher than 1.5°C, climate will be the dominant factor in regulating PFTC, according to our study. We also find that the individual contribution of temperature is generally greater than that of precipitation. The reason might be that boreal forests are more sensitive to global warming than other ecosystems (Gauthier et al. 2015) and because of their physiological and ecological characters. The influence of precipitation on vegetation is based on changing water balance of ecosystems, involving in multi-processes, including ET, which is also strongly dependent upon temperature. Other uncertainties, including model uncertainty (Jiang et al. 2012, Verheijen et al. 2015), climate data uncertainty (e.g., underling climate models have considerable disagreements in precipitation values, which may significantly impact the results) and RDA methods' uncertainty are not the focus of this study.

Conclusion

The vegetation distribution in the present-day boreal forest of Heilongjiang Province, China, is mainly determined by climate rather than fire disturbance. Climate can explain around 35%, while fire contributes 1–10.9% to the distribution of vegetation. Climate change, especially global warming, will have a strong impact on PFT coverages and fire regime, such as the burnt area. In addition, boreal forests will contract in the future in response to rising temperatures, while DBFs will progress rapidly until they become the dominant vegetation in Heilongjiang Province. Meanwhile, climate change will change the driving balance between climate and fires in local biome distribution. When temperature increases by 1.5°C, the local biome distribution will be determined by fires rather than climate (36.3% >29.3%). In other scenarios, temperature will be more important than fires in regulating vegetation distribution although other factors such as precipitation can also contribute. Spatial pattern has a significant impact on biome composition

(representing the largest part of the total variation) but will not change the driving balance between fires and climate in determining vegetation distribution. Our results are highly relevant for forest and wildfires' management.

Acknowledgments

This work was supported by the National Natural Science Foundation of China (31570475) and China Scholarship Council. We thank the two anonymous reviewers for providing insightful and valuable comments and suggestions, which significantly helped improve this paper. We thank Jun Yang for the suggestions of statistic methods. We also thank Yong Luo for providing the integrated climate data from CMIP5. NCEP Reanalysis data were provided by the NOAA/OAR/ESRL PSD, Boulder, Colorado, USA (http://www.esrl.noaa.gov/psd/). The authors declare no competing financial interests.

Literature Cited

Allen, C. D., and D. D. Breshears. 1998. Drought-induced shift of a forest–woodland ecotone: rapid landscape response to climate variation. Proceedings of the National Academy of Sciences USA 95:14839–14842.

Bachelet, D., K. Ferschweiler, T. J. Sheehan, B. M. Sleeter, and Z. L. Zhu. 2015. Projected carbon stocks in the conterminous USA with land use and variable fire regimes. Global Change Biology 21:4548–4560.

Bartholomé, E., and A. S. Belward. 2005. GLC2000: a new approach to global land cover mapping from Earth observation data. International Journal of Remote Sensing 26:1959–1977.

Bergeron, Y., and P.-R. Dansereau. 1993. Predicting the composition of Canadian southern boreal forest in different fire cycles. Journal of Vegetation Science 4:827–832.

Betts, R. A., P. M. Cox, and F. I. Woodward. 2000. Simulated responses of potential vegetation to doubled-CO_2 climate change and feedbacks on near-surface temperature. Global Ecology and Biogeography 9:171–180.

Bond, W. J. 2008. What limits trees in C_4 grasslands and savannas? Annual Review of Ecology, Evolution, and Systematics 39:641–659.

Bond, W. J., and J. E. Keeley. 2005. Fire as a global 'herbivore': the ecology and evolution of flammable ecosystems. Trends in Ecology & Evolution 20:387–394.

Bond, W. J., F. I. Woodward, and G. F. Midgley. 2005. The global distribution of ecosystems in a world without fire. New Phytologist 165:525–537.

Bond-Lamberty, B., S. D. Peckham, D. E. Ahl, and S. T. Gower. 2007. Fire as the dominant driver of central Canadian boreal forest carbon balance. Nature 450:89–92.

Borcard, D., F. Gillet, and P. Legendre. 2011a. Canonical ordination. Pages 153–225 in D. Borcard, F. Gillet, and P. Legendre, editors. Numerical ecology with R. Springer, New York, New York, USA.

Borcard, D., F. Gillet, and P. Legendre. 2011b. Spatial analysis of ecological data. Pages 227–292 in D. Borcard, F. Gillet, and P. Legendre, editors. Numerical ecology with R. Springer, New York, New York, USA.

Borcard, D., and P. Legendre. 2002. All-scale spatial analysis of ecological data by means of principal coordinates of neighbour matrices. Ecological Modelling 153:51–68.

Boschetti, L., et al. 2015. MODIS-Landsat fusion for large area 30 m burned area mapping. Remote Sensing of Environment 161:27–42.

Bowman, D., et al. 2009. Fire in the earth system. Science 324:481–484.

Bradshaw, C. J. A., and I. G. Warkentin. 2015. Global estimates of boreal forest carbon stocks and flux. Global and Planetary Change 128:24–30.

Brown, A. E., L. Zhang, T. A. McMahon, A. W. Western, and R. A. Vertessy. 2005. A review of paired catchment studies for determining changes in water yield resulting from alterations in vegetation. Journal of Hydrology 310:28–61.

Broxton, P. D., X. Zeng, D. Sulla-Menashe, and P. A. Troch. 2014. A global land cover climatology using MODIS data. Journal of Applied Meteorology and Climatology 53:1593–1605.

Chang, Y., H. S. He, Y. Hu, R. Bu, and X. Li. 2008. Historic and current fire regimes in the Great Xing'an Mountains, northeastern China: implications for long-term forest management. Forest Ecology and Management 254:445–453.

Chen, H., and P. C. Boutros. 2011. VennDiagram: a package for the generation of highly-customizable Venn and Euler diagrams in R. BMC Bioinformatics 12:35.

Chen, X., and W. Pan. 2002. Relationships among phenological growing season, time-integrated normalized difference vegetation index and climate forcing in the temperate region of eastern China. International Journal of Climatology 22:1781–1792.

Chen, Z., G. R. Yu, J. P. Ge, Q. F. Wang, X. J. Zhu, and Z. W. Xu. 2015. Roles of climate, vegetation and soil in regulating the spatial variations in ecosystem carbon dioxide fluxes in the Northern Hemisphere. PLoS ONE 10:e0125265.

Collatz, G. J., J. T. Ball, C. Grivet, and J. A. Berry. 1991. Physiological and environmental regulation of stomatal conductance, photosynthesis and transpiration: a model that includes a laminar boundary layer. Agricultural and Forest Meteorology 54:107–136.

Cramer, W., et al. 2001. Global response of terrestrial ecosystem structure and function to CO_2 and climate change: results from six dynamic global vegetation models. Global Change Biology 7:357–373.

Dale, V. H., et al. 2001. Climate change and forest disturbances. BioScience 51:723–734.

Dray, S., P. Legendre, and P. R. Peres-Neto. 2006. Spatial modelling: a comprehensive framework for principal coordinate analysis of neighbour matrices (PCNM). Ecological Modelling 196:483–493.

Duan, L., X. Man, L. B. Kurylyk, and T. Cai. 2017. Increasing winter baseflow in response to permafrost thaw and precipitation regime shifts in Northeastern China. Water 9:25.

Dybzinski, R., C. Farrior, A. Wolf, P. B. Reich, and S. W. Pacala. 2011. Evolutionarily stable strategy carbon allocation to foliage, wood, and fine roots in trees competing for light and nitrogen: an analytically tractable, individual-based model and quantitative comparisons to data. American Naturalist 177:153–166.

Ettinger, A., and J. HilleRisLambers. 2017. Competition and facilitation may lead to asymmetric range shift dynamics with climate change. Global Change Biology. https://doi.org/10.1111/gcb.13649

FAO. 1991. The digitized soil map of the world. Release 1.0. Food and Agriculture Organization of the United Nations, Rome, Italy.

Fyllas, N. M., et al. 2017. Solar radiation and functional traits explain the decline of forest primary productivity along a tropical elevation gradient. Ecology Letters 20:730–740.

Gauthier, S., P. Bernier, T. Kuuluvainen, A. Z. Shvidenko, and D. G. Schepaschenko. 2015. Boreal forest health and global change. Science 349:819–822.

Giglio, L., T. Loboda, D. P. Roy, B. Quayle, C. O. Justice, L. Giglio, J. T. Randerson, and G. R. van der Werf. 2009. An active-fire based burned area mapping algorithm for the MODIS sensor. Remote Sensing of Environment 113:408–420.

Giglio, L., J. T. Randerson, G. R. van der Werf, P. S. Kasibhatla, G. J. Collatz, D. C. Morton, and R. S. DeFries. 2010. Assessing variability and long-term trends in burned area by merging multiple satellite fire products. Biogeosciences 7:1171–1186.

Giglio, L., J. T. Randerson, G. R. van der Werf, J. T. Randerson, Y. Chen, G. R. van der Werf, B. M. Rogers, and D. C. Morton. 2013. Analysis of daily, monthly, and annual burned area using the fourth-generation global fire emissions database (GFED4). Journal of Geophysical Research: Biogeosciences 118:317–328.

Héon, J., D. Arseneault, and M.-A. Parisien. 2014. Resistance of the boreal forest to high burn rates. Proceedings of the National Academy of Sciences USA 111:13888–13893.

Hirota, M., M. Holmgren, E. H. Van Nes, and M. Scheffer. 2011. Global resilience of tropical forest and savanna to critical transitions. Science 334:232.

Holdridge, L. R. 1947. Determination of world plant formations from simple climatic data. Science 105:367.

Hulme, M. 2016. 1.5 °C and climate research after the Paris Agreement. Nature Climate Change 6:222–224.

Jackson, S. T. 2013. Natural, potential and actual vegetation in North America. Journal of Vegetation Science 24:772–776.

Jiafu, M., et al. 2015. Disentangling climatic and anthropogenic controls on global terrestrial evapotranspiration trends. Environmental Research Letters 10:094008.

Jiang, Y., Q. Zhuang, S. Schaphoff, S. Sitch, A. Sokolov, D. Kicklighter, and J. Melillo. 2012. Uncertainty analysis of vegetation distribution in the northern high latitudes during the 21st century with a dynamic vegetation model. Ecology and Evolution 2:593–614.

Jiang, Y., Q. Zhuang, S. Sitch, J. A. O'Donnell, D. Kicklighter, A. Sokolov, and J. Melillo. 2016. Importance of soil thermal regime in terrestrial ecosystem carbon dynamics in the circumpolar north. Global and Planetary Change 142:28–40.

Jolly, W. M., M. A. Cochrane, P. H. Freeborn, Z. A. Holden, T. J. Brown, G. J. Williamson, and D. M. J. S. Bowman. 2015. Climate-induced variations in global wildfire danger from 1979 to 2013. Nature Communications 6:1–11.

Josefsson, T., G. Hörnberg, and L. Östlund. 2009. Long-term human impact and vegetation changes in a boreal forest reserve: implications for the use of protected areas as ecological references. Ecosystems 12:1017–1036.

Justice, C. O., L. Giglio, S. Korontzi, J. Owens, J. T. Morisette, D. Roy, J. Descloitres, S. Alleaume, F. Petitcolin, and Y. Kaufman. 2002. The MODIS fire products. Remote Sensing of Environment 83: 244–262.

Kalnay, E., et al. 1996. The NCEP/NCAR 40-year reanalysis project. Bulletin of the American Meteorological Society 77:437–471.

Karhu, K., et al. 2010. Temperature sensitivity of soil carbon fractions in boreal forest soil. Ecology 91: 370–376.

Khvostikov, S., S. Venevsky, and S. Bartalev. 2015. Regional adaptation of a dynamic global vegetation model using a remote sensing data derived land

cover map of Russia. Environmental Research Letters 10:125007.

Knorr, W., T. Kaminski, A. Arneth, and U. Weber. 2014. Impact of human population density on fire frequency at the global scale. Biogeosciences 11:1085–1102.

Krawchuk, M. A., M. A. Moritz, M.-A. Parisien, J. Van Dorn, and K. Hayhoe. 2009. Global pyrogeography: the current and future distribution of wildfire. PLoS ONE 4:e5102.

Le Quéré, C., et al. 2015. Global carbon budget 2014. Earth System Science Data 7:47–85.

Lewis, S. L. 2006. Tropical forests and the changing earth system. Philosophical Transactions of the Royal Society B 361:195–210.

Liu, Z., J. Yang, Y. Chang, P. J. Weisberg, and H. S. He. 2012. Spatial patterns and drivers of fire occurrence and its future trend under climate change in a boreal forest of Northeast China. Global Change Biology 18:2041–2056.

Magnani, F., et al. 2007. The human footprint in the carbon cycle of temperate and boreal forests. Nature 447:848–850.

Mao, J. F., X. Y. Shi, P. E. Thornton, F. M. Hoffman, Z. C. Zhu, and R. B. Myneni. 2013. Global latitudinal-asymmetric vegetation growth trends and their driving mechanisms: 1982-2009. Remote Sensing 5:1484–1497.

McCullough, D. G., R. A. Werner, and D. Neumann. 1998. Fire and insects in northern and boreal forest ecosystems of North America. Annual Review of Entomology 43:107–127.

Meinshausen, M., et al. 2011. The RCP greenhouse gas concentrations and their extensions from 1765 to 2300. Climatic Change 109:213–241.

Mills, A. J., K. H. Rogers, M. Stalmans, and E. T. F. Witkowski. 2006. A framework for exploring the determinants of savanna and grassland distribution. BioScience 56:579–589.

Mitchell, R. M., J. D. Bakker, J. B. Vincent, and G. M. Davies. 2017. Relative importance of abiotic, biotic, and disturbance drivers of plant community structure in the sagebrush steppe. Ecological Applications 27:756–768.

Murphy, B. P., and D. M. J. S. Bowman. 2012. What controls the distribution of tropical forest and savanna? Ecology Letters 15:748–758.

Nemani, R. R., C. D. Keeling, H. Hashimoto, W. M. Jolly, S. C. Piper, C. J. Tucker, R. B. Myneni, and S. W. Running. 2003. Climate-driven increases in global terrestrial net primary production from 1982 to 1999. Science 300:1560–1563.

Oksanen, J., F. G. Blanchet, R. Kindt, P. Legendre, P. R. Minchin, R. B. O'Hara, G. L. Simpson, P. Solymos, M. H. H. Stevens, and H. Wagner. 2015. Vegan:

community ecology package. R package version 2.3-2. http://CRAN.R-project.org/package=vegan

Pan, Y., et al. 2011. A large and persistent carbon sink in the world's forests. Science 333:988–993.

Parmesan, C., T. L. Root, and M. R. Willig. 2000. Impacts of extreme weather and climate on terrestrial biota. Bulletin of the American Meteorological Society 81:443–450.

Parmesan, C., and G. Yohe. 2003. A globally coherent fingerprint of climate change impacts across natural systems. Nature 421:37–42.

Piao, S. L., et al. 2008. Net carbon dioxide losses of northern ecosystems in response to autumn warming. Nature 451:49–U43.

Poulter, B., P. Ciais, E. Hodson, H. Lischke, F. Maignan, S. Plummer, and N. E. Zimmermann. 2011. Plant functional type mapping for earth system models. Geoscientific Model Development 4:993–1010.

Prentice, I. C., and S. A. Cowling. 2013. Dynamic global vegetation models. Pages 607–689 in S. Levin, editor. Encyclopedia of biodiversity. Second edition. Elsevier, Amsterdam, The Netherlands.

Raich, J. W., and W. H. Schlesinger. 1992. The global carbon dioxide flux in soil respiration and its relationship to vegetation and climate. Tellus Series B-Chemical and Physical Meteorology 44:81–99.

Ramankutty, N., and J. A. Foley. 1999. Estimating historical changes in global land cover: croplands from 1700 to 1992. Global Biogeochemical Cycles 13:997–1027.

Ran, Y. H., X. Li, G. D. Cheng, T. J. Zhang, Q. B. Wu, H. J. Jin, and R. Jin. 2012. Distribution of permafrost in China: an overview of existing permafrost maps. Permafrost and Periglacial Processes 23:322–333.

Randerson, J. T., Y. Chen, G. R. van der Werf, B. M. Rogers, and D. C. Morton. 2012. Global burned area and biomass burning emissions from small fires. Journal of Geophysical Research: Biogeosciences 117:G04012.

Riahi, K., A. Grübler, and N. Nakicenovic. 2007. Scenarios of long-term socio-economic and environmental development under climate stabilization. Technological Forecasting and Social Change 74:887–935.

Rogers, B. M., A. J. Soja, M. L. Goulden, and J. T. Randerson. 2015. Influence of tree species on continental differences in boreal fires and climate feedbacks. Nature Geoscience 8:228–234.

Roy, D. P., L. Boschetti, C. O. Justice, and J. Ju. 2008. The collection 5 MODIS burned area product - Global evaluation by comparison with the MODIS active fire product. Remote Sensing of Environment 112:3690–3707.

RStudio. 2015. RStudio: integrated development environment for R. Version 0.99.489. RStudio, Boston, Massachusetts, USA.

Scheffer, M., M. Hirota, M. Holmgren, E. H. Van Nes, and F. S. Chapin. 2012. Thresholds for boreal biome transitions. Proceedings of the National Academy of Sciences USA 109:21384–21389.

Scheiter, S., and S. I. Higgins. 2009. Impacts of climate change on the vegetation of Africa: an adaptive dynamic vegetation modelling approach. Global Change Biology 15:2224–2246.

Scheiter, S., S. I. Higgins, J. Beringer, and L. B. Hutley. 2015. Climate change and long-term fire management impacts on Australian savannas. New Phytologist 205:1211–1226.

Scheiter, S., L. Langan, and S. I. Higgins. 2013. Next-generation dynamic global vegetation models: learning from community ecology. New Phytologist 198:957–969.

Seiler, C., R. W. A. Hutjes, B. Kruijt, J. Quispe, S. Añez, V. K. Arora, J. R. Melton, T. Hickler, and P. Kabat. 2014. Modeling forest dynamics along climate gradients in Bolivia. Journal of Geophysical Research: Biogeosciences 119:758–775.

Sitch, S., et al. 2003. Evaluation of ecosystem dynamics, plant geography and terrestrial carbon cycling in the LPJ dynamic global vegetation model. Global Change Biology 9:161–185.

Soja, A. J., N. M. Tchebakova, N. H. F. French, M. D. Flannigan, H. H. Shugart, B. J. Stocks, A. I. Sukhinin, E. I. Parfenova, F. S. Chapin III, and P. W. Stackhouse Jr. 2007. Climate-induced boreal forest change: predictions versus current observations. Global and Planetary Change 56:274–296.

Steffen, W., et al. 2015. Planetary boundaries: guiding human development on a changing planet. Science 347:1259855.

Stephenson, N. L. 1990. Climatic control of vegetation distribution: the role of the water balance. American Naturalist 135:649–670.

Syphard, A. D., V. C. Radeloff, J. E. Keeley, T. J. Hawbaker, M. K. Clayton, S. I. Stewart, and R. B. Hammer. 2007. Human influence on California fire regimes. Ecological Applications 17:1388–1402.

Tan, K., S. Piao, C. Peng, and J. Fang. 2007. Satellite-based estimation of biomass carbon stocks for northeast China's forests between 1982 and 1999. Forest Ecology and Management 240:114–121.

Tao, F., M. Yokozawa, Y. Xu, Y. Hayashi, and Z. Zhang. 2006. Climate changes and trends in phenology and yields of field crops in China, 1981–2000. Agricultural and Forest Meteorology 138:82–92.

Thonicke, K., S. Venevsky, S. Sitch, and W. Cramer. 2001. The role of fire disturbance for global vegetation dynamics: coupling fire into a dynamic global vegetation model. Global Ecology and Biogeography 10:661–677.

Tjoelker, M. G., J. Oleksyn, and P. B. Reich. 2001. Modelling respiration of vegetation: evidence for a general temperature-dependent Q_{10}. Global Change Biology 7:223–230.

Van der Werf, G. R., J. T. Randerson, L. Giglio, G. J. Collatz, P. S. Kasibhatla, and A. F. Arellano. 2006. Interannual variability in global biomass burning emissions from 1997 to 2004. Atmospheric Chemistry and Physics 6:3423–3441.

Venevsky, S., and S. Maksyutov. 2007. SEVER: a modification of the LPJ global dynamic vegetation model for daily time step and parallel computation. Environmental Modelling & Software 22:104–109.

Verheijen, L. M., R. Aerts, V. Brovkin, J. Cavender-Bares, J. H. C. Cornelissen, J. Kattge, and P. M. Van Bodegom. 2015. Inclusion of ecologically based trait variation in plant functional types reduces the projected land carbon sink in an earth system model. Global Change Biology 21:3074–3086.

Volney, W. J. A., and R. A. Fleming. 2000. Climate change and impacts of boreal forest insects. Agriculture, Ecosystems & Environment 82:283–294.

Weber, M. G., and M. D. Flannigan. 1997. Canadian boreal forest ecosystem structure and function in a changing climate: impact on fire regimes. Environmental Reviews 5:145–166.

Wu, M., W. Knorr, K. Thonicke, G. Schurgers, A. Camia, and A. Arneth. 2015. Sensitivity of burned area in Europe to climate change, atmospheric CO_2 levels, and demography: a comparison of two fire-vegetation models. Journal of Geophysical Research: Biogeosciences 120:2256–2272.

Xiao, X., S. Boles, J. Liu, D. Zhuang, and M. Liu. 2002. Characterization of forest types in Northeastern China, using multi-temporal SPOT-4 VEGETATION sensor data. Remote Sensing of Environment 82:335–348.

Xu, H., Z. Li, and Y. Qiu. 1997. Fire disturbance history in virgin forest in northern region of Daxinganling Mountains. Acta Ecologica Sinica 17:337–343.

Zhang, L. J., L. Q. Jiang, and X. Z. Zhang. 2015. Spatially precise reconstruction of cropland areas in Heilongjiang Province, northeast China during 1900-1910. Journal of Geographical Sciences 25:592–602.

Zhengqiu, Z., X. Yongkang, G. MacDonald, P. M. Cox, and G. J. Collatz. 2015. Investigation of North American vegetation variability under recent climate: a study using the SSiB4/TRIFFID biophysical/dynamic vegetation model. Journal of Geophysical Research: Atmospheres 120:1300–1321.

Zhu, Z., et al. 2016. Greening of the Earth and its drivers. Nature Climate Change 6:791–795.

Zielinski, W. J., J. M. Tucker, and K. M. Rennie. 2017. Niche overlap of competing carnivores across climatic gradients and the conservation implications of climate change at geographic range margins. Biological Conservation 209:533–545.

7

Effects of different nitrogen additions on soil microbial communities in different seasons in a boreal forest

Guoyong Yan,[1,2] Yajuan Xing,[1,3] Lijian Xu,[1] Jianyu Wang,[2] Xiongde Dong,[1] Wenjun Shan,[1] Liang Guo,[1] and Qinggui Wang[1,2,†]

[1]College of Agricultural Resource and Environment, Heilongjiang University, 74 Xuefu Road, Harbin 150080 China
[2]School of Forestry, Northeast Forestry University, 26 Hexing Road, Harbin 150040 China
[3]Institute of Forestry Science of Heilongjiang Province, 134 Haping Road, Harbin 150081 China

Abstract. In the global change scenario, nitrogen (N) deposition has the potential to affect the soil microbial communities that play critical roles in ecosystem functioning. Although the impacts of N deposition on soil microbial communities have been reasonably well studied, microorganism responses to the N addition combined with the seasonal change have rarely been reported. This study was conducted to evaluate the effects of different levels of N addition (0, 2.5, 5.0, and 7.5 g $N \cdot m^{-2} \cdot yr^{-1}$) on soil microbial communities in different seasons (spring, summer, and autumn) in a boreal forest. Our results showed that the soil physical–chemical properties were found to be changed by N addition and seasonal changes were correlated with microbial community structure. The N addition and seasonal changes significantly affect the diversity and abundance of the microbial community, while their interactions only affect the bacterial abundance and the fungal diversity. In addition, our results also provided clear evidence for specific responses of microorganisms to the different N additions and each season. Our findings suggest that the microbial community response to N deposition could be the seasonal change and will strongly correlate with environmental change.

Key words: boreal forest; nitrogen deposition; seasonal changes; soil microbial communities.

† E-mail: qgwang1970@163.com

Introduction

Atmospheric nitrogen (N) deposition from anthropogenic sources is currently estimated to be 30–50% greater than that from natural terrestrial sources (Canfield et al. 2010) and has increased approximately fourfold over the past century (IPCC 2013). It is predicted that the rate of N deposition will continuously increase, and by 2050, it may double (Phoenix et al. 2011). Especially in China, with the rapid economic growth, the average annual bulk deposition of N increased to 21.1 kg N/hm^2 in the 2000s from 13.2 kg N/hm^2 in the 1990s and atmospheric N deposition has increased by 8 kg N/hm^2 compared with the 1980s (Liu et al. 2013). Increased N deposition can have dramatic impacts on the structure and function of terrestrial ecosystem. Although attention has been paid to the effect of N deposition on ecosystem functioning, most studies focused on the response of the aboveground portion, such as aboveground plant production, plant community composition, plant diversity, and lignin in the leaf litter (Galloway et al. 2008, Xia and Wan 2008, Bleeker et al. 2011, Phoenix et al. 2011, Frey et al. 2014). The effects of N deposition on the soil microbial communities and their related ecological processes remain unclear, although they are also important in global change ecology (Hoover et al. 2012, Li et al. 2013a, b, Yao et al. 2014, Zhao et al. 2014, Xiong et al. 2016).

Soil microbial communities are directly related to soil biogeochemical processes, and they play an important role in the soil carbon cycle and soil N turnover (Bardgett et al. 2008, Xiong et al. 2014a, b, Zeng et al. 2016). The soil microorganisms can affect plant growth, soil organic matter decomposition and mineralization, nutrient cycling, and soil degradation (Klaubauf et al. 2010, Li et al. 2012, Graham et al. 2016). However, the community structure is considered to be a key determinant of the functions: Direct change in microbial community structure may alter microbial functions and soil N and C dynamics (Fierer et al. 2012, Chen et al. 2014, Matulich and Martiny 2014, Leff et al. 2015, Mannisto et al. 2016). Therefore, the shifts in the structure and composition of the microbial communities are strong indicators of soil biochemical processes and essential ecosystem functions (Edmeades 2003, Liu et al. 2006, Pardo et al. 2011).

Nitrogen deposition can affect structure and composition of soil microbial communities (Treseder 2008, Klaubauf et al. 2010, Tian and Niu 2015, Shi et al. 2016). Many previous experiments (including field and laboratory) have also identified the major shifts in the microbial community structure under N addition treatments (Ramirez et al. 2010a, b, Fierer et al. 2012, Ramirez et al. 2012, Koyama et al. 2014, Leff et al. 2015). Moreover, several meta-analyses have demonstrated that the responses of soil microbial communities to experimental N manipulations might alter soil ecosystem functions (Treseder 2008, Janssens et al. 2010, Zhou et al. 2014, Carey et al. 2016). Although previous studies suggested that N availability was an important determinant for the dominant life strategy of the soil microbial community (Fierer et al. 2012, Chen et al. 2014, Leff et al. 2015), the relationships between the different microbial taxa and N addition level are variable (Williams et al. 2013, Zhong et al. 2015), and the mechanisms underlying the soil microbial feedback in response to N deposition remain elusive (Zeng et al. 2016). Understanding the effect of N deposition on soil microbial communities is beneficial in the prediction of the manner in which soil

microorganisms will respond to environmental change in the future.

In addition to N deposition, soil microorganisms also have to face the seasonal variations in environmental conditions such as temperature and moisture (Lopez-Mondejar et al. 2015). These seasonal variations result in the seasonal change in microbial community (Aponte et al. 2010, Kaiser et al. 2010, Landesman and Dighton 2011, Kuffner et al. 2012, Zhang et al. 2013, Zhou et al. 2013, Voriskova et al. 2014, Siles et al. 2016). Thus, interaction between seasonal change and N deposition might have an impact on microbial communities. However, previous studies on the response of microbial communities to the N deposition have rarely focused on the seasonal changes.

Therefore, we designed an experiment to study the responses of the soil microbial community to N addition and seasonal changes. The average annual N deposition in China increased to 3.5 g $N \cdot m^{-2} \cdot yr^{-1}$ in 2012 from 1.3 g $N \cdot m^{-2} \cdot yr^{-1}$ in 1980 due to the increased anthropogenic activities and excessive application of fertilizer, and the current N deposition rate is 2.5 g $N \cdot m^{-2} \cdot yr^{-1}$ in northern China (Liu et al. 2013). Therefore, we investigated the soil microbial community in a controlled experiment plot that has received N addition at four different levels (0, 2.5, 5.0, and 7.5 g $N \cdot m^{-2} \cdot yr^{-1}$) for 4 yr in a boreal forest, northern China. The objective of this study was to examine the effects of N additions, seasonal changes, and their interaction on soil microbial community composition and diversity. We hypothesized that (1) N addition, seasonal changes, and their interaction will cause a change in the microbial community structure; (2) this change will be via modifying the soil properties and underground micro-environmental factors. The Illumina MiSeq sequencing analysis was used to determine the taxa of soil bacteria and fungi.

Materials and Methods

Site description

Our experiment was conducted in a boreal forest that is located in Nanwenghe National Natural Reserve (51°05′–51°39′ N, 125°07′–125°50′ E) in the Greater Khingan Mountains, China. The climate of the station is a typical cold temperate continental climate with a long, cold winter and a short, warm summer. Mean annual temperature is $-2.4°C$, ranging from $-26.3°C$ in January to $18.6°C$ in July. The mean annual precipitation is about 489 mm, mainly falling during short summer (July and August). The soil type of the studied areas is mainly stony to sandy loam, and the mean soil depth is 20 cm. The dominant and principal tree species is *Larix gmelinii* with a mean stand density of 2852 ± 99 trees/hm^2 and a mean diameter at breast height (1.3 m height, dbh) of 8.98 ± 0.32 cm.

Experimental design

To investigate the effects of N deposition, the N addition experiment was established in May 2011. This design was replicated in three random experimental blocks, each consisting of four plots measuring 20×20 m each, and the plots were separated by 10 m wide buffer strips to avoid disturbing nearby plots. In total, 12 plots were established. According to the current N deposition rate (2.5 g $N·m^{-2}·yr^{-1}$) in northern China (Liu et al. 2013), three N addition treatments—low level of N (LN, 2.5 g $N·m^{-2}·yr^{-1}$), medium level of N (MN, 5 g $N·m^{-2}·yr^{-1}$), high level of N (HN, 7.5 g $N·m^{-2}·yr^{-1}$)—and a control treatment (C, no N addition) were conducted randomly in four plots of each block. NH_4NO_3 was applied in this study as a form of N addition from May 2011 and was evenly distributed on five occasions during each growing season (from May to September) of local forests. The NH_4NO_3 was weighed (571.4, 1142.9, and 1714.3 g NH_4NO_3 for LN, MN, and HN in each plot in the 10th of each month, respectively) according to the N addition rate, mixed with 32 L of water. Then, a solution of NH_4NO_3 was sprayed evenly onto the forest floor of each plot by using a backpack sprayer. Each control plot received 32 L of N-free water.

Soil sampling

To determine the effect of seasonal changes, soil samples were collected in spring (21st May), summer (23rd July), and autumn (27th September) of 2015. Six soil columns (depth = 0–10 cm) were obtained at six random points on each plot using a stainless soil corer (5 cm inner diameter and 10 cm long). The six soil columns were then mixed to form a composite sample. All of the samples were sieved through a 2-mm soil sieve, and the roots, stones, and other debris were removed. Each sample was then divided into two sub-samples: One of them was immediately frozen in liquid N_2, put into the box containing dry ice, taken back to the laboratory, and stored at $-80°C$ for nucleic acid extraction, and the other one was packed in an icebox and transported to the laboratory for nutrient analysis.

Soil physicochemical properties

The soil temperature ($T_{5 cm}$, °C) and the soil moisture ($W_{5 cm}$, volumetric water content, m^3/m^3) at the 5 cm depth were simultaneously monitored by using a soil temperature probe (EM50; Decagon Devices, Pullman, Washington, USA) and the soil moisture probe (EC-5; Decagon Devices). The soil pH was measured by using a pH meter (Sartorius PB-10, Gottingen, Germany) with a 1:2.5 (soil:water, dry weight/volume) ratio of soil to water following shaking for 30 min. The soil total carbon and total N (TN) were measured by using a Multi N/C 3100 Analyzer (Analytik Jena, Thuringia, Germany). Dissolved organic carbon (DOC) and dissolved total N (DTN) were extracted from 10 g soil using 2 mol/L KCl extraction procedures. Dissolved total N and DOC were determined by using a Multi N/C 3100 Analyzer (Analytik Jena AG, Thuringia, Germany).

DNA extraction and PCR amplification

Microbial DNA in each sample was extracted from 0.5 g soil using the E.Z.N.A. Soil DNA Kit (Omega Bio-Tek, Norcross, Georgia, USA) according to the manufacturer's protocol. The protocol is available online (http://omegabiotek.com/store/wp-content/uploads/2013/04/D5625-Soil-DNA-Kit-Combo-Omega.pdf). The 16S ribosomal RNA gene was amplified by using a barcoded primer set 338F (5'-TGCTGCCTCCCGTAGGAGT-3')/806R (5'-GGACTACHVGGGTWTCTAAT-3'), and the fungal 18S ribosomal RNA genes were amplified by using the primer set ITS1F (5'-CTTG GTCATTTAGAGGAAGTAA-3')/2043R (5'-GCTG CGTTCTTCATCGATGC-3'). Three replicates of polymerase chain reaction were performed for each soil sample, and the products were pooled prior to purification. Amplicons were extracted from 2% agarose gels and purified using the Axy-Prep DNA Gel Extraction Kit (Axygen Biosciences, Union City, California, USA) according to the manufacturer's instructions and quantified with QuantiFluor-ST (Promega, Madison, Wisconsin,

USA). Then, the purified amplicons were pooled in equimolar concentrations and then sequenced on an Illumina MiSeq platform. The Illumina MiSeq sequencing was performed at Shanghai Majorbio Bio-Pharm Technology, Shanghai, China. The Illumina MiSeq sequencing was described in detail by Zhong et al. (2015).

Raw FASTQ files were used in the processing of the sequencing data. The raw FASTQ files were demultiplexed and quality-filtered using QIIME (version 1.17, http://qiime.org). Operational taxonomic units were clustered with a 97% similarity cutoff using UPARSE (version 7.1, http://drive5.com/uparse/), and the chimeric sequences were identified and removed using UCHIME (version 4.2.40, http://drive5.com/usearch/manual/uchime_algo.html). The phylogenetic affiliation of each 16S rRNA gene sequence was analyzed by the Ribosomal Database Project Classifier (version 2.2, http://sourceforge.net/projects/rdp-classifier/) against the SILVA (SSU117/119, https://www.arb-silva.de/) 16S rRNA and 18S rRNA database using a confidence threshold of 70% (Amato et al. 2013, Zhong et al. 2015). On average, 28,411 qualified 16S rRNA sequences and 30,334 qualified 18S rRNA sequences were obtained per sample. The MOTHUR software (version 1.30.1, http://www.mothur.org/wiki/Schloss_SOP#Alpha_diversity) was used to calculate alpha diversity and richness indexes. The calculation of diversity indexes included the Shannon index (http://www.mothur.org/wiki/Shannon), and that of richness indexes included the Chao1 estimator (http://www.mothur.org/wiki/Chao).

Statistical analyses

The soil factors in each sample were analyzed by analysis of variance (ANOVA) with Tukey's honestly significant difference (HSD) test to investigate differences in response to simulated N deposition, seasonal changes, and their interaction using R version 3.3.1 (R Development Core Team 2016, http://www.r-project.org). Moreover, general linear model and multivariate ANOVA with Tukey's HSD test were used to examine the main and interactive effects of N addition and seasonal changes on the richness and Shannon indexes for bacteria and fungi.

The characterization of microbial community under N addition and seasonal changes was described using the linear discriminant analysis (LDA) effect size (LEfSe) method (http://huttenhower.sph.harvard.edu/lefse/) for biomarker discovery. The LEfSe emphasizes both statistical significance and biological relevance (Segata et al. 2011). When the N addition treatments were used as the classes of the LEfSe analysis, the seasonal changes were used as subclasses of the LEfSe analysis, and vice versa. Both the significance alphas for the factorial Kruskal–Wallis test and the pairwise Wilcoxon test between subclasses were 0.05 and an effect size threshold for the logarithmic LDA score was 2 for all of the biomarkers evaluated in this study.

The redundancy analysis (RDA, post hoc permutation tests with 999 permutations) was used as the statistical criterion to assess multivariate changes in microbial community composition among treatments using the vegan package in R version 3.3.1, which is a multivariate direct gradient analysis method, was further implemented to test for significant associations between the soil microbial community and soil properties in the different treatments of N addition and seasonal changes.

The principal component analysis (PCA) based on the distance matrix was used to elucidate changes in soil microbial communities due to different treatments, and PERMANOVA was performed as described for the unweighted UniFrac distances. All statistical analyses were performed using the vegan package in R software version 3.3.1.

RESULTS

Effects of N addition and seasonal changes on soil properties

Whereas the values of $T_{5\ cm}$, $VWC_{5\ cm}$, DOC, DTN, and TN were not significantly different among the different N addition treatments, the soil pH and the C:N ratios were significantly different among the different N addition treatments (Tables 1, 2). And in different seasons, the effects of N addition on pH and C:N ratios were different (Tables 1, 2). In the spring, a significant decrease in pH was observed in MN and HN treatments, whereas no significant differences in soil pH were observed in control and LN treatments. In the summer, a significant decrease in soil pH was observed among the different N addition treatments. In the autumn, no

Table 1. Effects of nitrogen (N) addition on abiotic soil variables in different seasons (mean ± SD).

Treatments	pH	$T_{5\ cm}$ (°C)	$VWC_{5\ cm}$ (m³/m³)	DOC (mg/kg)	DTN (mg/kg)	TN (mg/g)	C:N
Spring
C	4.63 (0.03) a	6.46 (0.58) a	0.48 (0.02) a	0.55 (0.02) a	0.033 (0.002) a	0.96 (0.14) a	16.48 (1.18) a
LN	4.64 (0.02) a	5.49 (0.42) a	0.41 (0.05) a	0.56 (0.02) a	0.035 (0.001) a	0.86 (0.03) a	15.77 (0.83) b
MN	4.39 (0.02) b	5.58 (0.90) a	0.41 (0.06) a	0.55 (0.02) a	0.035 (0.002) a	0.82 (0.13) a	15.73 (5.70) b
HN	4.20 (0.03) b	7.07 (0.46) a	0.57 (0.11) a	0.56 (0.02) a	0.036 (0.002) a	0.90 (0.03) a	15.80 (1.23) c
Summer
C	5.37 (0.04) a	14.11 (0.074) a	0.51 (0.07) a	0.72 (0.04) a	0.041 (0.001) a	1.27 (0.46) a	15.83 (1.19) a
LN	5.07 (0.04) b	13.58 (0.57) a	0.42 (0.04) a	1.03 (0.33) a	0.046 (0.003) a	0.94 (0.09) a	16.61 (1.30) a
MN	4.88 (0.05) c	13.91 (0.63) a	0.52 (0.07) a	0.99 (0.20) a	0.046 (0.003) a	1.29 (0.01) a	21.29 (0.58) b
HN	5.06 (0.03) bd	13.58 (0.35) a	0.45 (0.01) a	0.93 (0.10) a	0.046 (0.002) a	0.81 (0.07) a	26.55 (1.92) a
Autumn
C	4.04 (0.03) b	12.37 (0.44) a	0.42 (0.02) a	0.43 (0.02) a	0.032 (0.000) ab	0.93 (0.11) a	13.06 (1.20) a
LN	4.21 (0.01) a	12.37 (0.46) a	0.38 (0.10) a	0.41 (0.02) a	0.031 (0.001) b	1.11 (0.34) a	11.64 (1.84) a
MN	3.98 (0.05) b	12.50 (0.46) a	0.38 (0.05) a	0.39 (0.02) a	0.033 (0.001) a	0.79 (0.03) a	12.12 (3.36) a
HN	4.05 (0.04) b	12.57 (0.44) a	0.49 (0.03) a	0.36 (0.02) a	0.031 (0.000) b	1.05 (0.12) a	13.69 (1.44) a

Notes: HN, high level of N; LN, low level of N; MN, medium level of N. The treatments C, LN, MN, and HN represent no N addition, 2.5 g N·m⁻²·yr⁻¹ N addition, 5 g N·m⁻²·yr⁻¹ N addition, and 7.5 g N·m⁻²·yr⁻¹ N addition, respectively. $T_{5\ cm}$ represents soil temperature at 5 cm soil depth; $VWC_{5\ cm}$ represents soil volume water content at 5 cm soil depth; DOC represents soil dissolved organic carbon; DTN represents soil dissolved total nitrogen; TN represents soil total nitrogen; C:N represents the ratio of soil dissolved carbon to soil dissolved nitrogen. The values in brackets represent standard deviation ($N = 3$). Lowercase letters for a given variable indicate significant difference ($P < 0.05$) among different N addition treatments in same season based on one-way ANOVA, followed by Tukey's HSD test.

significant differences were observed among the treatments for soil pH, except LN treatments. Nevertheless, responses of the C:N ratios to the N addition treatments were only significantly different in spring, and soil C:N ratios significantly decreased with the increase in N addition.

In contrast to N addition treatments, seasonal changes had significantly affected most of the measured soil properties, such as pH, $T_{5\ cm}$, DOC, DTN, and C:N ratios (Tables 1, 2). The changes in soil properties caused by seasonal changes had consistent patterns, and their values were highest in summer compared with the spring and autumn and significantly different among the seasons ($P < 0.05$). In addition, statistically significant interactions between N addition and seasonal changes influencing soil pH and $T_{5\ cm}$ were found in our study (Tables 1, 2).

Responses of soil microbial diversity and richness to N addition and seasonal changes

N addition had a significantly different impact on microbial (bacteria and fungi) richness and diversity in different seasons (Fig. 1). For bacteria, in spring, the Shannon index and the Chao index in the LN and MN treatments significantly increased compared with the control, while those significantly decreased in the HN treatment ($P < 0.05$). The Shannon index and the Chao1 index were found to be significantly decreased ($P < 0.05$) in the HN treatment compared with control in autumn. For fungi, the Chao1 index significantly increased in the LN and MN treatments and significantly decreased ($P < 0.05$) in the HN treatment compared with the control treatment. Additionally, the Shannon index significantly increased among N addition treatments in the

Table 2. The effects of nitrogen (N) addition, season, and their interaction on abiotic soil variables.

Parameters	pH	$T_{5\ cm}$ (°C)	$VWC_{5\ cm}$ (m³/m³)	DOC (mg/kg)	DTN (mg/kg)	TN (mg/g)	C:N
N addition	***	0.110	0.336	0.762	0.177	0.349	**
Season	***	***	0.256	***	***	0.842	*
N addition × Season	***	**	0.639	0.802	0.701	0.403	0.325

Notes: DOC, dissolved organic carbon; DTN, dissolved total nitrogen; $VWC_{5\ cm}$, volumetric water content. Two-way ANOVA was applied to indicate significant difference among variances.
*$P < 0.05$; **$P < 0.01$; ***$P < 0.001$.

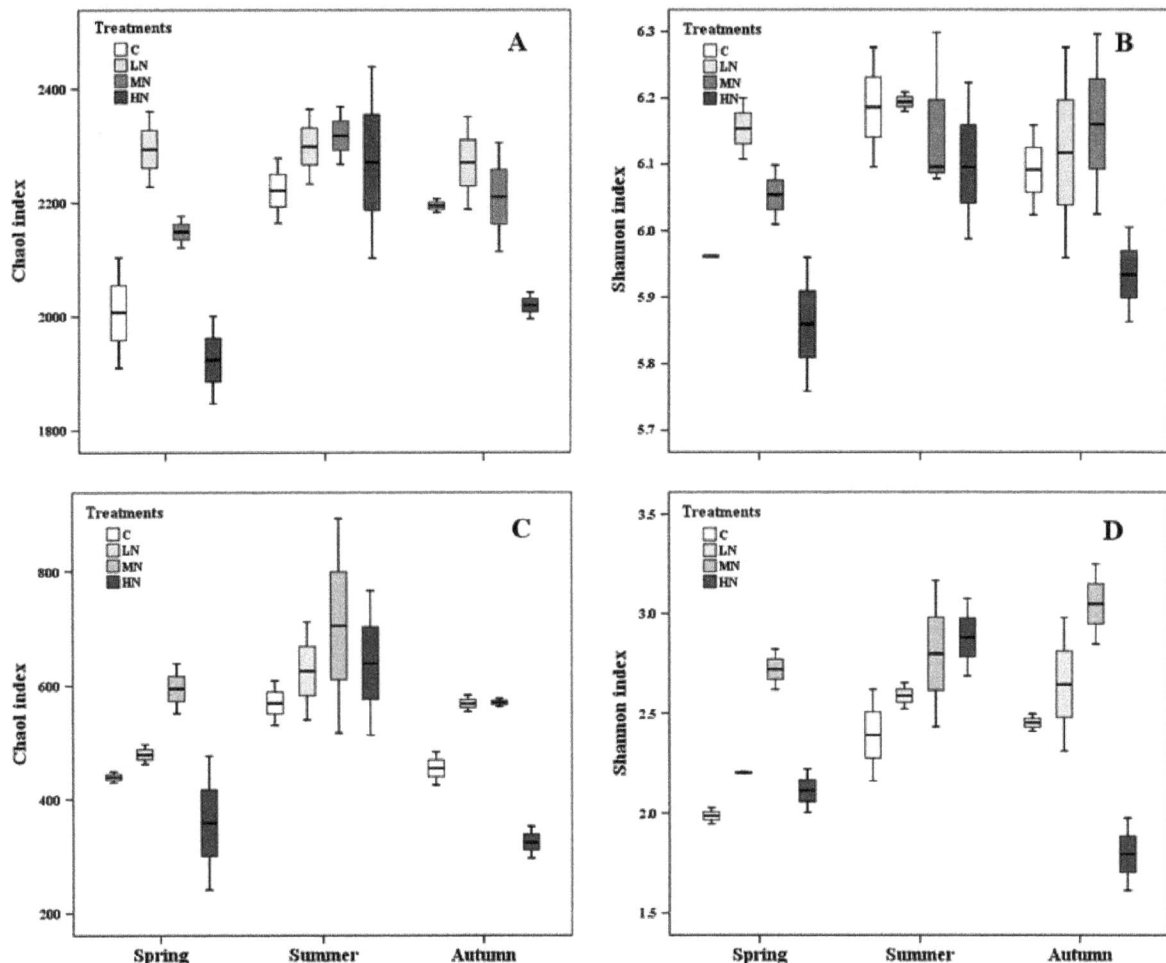

Fig. 1. Effects of N addition and season on alpha diversities of bacteria and fungi. Chao1 index (A) and Shannon index (B) of bacteria, Chao1 index (C) and Shannon index (D) of fungi.

spring and significantly increased ($P < 0.05$) in the MN treatment and significantly decreased in HN treatment in autumn. Moreover, in summer, the Shannon index significantly increased ($P < 0.05$) with the increase in N addition.

In addition to N addition treatments, seasonal changes also significantly affected diversity and richness of bacteria and fungi (Table 3). The diversity and richness of bacteria and fungi showed a different response to the seasonal changes (Fig. 1). The Chao1 index of bacteria and fungi significantly increased in summer compared with the spring and autumn, but the Shannon index of bacteria and fungi had no significant difference ($P < 0.05$) among different seasons. In general, the N addition and seasonal changes significantly affected diversity and

abundance of the microbial community, while their interactions only affected bacterial abundance and fungal diversity (Table 3).

Table 3. Results of mixed model evaluating effects of N treatments (N) and season (S) and their interaction on alpha diversities of bacteria and fungi.

Variables	N		S		N × S	
	F	P	F	P	F	P
Bacteria						
Chao1 index	12.89	***	16.19	***	3.08	*
Shannon index	7.06	**	8.03	**	1.03	0.373
Fungi						
Chao1 index	15.90	***	8.78	***	1.65	0.178
Shannon index	13.74	***	18.82	***	8.06	***

*$P < 0.05$; **$P < 0.01$; ***$P < 0.001$.

Effect of N addition on soil microbial community structure in different seasons

The simulated N deposition altered the relative abundance of bacteria and fungi at the phylum level as shown in Fig. 2. For fungi, Basidiomycota and Ascomycota were the dominant taxa in all the treatments, and Basidiomycota increased under N addition in spring and decreased in summer; however, the change in trend of Ascomycota was opposite to that of Basidiomycota in spring and summer

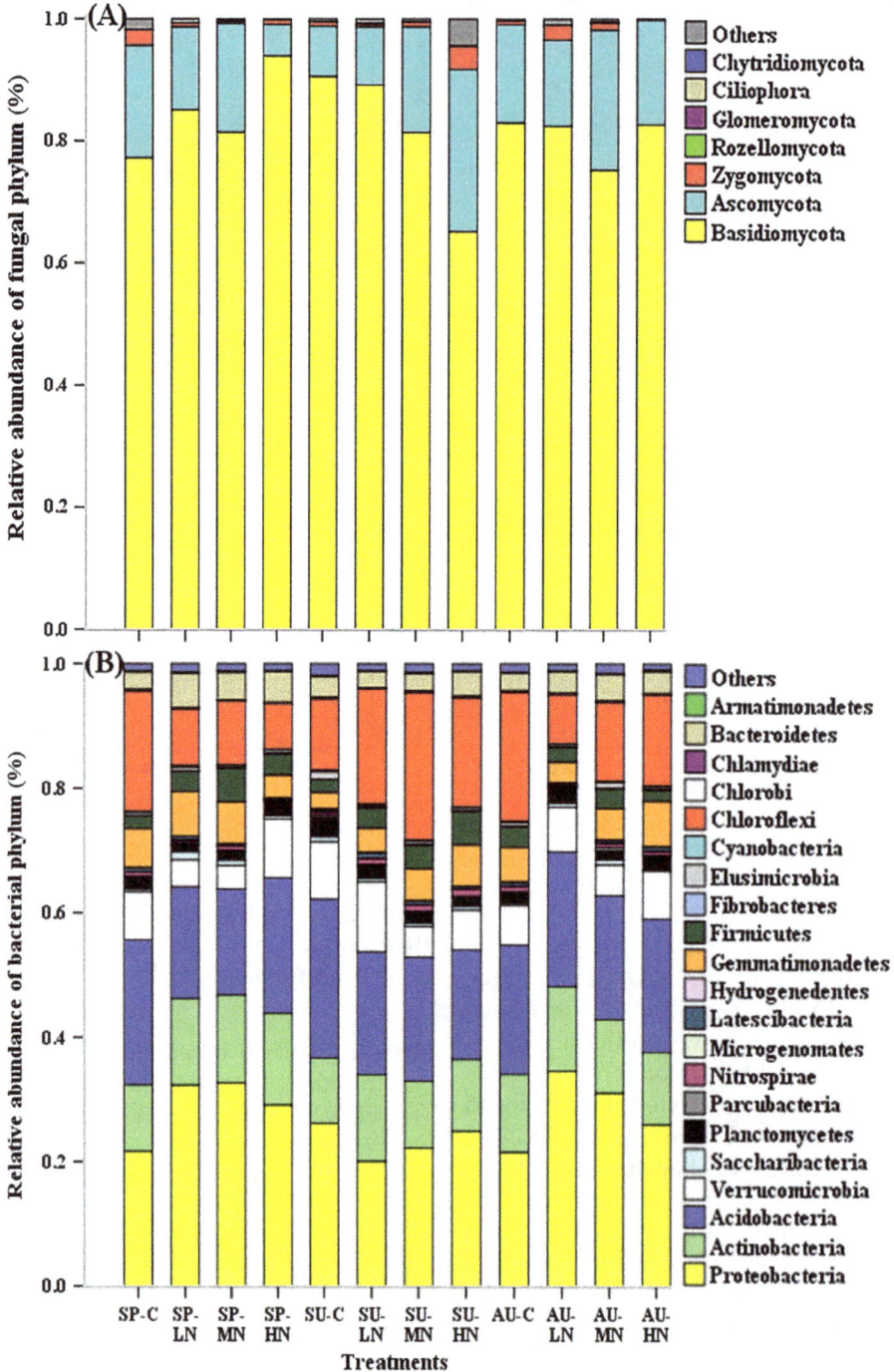

Fig. 2. The average of relative abundance (%) of fungal taxa (A) and bacterial taxa (B) at the phylum level (*n* = 3).

(Fig. 2A). For bacteria, Proteobacteria, Acidobacteria, Actinobacteria, and Chloroflexi were the dominant taxa in all the treatments. The Proteobacteria increased under the N addition treatments in spring and autumn, but decreased in summer, and the change in trend of Chloroflexi was opposite to that of Proteobacteria among all seasons.

Effects of different N addition treatments in different seasons on soil bacterial and fungal community were evaluated by principal component analysis (PCA). For bacteria, the first principal component (PC1) accounted for 31.31% and the second component (PC2) for 18.17% of the variation in the profiles (Fig. 3A). Under the same N addition level, the bacterial community structure in summer was significantly different from those of the spring and autumn ($P < 0.05$). For fungi, we also analyzed the first and second PCs, which explained 78.32% and 8.33% of variance in fungal community composition, respectively (Fig. 3B). The fungal community composition was significantly different between summer and autumn ($P < 0.05$). In addition, the N addition treatments also moderately altered the composition of the fungal and bacterial community. In general, seasonal variation, N deposition, and their interaction had significant effects on microbial community composition ($P < 0.05$).

Linear discriminant analysis effect size analyses were used to identify the statistical significance of differentially abundant taxa of bacteria and fungi in all of the treatments. The results of the LEfSe analyses for N addition and seasonal changes are shown in Figs. 4, 5, respectively. The LEfSe results at the genus level showed that *Pseudolabrys* (for LN), *OM27-clade* and *Elstera* (for MN), and *Ramlibacter* (HN) could be used as bacterial biomarkers for the N addition. And *Leptodontidium* (for control treatments), *Cystodendron* (for LN), and *Wilcoxina* (for MN) could be used as fungal biomarkers for the N addition at genus level. Besides those genera, *Thelephoraceae*, *Thelephorales*, *Pyronemataceae*, *Pezlzales*, and *Pezlzomycetes* could be fungal biomarkers for MN. Our results suggest that these microbial taxa could potentially be used as biomarkers for detecting different levels of N deposition in boreal forest soil.

By using the LEfSe analysis to identify bacterial changes caused by seasonal changes, we found that *Microbacterium* and *Epilithonimonas* primarily changed in the spring, whereas

Clostridium-sensu-stricto-13, *Clostridiaceae-1*, *Gemmatimonas*, *Gemmatimonadaceae*, *Geobacter*, and *Geobacteraceae* primarily changed in summer. There was no significant bacterial biomarker that could be used for autumn. With regard to fungi, the significant change in taxa was primarily found in autumn (*Paraphaeosphaeria*, *Stachybotrys*, and *hygrophorus*) and summer (*Xenopolyscytalum*, *Sarocladium*, *Cystofilobasidium*, *Guehomyces*, and *Xanthophyllomyces*). In addition, *Cystofilobasidiaceae*, *Cystofilobasidiales*, and *Agricales* were also used as biomarkers for the summer. These microbial species changes due to seasonal changes could be used as biomarkers for identifying different states of soils.

Relationships between the microbial community structure and soil properties

Redundancy analysis showed the relationship between soil properties and microbial (bacterial or fungal) community structure (Table 4). According to the post hoc permutation test, the soil bacterial community composition was significantly correlated with DOC ($R^2 = 0.34$, $P < 0.001$), DTN ($R^2 = 0.39$, $P < 0.001$), TN ($R^2 = 0.18$, $P < 0.028$), C:N ($R^2 = 0.32$, $P = 0.002$), and pH ($R^2 = 0.18$, $P = 0.047$), but not $T_{5\ cm}$ or $VWC_{5\ cm}$. The change in DOC explained 73.29% of the total variability of the bacterial community composition. The soil fungal community was only marginally correlated with $VWC_{5\ cm}$ ($R^2 = 0.31$, $P = 0.006$) and explained 29.23% of the total variability of the fungal community composition. The soil property indices explained 31.94% of the variation in the bacterial community and 29.11% of the variation in the fungal community. In general, seasonally induced variability in microbial community composition was larger than N treatment-induced variability.

DISCUSSION

Changes in soil bacterial community composition associated with changes in quantity and quality of soil organic matter are likely to subsequently influence soil C storage (Cusack et al. 2011, Zeng et al. 2016), and changes in soil fungal community composition associated with changes in decomposers of organic matter and symbiosis of plants are likely to affect ecosystem C and N cycles (Allen et al. 2010, Pardo et al.

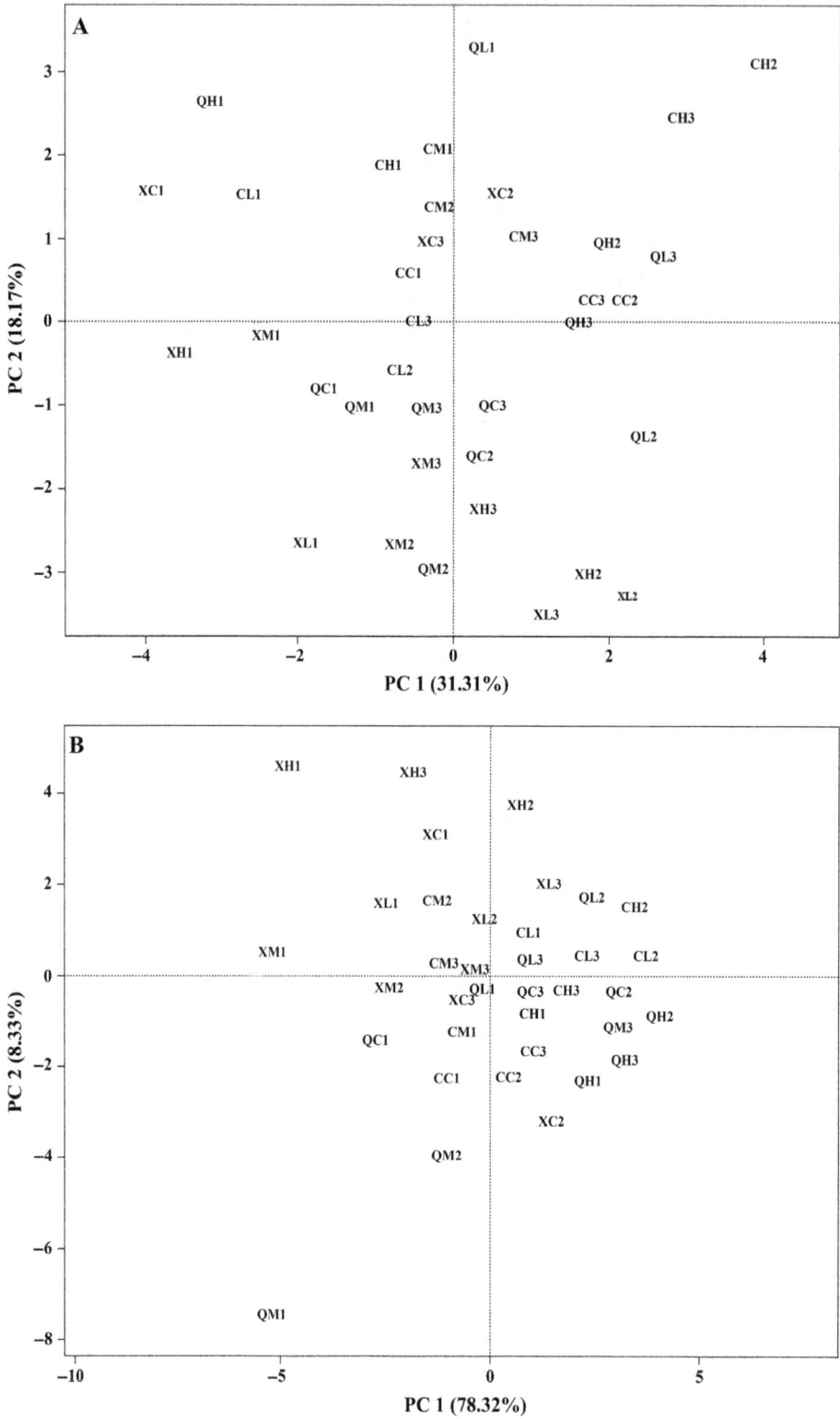

Fig. 3. Principal component analysis of microbial community data by permutation-based analysis of variance

(Fig. 3. *Continued*)

illustrating significant separation of the microbial community in N addition and season treatments. CC1, CC2, and CC3 represent control treatments; CL1, CL2, and CL3 represent low level of N (LN) treatments; CM1, CM2, and CM3 represent medium level of N (MN) treatments; CH1, CH2, and CH3 represent high level of N (HN) treatments; XC1, XC2, and XC3 represent control treatments; XL1, XL2, and XL3 represent LN treatments; XM1, XM2, and XM3 represent MN treatments; XH1, XH2, and XH3 represent HN treatments; QC1, QC2, and QC3 represent control treatments; QL1, QL2, and QL3 represent LN treatments; QM1, QM2, and QM3 represent MN treatments; QH1, QH2, and QH3 represent HN treatments.

2011, Fernandez-Fueyo et al. 2012, Schneider et al. 2012, Morrison et al. 2016). In addition, the loss of bacterial and fungal species diversity following N enrichment may threaten ecosystem stability (Tilman et al. 2006). Therefore, in studying effects of N deposition on forest ecosystem functions, it is important to take into account shifts of soil microbial community composition and diversity. The soil microbial community composition should be more copiotrophic in microcosms with higher soil N concentrations and stronger seasonal fluctuations (Mannisto et al. 2016). However, the studies of effects of N addition on soil bacterial and fungal communities have rarely focused on seasonal changes, neglecting interactions between N deposition and seasonal variation. In this study, we demonstrated that N addition and seasonal changes caused fundamental changes in the diversity and community composition of soil bacteria and fungi in a boreal forest. In addition, we found that seasonal changes might mediate effects of N addition on soil microbial communities by affecting soil properties.

Effects of seasonal changes on soil microbial diversity response to N addition

Boreal forests are generally N-limited. Therefore, the soil microorganisms in boreal forests might be N-limited as well (Tamm 1991, Sistla et al. 2012, Koyama et al. 2013, Stark et al. 2014). Previous studies have shown that low-level addition of N can stimulate soil microbial growth (Zhang et al. 2008, Bai et al. 2010), which is consistent with our results in LN and MN treatments in spring and autumn. This suggests that N might be a limiting factor for soil microbial growth in our study site, and a LN addition improved soil microbial activity and increased microbial diversity. Nevertheless, the HN addition might inhibit soil microorganisms

(Freedman et al. 2015, Zeng et al. 2016). Our results showed that the HN addition resulted in the decrease in microbial diversity in spring and autumn. It is consistent with other previous studies that showed a decline in microbial diversity following HN addition treatments (Burke et al. 2006, Campbell et al. 2010, Janssens et al. 2010, Ramirez et al. 2010a, b, Freedman et al. 2015, Zeng et al. 2016). This result suggested that the HN addition might have reached or exceeded the critical value of soil N for microorganisms. According to the related references, we summarized three potential mechanisms that have been proposed to clarify the HN-induced decline of microbial diversity: (1) N accumulation caused a decline in biodiversity by the expansion of nitrophilous species and competitive exclusion of others (Bobbink et al. 2010); (2) microbial responses to N addition were plant species specific, and direct changes in plant composition due to HN addition may affect microbial diversity by altering the quality of plant-derived C (Sagova-Mareckova et al. 2011, Zeng et al. 2016); and (3) the decrease in microbial diversity may be due to changes in soil properties (i.e., soil water and P status) caused by the N addition (Geisseler and Scow 2014, Zhong et al. 2015, Zeng et al. 2016). For example, a HN addition can induce soil acidification, exerting deleterious effects on microbial growth (Vitousek et al. 1997, Wei et al. 2013). In our study, the soil pH also showed a significant decrease in HN treatments, and decrease in soil pH usually could significantly reduce the microbial diversity with N addition (Feng et al. 2014, Liu et al. 2014, Zhou et al. 2016). In addition, Allen et al. (2010) found that at high levels of N application, plants rejected ectomycorrhizal fungi, thereby setting the stage for water and P limits to growth, which might result in a decrease in fungal diversity. Thus, N has not just a direct impact on soil microbes, but also an

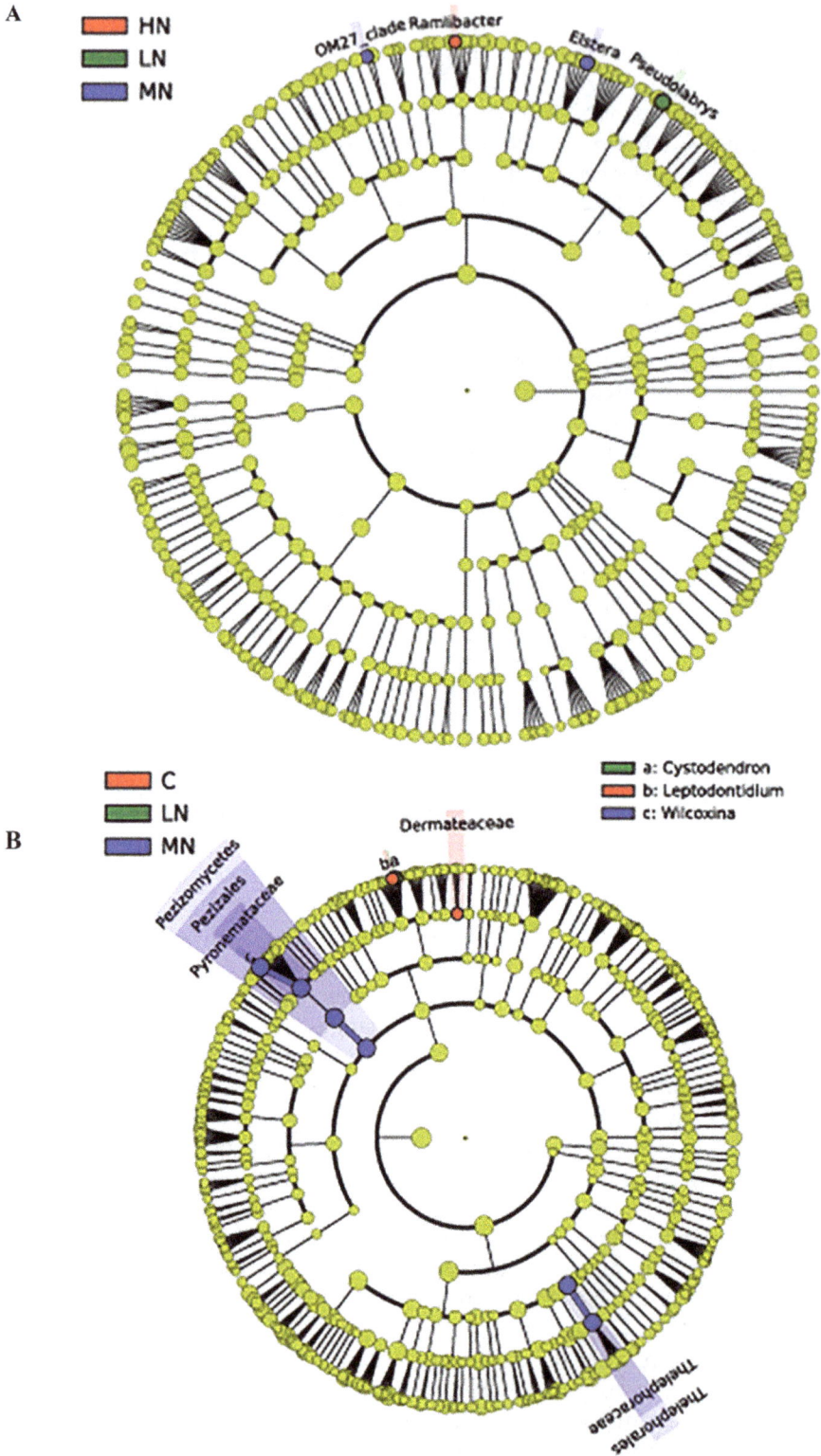

Fig. 4. A linear discriminant analysis effect size method identifies the significantly different abundant taxa of

(Fig. 4. *Continued*)

bacteria (A) and fungi (B) in all of the N treatments. The taxa with significantly different abundance among treatments are represented by colored dots, and from the center outward, they represent the kingdom, phylum, class, order, family, and genus levels. The colored shadows represent trends of the significantly differed taxa. Only taxa meeting an linear discriminant analysis significance threshold of >2 are shown. The treatments C, low level of N (LN), medium level of N (MN), and high level of N (HN) represent no N addition, 2.5 g $N \cdot m^{-2} \cdot yr^{-1}$ N addition, 5 g $N \cdot m^{-2} \cdot yr^{-1}$ N addition, and 7.5 g $N \cdot m^{-2} \cdot yr^{-1}$ N addition, respectively. All the season treatments were included in this analysis.

indirect impact mediated through the plant responses. Although we did not measure plant physiological data, we found that fine root biomass significantly decreased in the high levels of N treatment (Yan et al. 2017); therefore, we also think that a HN treatment may change the microbial community by indirectly mediating the response of plants.

Vorískova et al. (2014) found that the fungal community experiences significant seasonal changes in activity, biomass content, composition, and relative abundance of different groups, which is consistent with our results. However, the effects of N addition on soil microbial diversity were weak in summer compared with spring and autumn in our site. It seems that the microbial response to N deposition might be seasonally or temporally dependent (Freedman et al. 2015). In addition, many previous studies also suggest that both bacterial and fungal communities exhibit seasonal community dynamics (Lipson and Schmidt 2004, Bevivino et al. 2014, Vorískova et al. 2014, Matulich et al. 2015). In our results, seasonal variations significantly changed soil moisture, soil temperature, DOC, and DTN, consistent with another previous report (Zhang et al. 2015). Moreover, we speculated that soil microbial diversity responses to N deposition might be modified by seasonal changes that altered soil properties (including soil temperature, soil moisture, and soil nutrients; Zhang and Han 2012, Freedman et al. 2015). To achieve a deeper understanding of the microbial response to future rates of N deposition, we must gain a greater understanding of the seasonal and temporal robustness of soil microbial response to this pervasive agent of climate change.

Responses of soil microbial community composition to N addition and seasonal changes

The changes in soil microbial community composition are often associated with changes in the

ecosystem functions they mediate (Allison et al. 2013). Therefore, the mechanisms that affect microbial communities can be critical for understanding how ecosystem processes respond to environmental changes (Freedman et al. 2015). In our study, N addition and seasonal changes significantly altered soil bacterial and fungal community composition (Fig. 3), consistent with our hypothesis. Many previous studies also have demonstrated that shifts in soil bacterial and fungal community composition might occur as a result of N addition (Clegg et al. 2006, Singh et al. 2010, Li et al. 2013a, b, Morrison et al. 2016). However, in contrast, some studies also found that N addition did not alter microbial community composition (Frey et al. 2004, 2008, Bradford et al. 2008, Contosta et al. 2015). These different results suggest that effects of N addition on soil microbial composition could depend on the length of treatments and the type of N addition (Dong et al. 2014, Contosta et al. 2015). In addition, other studies have reported that seasonal variation in microbial community may be driven by shifts in climate and resource availability (Bohlen et al. 2001, Waldrop and Firestone 2006, Vorískova et al. 2014), which is consistent with our results.

The LEfSe analyses found that microbial taxa (*Pseudolabrys, OM27-clade, Elstera, Ramlibacter, Leptodontidium, Cystodendron,* and *Wilcoxina*) might be used as biomarkers of microbial response to N addition (Fig. 4). These results indicate that different levels of N addition had a significant effect on certain microbial species. *Ramlibacter* belongs to *Proteobacteria*, and field experiments have shown increased abundance of *Proteobacteria* in N addition plots (Ramirez et al. 2010a, b, Koyama et al. 2014), which is consistent with our results in spring (Fig. 2). *Elstera* and *Pseudolabrys* also belong to *Proteobacteria*. In contrast to what we observed in this experiment,

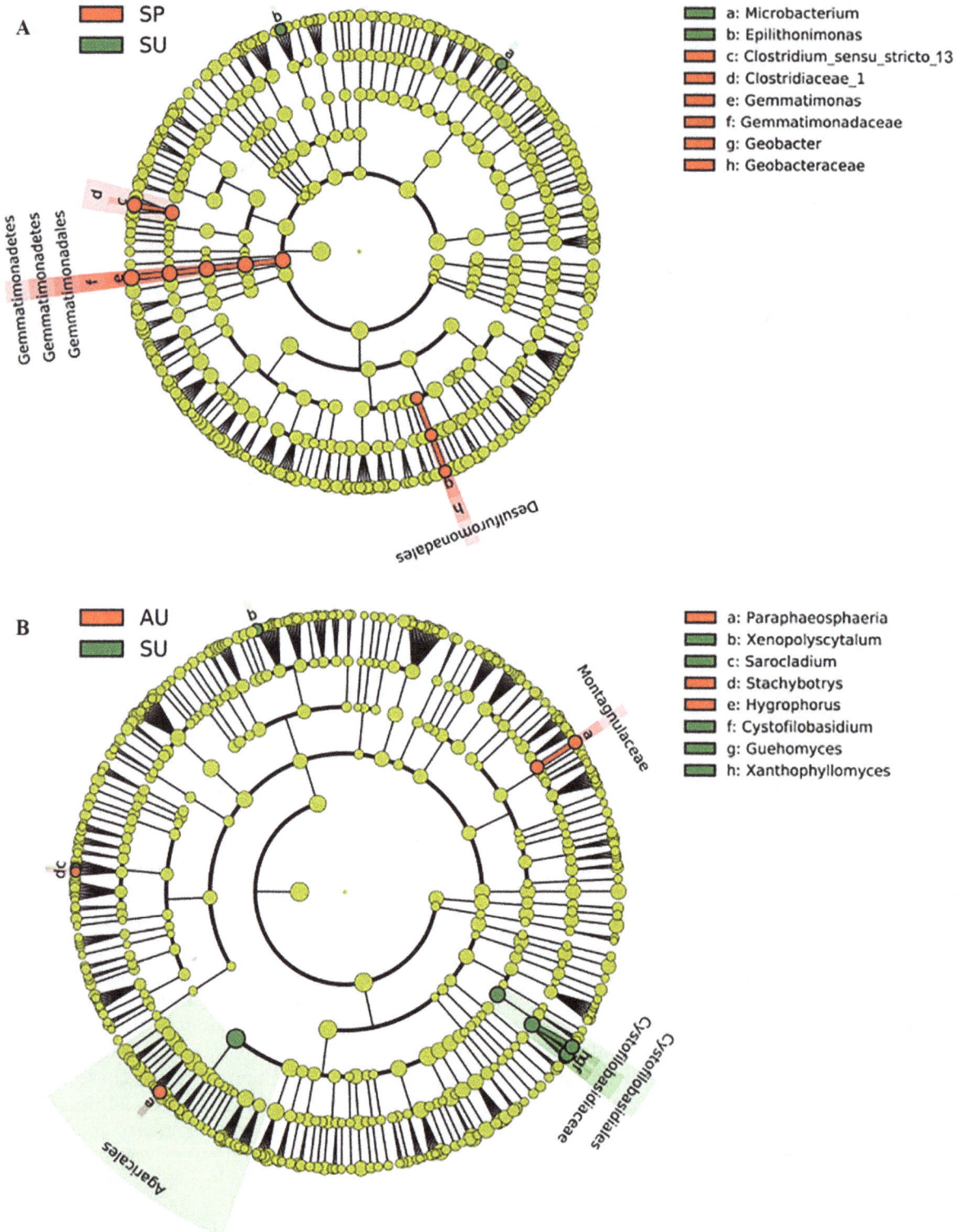

Fig. 5. A linear discriminant analysis effect size method identifies the significantly different abundant taxa of bacteria (A) and fungi (B) in all of the season changes. The taxa with significantly different abundance among

(Fig. 5. *Continued*)

treatments are represented by colored dots, and from the center outward, they represent the kingdom, phylum, class, order, family, and genus levels. The colored shadows represent trends of the significantly differed taxa. Only taxa meeting an linear discriminant analysis significance threshold of >2 are shown. The treatments SP, SM, and AU represent spring season, summer season, and autumn season, respectively. All the treatments of N addition treatments were included in this analysis.

Mannisto et al. (2016) have shown decreased abundance of *Alphaproteobacteria* in N addition plots, which is consistent with our results in summer and autumn (Fig. 2). These different results suggest that the season may modulate the response of *Proteobacteria* to N addition. Moreover, we found that the relative abundance of *Ascomycota* significantly increased in control, LN, and MN treatments, respectively. Increase in its abundance may be improving the rates of soil C decomposition in low-level N treatments (LN and MN). However, Wang et al. (2015) found that members of the *Ascomycota* are particularly vulnerable to high nutrient levels and explained why there is no change in the HN treatment. The shift in the relative abundance of *Ascomycota* may in turn dramatically affect the soil C decomposition (Xiong et al. 2014*a*, *b*). By using the LEfSe analysis to identify microbial changes caused by seasonal changes (Fig. 5), we found that specific bacterial taxa (*Microbacterium* and *Epilithonimonas*, spring; *Clostridium-sensu-stricto-13*, *Clostridiaceae-1*, *Gemmatimonas*, *Gemmatimonadaceae*, *Geobacter*, and *Geobacteraceae*, summer) and fungal taxa (*Paraphaeosphaeria*, *Stachybotrys*, and *hygrophorus*, autumn; *Xenopolyscytalum*, *Sarocladium*, *Cystofilobasidium*, *Guehomyces*, and *Xanthophyllomyces*, summer) were found to be

significantly changed. The significant change in bacterial and fungal taxa predicted by the LEfSe analysis was primarily observed in summer, which may indicate that changes in bacterial and fungal taxa in different seasons correlate with ecosystem functions and processes. For example, one previous study suggested that the fungal biomass increased approximately threefold from spring to summer, which corresponded to the expected increase in photosynthate allocation below the ground (Vorískova et al. 2014). However, N treatments and seasonal changes caused a significant increase or decrease in the communities of specific microbial groups (Dong et al. 2014). Our results suggest that these microbial species could potentially be used as biomarkers for detecting different levels of N and for identifying seasonal responses of microorganisms in boreal forests.

The RDA illustrated that these changes in microbial community composition were correlated with soil chemical properties (Table 4). Bacterial community composition was found to be significantly correlated with DTN and TN. Our result is consistent with those of Fontaine and Barot (2005) and Fierer et al. (2007), who showed that the shift in dominant life-history strategies may be a result of the increase in N availability. Li

Table 4. The redundancy analysis used to explore the relationships between microbial community and selected soil properties for bacteria and fungi.

Variables	Bacteria				Fungi			
	%Var	R^2	P	Cum%	%Var	R^2	P	Cum%
$T_{5\ cm}$	4.03	0.07	0.304	4.03	5.07	0.18	0.058	5.07
$VWC_{5\ cm}$	4.50	0.14	0.083	8.53	8.51	0.31	**	13.58
DOC	5.04	0.34	***	13.57	2.57	0.03	0.613	16.15
DTN	5.72	0.39	***	19.29	6.59	0.08	0.263	22.74
TN	4.28	0.18	**	23.57	2.91	0.01	0.924	25.65
C:N	4.70	0.32	**	28.27	2.07	0.03	0.595	27.72
pH	3.67	0.18	*	31.94	1.39	0.01	0.868	29.11

Notes: DOC, dissolved organic carbon; DTN, dissolved total nitrogen. %Var represents percentage variance in our data explained by that variable; Cum% represents the sum percentage of variance explained.
$^*P < 0.05$; $^{**}P < 0.01$; $^{***}P < 0.001$.

et al. (2014) also reported that soil TN had a great impact on the microbial community structure. The DOC and pH were also related to bacterial distribution in our study. Few studies focused on the relationship between DOC and bacterial community composition, but Sul et al. (2013) reported that SOC (soil organic carbon) was the most important factor explaining the differences in soil bacterial community structure. Moreover, soil pH is a major factor influencing the structure of the soil microbial community (Nilsson et al. 2007, Lauber et al. 2009, Feng et al. 2014, Liu et al. 2014, Zhong et al. 2015). Soil pH has been shown to decrease with N addition (Zhong et al. 2015), which is consistent with our results. And decreased pH caused a shift in the soil microbial community (Feng et al. 2014, Liu et al. 2014). In addition, the results of the RDA also implied that fungi are more resistant to environmental changes than bacteria (Hawkes et al. 2011, Jin et al. 2011), and fungal community composition only correlated with $VWC_{5\ cm}$ in our study. Previous studies show evidence that the soil microbial community is significantly correlated with soil moisture (Bi et al. 2012, Zhang et al. 2013).

Conclusion

After a four-year experiment in a boreal forest, N addition resulted in significant changes in soil bacterial and fungal community composition and these changes exhibited significant differences in spring, summer, and autumn, respectively, in a boreal forest. Moreover, some of specific microbial taxa could be used as biomarkers for microbial responses to the N addition and seasonal changes. The $VWC_{5\ cm}$, DOC, DTN, TN, and pH contents were obviously correlated with the soil microbial community composition. It indicated that they were the key variables for maintaining the soil microbial community. In addition, we found that seasonal changes might change the effects of N addition on soil microbial communities by affecting soil properties. Taken together, the responses of microbial responses to N addition are significantly different in different seasons.

Acknowledgments

This research was supported by grants from the National Key Research and Development Program of China (2016YFA0600800), National Natural Science Foundation of China (41575137, 31370494, 31170421, 31070406), and the Key Projects of Heilongjiang Province Natural Science Foundation (ZD201406). G. Yan and Y. Xing contributed equally to this work.

Literature Cited

Allen, M. F., E. B. Allen, J. L. Lansing, K. S. Pregitzer, R. L. Hendrick, R. W. Ruess, and S. L. Collins. 2010. Responses to chronic N fertilization of ectomycorrhizal pinon but not arbuscular mycorrhizal juniper in a pinon-juniper woodland. Journal of Arid Environments 74:1170–1176.

Allison, S. D., Y. Lu, C. Weihe, M. L. Goulden, A. C. Martiny, K. K. Treseder, and J. B. Martiny. 2013. Microbial abundance and composition influence litter decomposition response to environmental change. Ecology 94:714–725.

Amato, K. R., C. J. Yeoman, A. Kent, N. Righini, F. Carbonero, A. Estrad, H. R. Gaskins, R. M. Stumpf, S. Yildirim, and M. Torralba. 2013. Habitat degradation impacts black howler monkey (*Alouatta pigra*) gastrointestinal microbiomes. ISME Journal 7:1344–1353.

Aponte, C., L. V. Garcıa, T. Maranon, and M. Gardes. 2010. Indirect host effect on ectomycorrhizal fungi: Leaf fall and litter quality explain changes in fungal communities on the roots of co-occurring Mediterranean oaks. Soil Biology and Biochemistry 42:788–796.

Bai, Y. F., J. G. Wu, C. M. Clark, S. Naeem, Q. M. Pan, J. H. Huang, L. X. Zhang, and X. G. Han. 2010. Tradeoffs and thresholds in the effects of nitrogen addition on biodiversity and ecosystem functioning: evidence from inner Mongolia Grasslands. Global Change Biology 16:358–372.

Bardgett, R. D., C. Freeman, and N. J. Ostle. 2008. Microbial contributions to climate change through carbon cycle feedbacks. International Society for Microbial Ecology Journal 2:805–814.

Bevivino, A., et al. 2014. Soil bacterial community response to differences in agricultural management along with seasonal changes in a Mediterranean region. PLoS ONE 9:e105515.

Bi, J., N. L. Zhang, Y. Liang, L. Cai, and K. P. Ma. 2012. Impacts of increased N use and precipitation on microbial C utilization potential in the semiarid grassland of Inner Mongolia. Chinese Journal of Eco-Agriculture 20:1586–1593.

Bleeker, A., W. Hicks, F. Dentener, J. Galloway, and J. Erisman. 2011. N deposition as a threat to the World's protected areas under the Convention on Biological Diversity. Environmental Pollution 159:2280–2288.

Bobbink, R., et al. 2010. Global assessment of nitrogen deposition effects on terrestrial plant diversity: a synthesis. Ecological Applications 20:30–59.

Bohlen, P. J., P. M. Groffman, C. T. Driscoll, T. J. Fahey, and T. G. Siccama. 2001. Plant-soil-microbial interactions in a northern hardwood forest. Ecology 82:965–978.

Bradford, M. A., C. A. Davies, S. D. Frey, T. R. Maddox, J. M. Melillo, J. E. Mohen, J. F. Reynolds, K. K. Treseder, and M. D. Wallenstein. 2008. Thermal adaptation of soil microbial respiration to elevated temperature. Ecology Letters 11:1316–1327.

Burke, D. J., A. M. Kretzer, P. T. Rygiewicz, and M. A. Topa. 2006. Soil bacterial diversity in a loblolly pine plantation: influence of ectomycorrhizas and fertilization. FEMS Microbiology Ecology 57:409–419.

Campbell, B. J., S. W. Polson, T. E. Hanson, M. C. Mack, and E. A. G. Schuur. 2010. The effect of nutrient deposition on bacterial communities in Arctic tundra soil. Environmental Microbiology 12:1842–1854.

Canfield, D. E., A. N. Glazer, and P. G. Falkowski. 2010. The evolution and future of earth's nitrogen cycle. Science 330:192–196.

Carey, C. J., N. C. Dove, J. M. Beman, S. C. Hart, and E. L. Aronson. 2016. Meta-analysis reveals ammonia-oxidizing bacteria respond more strongly to nitrogen addition than ammonia-oxidizing archaea. Soil Biology and Biochemistry 99:158–166.

Chen, R., M. Senbayram, S. Blagodatsky, O. Myachina, K. Dittert, X. Lin, E. Blagodatskaya, and Y. Kuzyakov. 2014. Soil C and N availability determine the priming effect: microbial N mining and stoichiometric decomposition theories. Global Change Biology 20:2356–2367.

Clegg, C. D., R. D. L. Lovell, and P. J. Hobbs. 2006. The impact of grassland management regime on the community structure of selected bacterial groups in soils. FEMS Microbiology Ecology 43:263–270.

Contosta, A. R., S. D. Frey, and A. B. Cooper. 2015. Soil microbial communities vary as much over time as with chronic warming and nitrogen additions. Soil Biology and Biochemistry 88:19–24.

Cusack, D. F., W. L. Silver, M. S. Torn, S. D. Burton, and M. K. Firestone. 2011. Changes in microbial community characteristics and soil organic matter with nitrogen additions in two tropical forests. Ecology 92:621–632.

Dong, W., X. Zhang, X. Dai, X. Fu, F. Yang, X. Liu, X. Sun, X. Wen, and S. Schaeffer. 2014. Changes in soil microbial community composition in response to fertilization of paddy soils in subtropical China. Applied Soil Ecology 84:140–147.

Edmeades, D. 2003. The long-term effects of manures and fertilisers on soil productivity and quality: a review. Nutrient Cycling in Agroecosystems 66:165–180.

Feng, Y., P. Grogan, J. G. Caporaso, H. Zhang, X. Lin, R. Knight, and H. Chu. 2014. pH is a good predictor of the distribution of anoxygenic purple phototrophic bacteria in arctic soils. Soil Biology and Biochemistry 74:193–200.

Fernandez-Fueyo, E., et al. 2012. Comparative genomics of Ceriporiopsis subvermispora and Phanerochaete chrysosporium provide insight into selective ligninolysis. Proceedings of the National Academy of Sciences USA 109:5458–5463.

Fierer, N., M. A. Bradford, and R. B. Jackson. 2007. Toward an ecological classification of soil bacteria. Ecology 88:1354–1364.

Fierer, N., C. L. Lauber, K. S. Ramirez, J. Zaneveld, M. A. Bradford, and R. Knight. 2012. Comparative metagenomic, phylogenetic and physiological analyses of soil microbial communities across nitrogen gradients. ISME Journal 6:1007–1017.

Fontaine, S., and S. Barot. 2005. Size and functional diversity of microbe populations control plant persistence and long-term soil carbon accumulation. Ecology Letters 8:1075–1087.

Freedman, Z. B., K. J. Romanowicz, R. A. Upchurch, and D. R. Zak. 2015. Differential responses of total and active soil microbial communities to long-term experimental N deposition. Soil Biology and Biochemistry 90:275–282.

Frey, S. D., R. Drijber, H. Smith, and J. Melillo. 2008. Microbial biomass, functional capacity, and community structure after 12 years of soil warming. Soil Biology and Biochemistry 40:2904–2907.

Frey, S. D., M. Knorr, J. L. Parrent, and R. T. Simpson. 2004. Chronic nitrogen enrichment affects the structure and function of the soil microbial community in temperate hardwood and pine forests. Forest Ecology and Management 196:159–171.

Frey, S. D., et al. 2014. Chronic nitrogen additions suppress decomposition and sequester soil carbon in temperate forests. Biogeochemistry 121:305–316.

Galloway, J. N., A. R. Townsend, J. W. Erisman, M. Bekunda, Z. Cai, J. R. Freney, L. A. Martinelli, S. P. Seitzinger, and M. A. Sutton. 2008. Transformation of the nitrogen cycle: recent trends, questions, and potential solutions. Science 320:889–892.

Geisseler, D., and K. M. Scow. 2014. Long-term effects of mineral fertilizers on soil microorganisms: a review. Soil Biology and Biochemistry 75:54–63.

Graham, E. B., et al. 2016. Microbes as engines of ecosystem function: When does community structure enhance predictions of ecosystem processes? Frontiers in Microbiology 7:214.

Hawkes, C. V., S. N. Kivlin, J. D. Rocca, V. Huguet, M. A. Thomsen, and K. B. Suttle. 2011. Fungal community responses to precipitation. Global Change Biology 17:1637–1645.

Hoover, S. E., J. J. Ladley, A. A. Shchepetkina, M. Tisch, S. P. Gieseg, and J. M. Tylianakis. 2012. Warming, CO_2, and nitrogen deposition interactively affect a plant-pollinator mutualism. Ecology Letters 15:227–234.

IPCC. 2013. Climate change 2013: the physical science basis. Pages 466–570 in T. F. Stocker, et al., editors. Contribution of Working Group I to the Fifth Assessment Report of the Intergovernmental Panel on Climate Change. Cambridge University Press, Cambridge, UK.

Janssens, I., W. Dieleman, S. Luyssaert, J. A. Subke, M. Reichstein, R. Ceulemans, P. Ciais, A. J. Dolman, J. Grace, and G. Matteucci. 2010. Reduction of forest soil respiration in response to nitrogen deposition. Nature Geoscience 3:315–322.

Jin, V. L., S. M. Schaeffer, S. E. Ziegler, and R. D. Evans. 2011. Soil water availability and microsite mediate fungal and bacterial phospholipid fatty acid biomarker abundances in Mojave Desert soils exposed to elevated atmospheric CO_2. Journal of Geophysical Research Biogeoscience 116:G02001.

Kaiser, C., M. Koranda, B. Kitzler, L. Fuchslueger, J. Schnecker, P. Schweiger, F. Rasche, S. Zechmeister-Boltenstern, A. Sessitsch, and A. Richter. 2010. Belowground carbon allocation by trees drives seasonal patterns of extracellular enzyme activities by altering microbial community composition in a beech forest soil. New Phytologist 187:843–858.

Klaubauf, S., E. Inselsbacher, S. Zechmeister-Boltenstern, W. Wanek, R. Gottsberger, J. Strauss, and M. Gorfer. 2010. Molecular diversity of fungal communities in agricultural soils from Lower Austria. Fungal Diversity 44:65–75.

Koyama, A., M. D. Wallenstein, R. T. Simpson, and J. C. Moore. 2013. Carbon-degrading enzyme activities stimulated by increased nutrient availability in arctic tundra soils. PLoS ONE 8:e77212.

Koyama, A., M. D. Wallenstein, R. T. Simpson, and J. C. Moore. 2014. Soil bacterial community composition altered by increased nutrient availability in arctic tundra soils. Frontiers in Microbiology 5:516.

Kuffner, M., B. Hai, T. Rattei, C. Melodelima, M. Schloter, S. Zechmeister-Boltenstern, R. Jandl, A. Schindlbacher, and A. Sessitsch. 2012. Effects of season and experimental warming on the bacterial community in a temperate mountain forest soil assessed by 16S rRNA gene pyrosequencing. FEMS Microbiology Ecology 82:551–562.

Landesman, W., and J. Dighton. 2011. Shifts in microbial biomass and the bacteria: fungi ratio occur under field conditions within 3 h after rainfall. Microbial Ecology 62:228–236.

Lauber, C. L., M. Hamady, R. Knight, and N. Fierer. 2009. Pyrosequencing-based assessment of soil pH as a predictor of soil bacterial community structure at the continental scale. Applied and Environmental Microbiology 75:5111–5120.

Leff, J. W., et al. 2015. Consistent responses of soil microbial communities to elevated nutrient inputs in grasslands across the globe. Proceedings of the National Academy of Sciences USA 112:10967–10972.

Li, Q., H. Bai, W. Liang, J. Xia, S. Wan, and W. H. van der Putten. 2013a. Nitrogen addition and warming independently influence the belowground microfood web in a temperate steppe. PLoS ONE 8: e60441.

Li, Y., Y. L. Chen, M. Li, X. G. Lin, and R. J. Liu. 2012. Effects of arbuscular mycorrhizal fungi communities on soil quality and the growth of cucumber seedlings in a greenhouse soil of continuously planting cucumber. Pedosphere 22:79–87.

Li, F. L., M. Liu, Z. P. Li, C. Y. Jiang, F. X. Han, and Y. P. Che. 2013b. Changes in soil microbial biomass and functional diversity with a nitrogen gradient in soil columns. Applied Soil Ecology 64:1–6.

Li, C., K. Yan, L. Tang, Z. Jia, and Y. Li. 2014. Change in deep soil microbial communities due to long-term fertilization. Soil Biology and Biochemistry 75:264–272.

Lipson, D. A., and S. K. Schmidt. 2004. Seasonal changes in an alpine soil bacterial community in the Colorado Rocky Mountains. Applied and Environmental Microbiology 70:2867–2879.

Liu, P., J. H. Huang, X. G. Han, O. J. Sun, and Z. Y. Zhou. 2006. Differential responses of litter decomposition to increased soil nutrient and water between two contrasting grassland plant species of Inner Mongolia, China. Applied Soil Ecology 34:266–275.

Liu, J., Y. Sui, Z. Yu, Y. Shi, H. Chu, J. Jin, X. Liu, and G. Wang. 2014. High throughput sequencing analysis of biogeographical distribution of bacterial communities in the black soils of northeast China. Soil Biology and Biochemistry 70:113–122.

Liu, X., et al. 2013. Enhanced nitrogen deposition over China. Nature 494:459–462.

Lopez-Mondejar, R., J. Voriskova, T. Vetrovsky, and P. Baldrian. 2015. The bacterial community inhabiting temperate deciduous forests is vertically stratified and undergoes seasonal dynamics. Soil Biology Biochemistry 87:43–50.

Mannisto, M., L. Ganzert, M. Tiirola, M. M. Haggblom, and S. Stark. 2016. Do shifts in life strategies explain microbial community responses to increasing nitrogen in tundra soil? Soil Biology and Biochemistry 96:216–228.

Matulich, K. L., and J. B. H. Martiny. 2014. Microbial composition alters the response of litter decomposition to environmental change. Ecology 96:154–163.

Matulich, K. L., C. Weihe, S. D. Allison, A. S. Amend, R. Berlemont, M. L. Goulden, S. Kimball, A. C. Martiny, and J. B. H. Martiny. 2015. Temporal variation overshadows the response of leaf litter microbial communities to simulated global change. ISME Journal 9:2477–2489.

Morrison, E. W., S. D. Frey, J. J. Sadowsky, L. T. A. Van Diepen, W. K. Thomas, and A. Pringle. 2016. Chronic nitrogen additions fundamentally restructure the soil fungal community in a temperate forest. Fungal Ecology 23:48–57.

Nilsson, L. O., E. Bååth, U. Falkengren-Grerup, and H. Wallander. 2007. Growth of ectomycorrhizal mycelia and composition of soil microbial communities in oak forest soils along a nitrogen deposition gradient. Oecologia 153:375–384.

Pardo, L. H., et al. 2011. Effects of nitrogen deposition and empirical nitrogen critical loads for ecoregions of the United States. Ecological Applications 21: 3049–3082.

Phoenix, G. K., et al. 2011. Impacts of atmospheric nitrogen deposition: responses of multiple plant and soil parameters across contrasting ecosystems in long-term field experiments. Global Change Biology 18:1197–1215.

Ramirez, K. S., J. M. Craine, and N. Fierer. 2010a. Nitrogen fertilization inhibits soil microbial respiration regardless of the form of nitrogen applied. Soil Biology and Biochemistry 42:2336–2338.

Ramirez, K. S., J. M. Craine, and N. Fierer. 2012. Consistent effects of nitrogen amendments on soil microbial communities and processes across biomes. Global Change Biology 18:1918–1927.

Ramirez, K. S., C. L. Lauber, R. Knight, M. A. Bradford, and N. Fierer. 2010b. Consistent effects of nitrogen fertilization on soil bacterial communities in contrasting systems. Ecology 91:3463–3470.

Sagova-Mareckova, M., M. Omelka, L. Cermak, Z. Kamenik, J. Olsovska, E. Hackl, J. Kopecky, and F. Hadacek. 2011. Microbial communities show parallels at sites with distinct litter and soil characteristics. Applied and Environmental Microbiology 77:7560–7567.

Schneider, T., K. M. Keiblinger, E. Schmid, K. Sterflinger-Gleixner, G. Ellersdorfer, B. Roschitzki, A. Richter, L. Eberl, S. Zechmeister-Boltenstern, and K. Riedel. 2012. Who is who in litter decomposition? Metaproteomics reveals major microbial players and their biogeochemical functions. ISME Journal 6:1749–1762.

Segata, N., J. Izard, L. Waldron, D. Gevers, L. Miropolsky, W. S. Garrett, and C. Huttenhower. 2011. Metagenomic biomarker discovery and explanation. Genome Biology 12:R60.

Shi, L., et al. 2016. Consistent effects of canopy vs. understory nitrogen addition on the soil exchangeable cations and microbial community in two contrasting forests. Science of the Total Environment 553:349–357.

Siles, J. A., T. Cajthaml, S. Minerbi, and R. Margesin. 2016. Effect of altitude and season on microbial activity, abundance and community structure in Alpine forest soils. FEMS Microbiology Ecology, 92:fw008.

Singh, B. K., R. D. Bardgett, P. Smith, and D. S. Reay. 2010. Microorganisms and climate change: terrestrial feedbacks and mitigation options. Nature Reviews Microbiology 8:779–790.

Sistla, S. A., S. Asao, and J. P. Schimel. 2012. Detecting microbial N-limitation in tussock tundra soil: implications for arctic soil organic carbon cycling. Soil Biology and Biochemistry 55:78–84.

Stark, S., M. Mannisto, and A. Eskelinen. 2014. Nutrient availability and pH jointly constrain microbial extracellular enzyme activities in nutrient-poor tundra soils. Plant and Soil 383:373–385.

Sul, W. J., S. Asuming-Brempong, Q. Wang, D. M. Tourlousse, C. R. Penton, Y. Deng, J. L. Rodrigues, S. G. Adiku, J. W. Jones, and J. Zhou. 2013. Tropical agricultural land management influences on soil microbial communities through its effect on soil organic carbon. Soil Biology and Biochemistry 65:33–38.

Tamm, C. O. 1991. Nitrogen in terrestrial ecosystems: questions of productivity, vegetational changes, and ecosystem stability. Springer, Berlin, Germany.

Tian, D., and S. Niu. 2015. A global analysis of soil acidification caused by nitrogen addition. Environmental Research Letters 10:024019.

Tilman, D., P. B. Reich, and J. M. H. Knops. 2006. Biodiversity and ecosystem stability in a decade-long grassland experiment. Nature 441:629–632.

Treseder, K. K. 2008. Nitrogen additions and microbial biomass: a meta-analysis of ecosystem studies. Ecology Letters 11:1111–1120.

Vitousek, P. M., J. D. Aber, R. W. Howarth, G. E. Likens, P. A. Matson, D. W. Schindler, W. H. Schlesinger, and D. Tilman. 1997. Human alteration of the global nitrogen cycle: sources and consequences. Ecological Applications 7:737–750.

Voriskova, J., V. Brabcova, T. Cajthaml, and P. Baldrian. 2014. Seasonal dynamics of fungal communities in a temperate oak forest soil. New Phytologist 201:269–278.

Waldrop, M. P., and M. K. Firestone. 2006. Response of microbial community composition and function to soil climate change. Microbial Ecology 52:716–724.

Wang, J. T., Y. M. Zheng, H. W. Hu, L. M. Zhang, J. Li, and J. Z. He. 2015. Soil pH determines the alpha diversity but not beta diversity of soil fungal community along altitude in a typical Tibetan forest ecosystem. Journal Soils and Sediments 15:1224–1232.

Wei, C., Q. Yu, E. Bai, X. Lü, Q. Li, J. Xia, P. Kardol, W. Liang, Z. Wang, and X. Han. 2013. Nitrogen deposition weakens plant–microbe interactions in grassland ecosystems. Global Change Biology 19: 3688–3697.

Williams, A., G. Borjesson, and K. Hedlund. 2013. The effects of 55 years of different inorganic fertiliser regimes on soil properties and microbial community composition. Soil Biology and Biochemistry 67:41–46.

Xia, J., and S. Wan. 2008. Global response patterns of terrestrial plant species to nitrogen addition. New Phytologist 179:428–439.

Xiong, Q., K. Pan, L. Zhang, Y. Wang, W. Li, X. He, and H. Luo. 2016. Warming and nitrogen deposition are interactive in shaping surface soil microbial communities near the alpine timberline zone on the eastern Qinghai-Tibet Plateau, southwestern China. Applied Soil Ecology 101:72–83.

Xiong, J., F. Peng, H. Sun, X. Xue, and H. Chu. 2014a. Divergent responses of soil fungi functional groups to short-term warming. Microbial Ecology 68:708–715.

Xiong, J., H. Sun, F. Peng, H. Zhang, X. Xue, S. M. Gibbons, J. A. Gilbert, and H. Chu. 2014b. Characterizing changes in soil bacterial community structure in response to short-term warming. FEMS Microbiology Ecology 89:281–292.

Yan, G., F. Chen, X. Zhang, J. Wang, S. Han, Y. Xing, and Q. Wang. 2017. Spatial and temporal effects of nitrogen addition on root morphology and growth in a boreal forest. Geoderma 303:178–187.

Yao, M., et al. 2014. Rate-specific responses of prokaryotic diversity and structure to nitrogen deposition in the Leymus chinensis steppe. Soil Biology and Biochemistry 79:81–90.

Zeng, J., X. Liu, L. Song, X. Lin, H. Zhang, C. Shen, and H. Chu. 2016. Nitrogen fertilization directly affects soil bacterial diversity and indirectly affects bacterial community composition. Soil Biology and Biochemistry 92:41–49.

Zhang, X., and X. Han. 2012. Nitrogen deposition alters soil chemical properties and bacterial communities in the Inner Mongolia grassland. Journal of Environmental Sciences 24:1483–1491.

Zhang, N. L., W. X. Liu, H. J. Yang, X. J. Yu, J. L. M. Gutknecht, Z. Zhang, S. Q. Wan, and K. P. Ma. 2013. Soil microbial responses to warming and increased precipitation and their implications for ecosystem C cycling. Oecologia 173:1125–1142.

Zhang, Y., L. X. Zheng, X. J. Liu, T. Jickells, J. N. Cape, K. Goulding, A. Fangmeier, and F. S. Zhang. 2008. Evidence for organic N deposition and its anthropogenic sources in China. Atmospheric Environment 42:1035–1041.

Zhang, N., et al. 2015. Precipitation modifies the effects of warming and nitrogen addition on soil microbial communities in northern Chinese grasslands. Soil Biology and Biochemistry 89:12–23.

Zhao, C., L. Zhu, J. Liang, H. Yin, C. Yin, D. Li, N. Zhang, and Q. Liu. 2014. Effects of experimental warming and nitrogen fertilization on soil microbial communities and processes of two subalpine coniferous species in Eastern Tibetan Plateau, China. Plant and Soil 382:189–201.

Zhong, Y., W. Yan, and Z. Shangguan. 2015. Impact of long-term N additions upon coupling between soil microbial community structure and activity, and nutrient-use efficiencies. Soil Biology and Biochemistry 91:151–159.

Zhou, X. Q., C. R. Chen, Y. F. Wang, Z. H. Xu, J. C. Duan, Y. B. Hao, and S. Smaill. 2013. Soil extractable carbon and nitrogen, microbial biomass and microbial metabolic activity in response to warming and increased precipitation in a semiarid Inner Mongolian grassland. Geoderma 206: 24–31.

Zhou, L., X. Zhou, B. Zhang, M. Lu, Y. Luo, L. Liu, and B. Li. 2014. Different responses of soil respiration and its components to nitrogen addition among biomes: a meta-analysis. Global Change Biology 20:2332–2343.

Zhou, J., et al. 2016. Thirty four years of nitrogen fertilization decreases fungal diversity and alters fungal community composition in black soil in northeast China. Soil Biology Biochemistry 95: 135–143.

8

Plant diversity increases predation by ground-dwelling invertebrate predators

Lionel R. Hertzog,[1,3,†] Anne Ebeling,[2] Wolfgang W. Weisser,[1] and Sebastian T. Meyer[1]

[1]Terrestrial Ecology Research Group, Department of Ecology and Ecosystem Management, Center for Food and Life Sciences Weihenstephan, Technische Universität München, Hans-Carl-von-Carlowitz-Platz 2, DE-85354 Freising Germany
[2]Institute for Ecology, Friedrich-Schiller University Jena, Dornburger Strasse 159, DE-07743 Jena Germany

Abstract. Global declines in biodiversity have raised concerns over the implications of diversity loss for the functioning of ecosystems. Plant diversity loss has impacts throughout food webs affecting both consumer communities and ecosystem functions mediated by consumers. Effects of plant diversity loss on communities of invertebrate predators have been documented, yet little is known about how these translate into variations in predation rates. We measured predation rates along two plant diversity gradients in grassland experiments manipulating species richness and functional diversity. Measurements were conducted at two different heights (ground and vegetation) and in two different seasons (spring and summer), using three different types of baits. Our results show that overall predation rates increase with plant species richness, but effects are seasonally variable and are much more pronounced on the ground than in the vegetation. Plant functional diversity did not consistently affect predation rates in our experiments. Potential mechanistic explanations for an effect of plant diversity on predation include higher complementarity between predator species or reduced intraguild predation with increasing structural complexity at higher plant diversity. These results underline the importance of high local plant diversity for natural pest control.

Key words: arthropods; biodiversity; ecosystem function; Jena Experiment; Rapid Ecosystem Function Assessment (REFA).

† E-mail: lionel.hertzog@ugent.be

INTRODUCTION

Positive effects of local plant diversity on ecosystem functions such as plant productivity, nutrient cycling, or decomposition are a consensus among the scientific community (Hooper et al. 2005). Furthermore, changes in plant diversity can affect the structure, composition, and stability of whole food webs (Scherber et al. 2010, Haddad et al. 2011), consequently also affecting related ecosystem functions. Understanding how diversity affects ecosystem functioning within but also between trophic levels is a prerequisite to fully evaluate how diversity loss affects multiple ecosystem functions (Hines et al. 2015).

Linking shifts in consumer communities and the ecosystem function they mediate to changes in plant diversity is more complex than for directly plant-associated functions (Duffy 2002, Ives et al. 2005). Therefore, these relationships have been less intensively studied (Cardinale et al. 2012). Yet, there is growing evidence for higher consumer-related functioning with higher plant diversity for herbivory (Meyer et al. 2017) and pollination (Ebeling et al. 2008). In contrast, no study has so far measured predation rates caused by diverse predator communities (i.e., hundreds of species) along plant diversity gradients. That is surprising because changes in predation rates with plant diversity are likely. Numerous studies have reported changes in predator density and diversity along gradients of plant diversity (Haddad et al. 2009, Letourneau et al. 2011, Hertzog et al. 2016), and any change in consumer communities can be a mechanistic cause of changes in predation rates (Ebeling et al. 2014a). Several potentially counteracting mechanisms have been proposed to link plant diversity, predator communities, and predation rates. Consequently, predictions of the direction and magnitude of the effect of plant diversity on predation rates are difficult. Predation may increase or decrease depending on whether interactions between predator species are complementary, synergistic, or antagonistic (Letourneau et al. 2009). The link between predator diversity and top-down control has been intensively studied due to its importance in natural pest control; however, different meta-analyses on this topic found contrasting results (Letourneau et al. 2009, Griffin et al. 2013, Katano et al. 2015). The prevailing lack of empirical data diminishes the chance of successful management aiming at increasing pest control via natural enemies (Landis et al. 2000).

Predation rates have been shown to vary (1) globally, along a latitudinal gradient following the shifts in organismic diversity (Roslin et al. 2017); (2) seasonally (Winder et al. 1994), as seasonal shifts in predator identity (Douglass et al. 2008), prey community composition (Wilby and Thomas 2002), and vegetation structure (Finke and Denno 2002) affect predation rates; and (3) spatially, as predation rates may differ between the ground and the vegetation layer (strata), for example, because of the variation in the abundance of some important and voracious predator groups such as ground beetles that only forage on or near the ground (Miller et al. 2014). Therefore, when studying the determinants of predation rates, an important element is to take these variations and context dependency into account (Tylianakis and Romo 2010).

Predation rates under field conditions can be assessed using different methods (Letourneau et al. 2009). Many biological control studies have extensively studied predator–prey interactions using cages to set up treatments with various predator or prey communities (Cardinale et al.

2003), but the systems are often reduced to a few dominant interacting species (see, e.g., Table 1 in Janssen et al. [2006]). However, natural communities are highly diverse with potentially hundreds of predator species and as a consequence much higher potential for direct or indirect interactions affecting ecosystem functioning (Snyder et al. 2005). For a high level of replication in complex and diverse systems, the Rapid Ecosystem Function Assessment has been proposed as a toolbox of methods that measure proxies for ecosystem functions in an easy-to-use, simple, and cost-effective way (Meyer et al. 2015). For predation, sentinel prey methods are frequently used (Meyer et al. 2015, Lövei and Ferrante 2017). These methods fix live, dead, or even artificial prey items (baits) with glue or needles to sampling locations. Because prey is fixed, it cannot run away reducing the need for permanent observation. After exposure, sampling locations are surveyed and complete or partial removal of the prey items is noted as an indication of predation events. In case of artificial prey items, marks left during predation attempts can be evaluated during checking to determine groups of predators (Meyer et al. 2015).

Different prey items may show different predation rates (Lövei and Ferrante 2017) and be attractive to different groups of predators. This potential bias can be used to an advantage when estimating predation rates using different types of baits as sentinel prey to record predation rates from a broad range of potential predators. We used three different types of commonly used baits: pea aphids (Östman et al. 2003), mealworms (Rouabah et al. 2014), and dummy caterpillars made from plasticine (Sam et al. 2015), all being common sentinel prey items used to assess predation rates (Lövei and Ferrante 2017).

Here, we measured predation rates in two experiments manipulating both taxonomic diversity (species richness) and functional diversity. Using different diversity experiments, we could test whether the observed patterns are robust to variation in the length of the diversity gradient and the age of the gradient. To test for additional context dependency, we measured predation rates in two different seasons (spring and summer) and in two different strata (on the ground and in the vegetation). Specifically, we tested (1) whether predation rates increase at higher plant species richness and/or plant functional diversity and (2) how strongly the plant diversity–predation relationship depends on the context, concerning season, stratum, prey type, and age of the plant community.

MATERIALS AND METHODS

Experimental field site

The study was conducted in 2014 in a grassland biodiversity experiment (The Jena Experiment) situated in the floodplain of the river Saale in Jena (Thuringia, Germany, 50°55′ N, 11°35′ E, 130 m above sea level). The mean annual air temperature at the site is 9.9°C, and yearly precipitation is 610 mm (Hoffmann et al. 2014). Two plant diversity experiments have been established at the field site (see Appendix S1: Fig. S1 for a representation of the spatial distribution of the plots). The first one, called the Main Experiment, was established in 2002 and included at the time of the study 80 plots of 6 × 7 m (Roscher et al. 2004). The plots were sown with combinations of grassland plant species from a species pool of 60 species commonly found in Molinio-Arrhenatherea meadows that naturally occur in the area of the field site. The plots formed a gradient of species richness with 1, 2, 4, 8, 16, and 60 sown species, each level being replicated 16 times except for the 16-species (14 replicates) and the 60-species mixtures (four replicates). Two monocultures were abandoned in 2009 due to poor performance, reducing the number of plots from 82 to 80. The experiment also manipulated functional diversity of the plant communities by varying the number of plant functional groups (grasses, small herbs, tall herbs, and legumes) sown into the plots in a full-factorial design. Limitations of the design were that monocultures can contain only one functional group, two-species mixtures contained only one or two functional groups, and all 60-species mixtures contained all four functional groups because these mixtures were comprised of the complete species pool (Roscher et al. 2004). Species composition was randomly drawn for each plot constrained by plant species number and the number and identity of the plant functional groups. To control for the effect of varying soil texture at the field site, a block design was implemented where four blocks were established parallel to the river. The second plant diversity gradient, called the Trait-Based Experiment (TBE), was established in 2010 and consisted of 138 plots of 3.5 × 3.5 m (Ebeling et al. 2014b). The aim of this experimental gradient was to directly manipulate the trait composition of the plant community to better understand the links between biodiversity and ecosystem processes. In this experiment, six plant traits (maximum height, leaf area, rooting depth, root density, growth start, and flowering start) were used to characterize the plants of the species pool of The Jena Experiment using a Principal Component Analysis. Three pools of eight species each (excluding legumes) were chosen from the PCA in which the first two axes represented traits related to spatial and temporal resource acquisition, respectively. Species pool 1 manipulated functional diversity based on the first axis (spatial resource acquisition), species pool 2 selected species along the second axis (temporal resource acquisition), and species pool 3 maximized functional diversity along both axes, covering the four corners of the trait space. Within each of these species pools, functional diversity was defined as the distance between the species grouped in sectors along the PCA axes and ranged between 1 and 4 (Ebeling et al. 2014b). In addition to the gradient in functional diversity, the TBE also includes a gradient in plant species richness ranging from 1 to 8 sown species. As in the Main Experiment, blocks were established to control for variation in environmental conditions. All plots in both experiments were manually weeded three times per year to maintain the species mixtures. Also, the field site was mown twice a year in late spring and summer as is the common practice for unfertilized meadows in the region. The realized species richness was tightly correlated with sown species richness (see Table 2 in Marquard et al. [2009]). The matrix between all plots consisted of frequently mowed standard meadow.

Predation assessment

We measured predation rates using three different types of baits: pea aphids, mealworms, and plasticine dummies. Data were collected in two seasons (spring and summer of 2014) before the two peaks in standing biomass. On seven days during each season (between 12 and 23 May and between 11 and 22 August), baits were exposed at ten specific positions per plot with a distance of 50 cm between them (see Appendix S1: Fig. S3). Baits were exposed in the morning between 9 am and noon and recovered the next day after 24 h of exposure. The type of bait placed at each position was randomized and different on each day. For example, on the first day of the measurement, the bait at position 3 could have been a mealworm, while on the second day the bait would have been a dummy. At the end of each season, a total of 10 aphids, 20 dummies, and 40 mealworms were exposed per plot (see Appendix S1: Table S1, Fig. S1). Dummies and mealworms were placed in two strata, both on the ground and in the vegetation; aphids were placed only on the ground because of a limited number of aphids available. Baits exposed in the vegetation were put on a plant shoot at approximately half the maximum height of the selected shoot. As baits were fixed and could not move (i.e., sentinel prey), opportunistic scavengers such as slugs may attack or consume baits in our experiments in addition to actively hunting predators.

Pea aphid baits.—Pea aphids (*Acyrthosiphon pisum* (Harris, 1776), hereafter aphids) were cultivated on broad beans (*Vicia faba*) in a climate chamber at 20°C with 50% air humidity. Aphid colonies were reared continuously throughout the experiments by frequently transferring winged adults to new un-colonized plants. Plants were covered with a transparent plastic foil to avoid the dispersal of individuals. Every day in the morning, we glued fourth-instar or adult individuals (largest individuals in the colonies of at least 1 mm in size) to white labels using waterproof glue (Pattex 100% Kleber; Henkel & Cie, CH-4133 Pratteln). After exposure, we recorded whether aphids were removed from the label or not. Partially remaining aphids were counted as removed. In total, we exposed 4400 aphid baits.

Mealworms baits.—Mealworms (larvae of the beetle *Tenebrio molitor* (Linnaeus 1775)) were bought from a nearby pet shop and stored at 7°C in a fridge for a maximum of 15 d to keep mealworms in a larval stage. We used only medium- and large-sized mealworms (minimum size 3 cm, maximum size 5 cm). The mealworms were pinned with insect needles (0.35 × 38 mm; Bioform, Nürnberg, Germany) either to the ground or to plant shoots. As for the aphids, we recorded whether mealworms were removed or not. In cases where mealworms were only partly consumed, we counted this as removed for the binary analysis. In total, we exposed 17,600 mealworm baits.

Plasticine dummy baits.—We used green plasticine (Staedtler Noris Club, Nürnberg, Germany) to form dummies: plain cylinders of 0.6 × 2 cm in size to vaguely resemble lepidopteran larvae (Meyer et al. 2015). Dummies were pinned either on the ground or in the vegetation using insect needles (0.35 × 38 mm). After exposure dummies were checked for predation marks using a stereo microscope with a five times magnification. Predation marks were classified into broad categories of predators based on the large collection of photographs from Low et al. (2014) and our photographs that were taken at the field site. We used five categories of marks (Appendix S1: Fig. S4): (1) rasping marks (gastropods), (2) mandibular marks (biting insects), (3) teeth marks (rodent), (4) beak marks (bird), and (5) stylet (predatory bugs) or ovipositor marks (e.g., by parasitic Hymenoptera). In total, we exposed 8800 dummy baits. Each dummy was scored for each predator type independently as showing any marks or not (binary variables).

Additional cage experiment.—We observed many birds on the field site in May 2014 together with exceptionally high predation rates on the exposed mealworms. As the effect of the diversity gradient is certainly different between a large mobile vertebrate predator and invertebrate predators, we conducted an additional cage experiment to exclude vertebrates as potential predators. In 2015, between 18 and 22 May, we selected a subset of 16 plots across the diversity gradient in the Main Experiment with 1, 4, 16, and 60 sown plant species (four replicates each) to expose ten mealworms per plot each day. Half of the mealworms were individually protected from birds by a cage made of green plastic mesh (mesh size 10 × 10 mm) covered with a transparent plastic roof. The cages were fixed to the ground. Baits were replaced every day between 09:00 and 11:00 am, and after 24 h of exposure, removal rates were assessed. The position of the baits with cages was randomized between the days of the experiment. For example, on the second day, the bait at position number 5 was caged, and on the fourth day, the bait at the same position was uncaged.

Data analysis

All mealworms and dummies for which not even the needle was found after one day of exposure were considered to be lost. These baits were not included in the calculation of the respective proportions of predated baits. In total, 13 dummies (<1% of the total) and 2299 mealworms (13% of the total) have been lost. The latter were likely due to heavy bird predation in spring (see *Additional cage experiment*) when occasionally birds might have removed also the needles with which mealworms were pinned. For aphids, no plastic labels were lost so that for all exposed aphids their status could be determined. To allow comparability in the results between the dummies and the other two types of baits, we combined all types of predation marks into an indicator of total predation (i.e., marks of any type as a binary variable; 0/1). In addition, we investigated the potential effect of including opportunistic scavengers on our results by analyzing predation on dummies excluding rasping marks caused by gastropods.

In both plant diversity experiments, data were aggregated across the replicates within plots separately for each type of bait, season, and stratum (ground vs. vegetation). The number of removed or attacked baits was computed. This number was divided by the number of recovered baits to derive the proportion of removed or attacked baits. This proportion, weighted by the number of recovered baits, was used as the response variable in our models. The effect of plant species richness and plant functional diversity on the proportion of predated baits was tested using generalized linear mixed models for the two experiments and the two strata separately. Models were fitted using the function *glmer* implemented in the library lme4 v1.1 (Bates et al. 2015) in R v3.2 (R Core Team 2015). A binomial distribution with a logit link was used. Models included a random effect for plots nested within blocks to account for any block- and plot-level random deviations. In addition, we included an observation-level random effect to account for overdispersion (Harrison 2014). For the Main Experiment, the models contained as fixed effects the type of baits (aphids, dummies, or mealworms), the season, plant species richness, and plant functional diversity (i.e., the number of plant functional groups) in that order. Three-way interactions between the two measures of plant diversity each and type of baits and season were included. For the TBE, equivalent models were fit including as fixed effects the type of bait (aphids, dummies, or mealworms), the season, the pool,

plant species richness, and plant functional diversity (i.e., FDjena) in that order. A three-way interaction between plant species richness and type of bait and season was included as well as an additional four-way interaction between FDjena, type of bait, season, and pool as the effect of functional diversity might differ between the different plant species pools. Plant species richness was log-transformed to improve model fit. To assess the significance of individual fixed-effect terms, we sequentially reduced the models starting from the most complex interaction terms up to single main effects and computed at each step a likelihood ratio test (LRT) statistic. This value represents the increase in model deviance as the model is gradually reduced. The significance of the LRT statistics was assessed using a chi-square distribution. The sequence in which terms were dropped from models was based on the order in which the variables were included in the models starting with the most complex interactions. We chose this method because it is an extension of classical analysis of deviance used for generalized linear models also using variation in deviance and chi-square tests to assess the significance of terms (Nelder and Baker 1972). Marginal and conditional R^2 of the models were computed using the approach described in Nakagawa and Schielzeth (2013). As the logit transformation used to model the predation rates is a nonlinear transformation, we compute the effect sizes by comparing the change in the back-transformed predicted values at the endpoints of the diversity gradients. For example, if the model predicts predation rates of 30% for the monocultures and 70% for the 60-species mixtures, we report an increase of 40 percentage points (pp) in the predation rates along the species richness gradient.

For the caging experiment, we aggregated all observations per plot and similarly computed the proportion of removed mealworm. To analyze the effect of caging on the relationship between plant species richness and predation rates, we used a generalized linear model with a binomial distribution and a logit link. The variables in the model were the caging status of the baits (caged/uncaged), plant species richness, and the interaction between these variables. We assessed the significance of the terms using an analysis of deviance with a chi-square test (Nelder and Baker 1972) as above.

RESULTS

Average predation rates and effects of season and heights

In the Main Experiment, 99% of aphid, 90% of mealworm, and 100% of dummy baits could be assessed after one day of exposure. On average, predation rates on aphids were highest with 60%, followed by predation rates on mealworms (53%) and dummies (19%). In the TBE, 99% of aphids, 86% of mealworms, and 99% of dummies were recovered, and average predation rates showed similar values to the Main Experiment (63% on aphids, 56% on mealworms, and 16% on dummies). Predations rates strongly differed between seasons. Predation rates on aphids and mealworms were significantly higher in May than in August. They decreased on average by 24 pp from 72% to 48% and by 35 pp (from 71% to 36%), respectively, from May to August in the Main Experiment (Appendix S1: Table S2), and by 33 pp (from 79% to 49%) and 48 pp (from 82% to 34%) in the TBE (Appendix S1: Table S2). Predation rates on dummies increased slightly by 7 pp (from 15% to 22%) from May to August in the Main Experiment, while they stayed unchanged at 15% in the TBE. Predation rates also differed between strata. Predation rates on mealworms and dummies in the Main Experiment were 36 pp (70% vs. 34%) and 21 pp (30% vs. 9%) higher on the ground than in the vegetation (Appendix S1: Table S2). Similarly, in the TBE, predation rates were 33 pp (71% vs. 38%) and 17 pp (24% vs. 7%) higher on the ground than in the vegetation for mealworms and dummies, respectively. Aphid baits were only exposed on the ground in both experiments. Regarding the identity of predator groups that we could identify based on the attack marks left in the dummies, biting insects and gastropods were the main group of predators with 52% and 46% of the dummies with attack marks showing marks from these two groups in the Main Experiment and 53% and 35% in the TBE, respectively. Intermediate proportions of marks were caused by rodents (Main: 10%; Trait-based: 8%). Less than 0.1% of the total observed predation marks were attributed to the other types of predators: birds, predatory bugs, and parasitic Hymenoptera.

Plant species richness effects on predation rates

Our models explained between 26% and 43% of the observed variation in predation rates (Appendix S1: Table S5), which is a relatively large fraction of the variance given that our models did not include any variables describing the predator communities. While baits exposed in the vegetation showed generally very low lower predation rates (see *Average predation rates and effects of season and heights*) without significant main effects of plant species richness, plant species richness had significant main effects on predation rates across all three types of baits and in both diversity experiments for baits exposed on the ground while for baits in the vegetation no significant main effects were found (Fig. 1, Table 1). Effects of plant species richness for baits exposed on the ground interacted with the type of dummy used for the experiment and with season. In summer, The plant species richness effects were consistently positive across all types of baits and both experiments ranging from an increase of 10 pp (from 18% in monocultures to 28% in eight-species mixtures) in the TBE for dummies to 29 pp from 44% in monocultures to 73% in the 60-species plots for mealworms in the Main Experiment. In contrast, in spring differences in strength and direction of plant species richness effects were observed (Fig. 1, Table 1). For aphids and dummies, predation rates increased with plant species richness also in May by 48 pp (from 48% in the monocultures to 96% in the 60-species mixtures) and 20 pp (from 37% in monocultures to 57% in eight-species mixtures) for aphids and by 20 pp (from 19% to 39%) and 10 pp (from 18% to 28%) for dummies and for the Main Experiment and the TBE, respectively. In the Main Experiment, plant diversity effects increased predation rates on aphids by 48 pp in May (from 48% in the monocultures to 96% in the 60-species mixtures) and 28 pp in August (from 38% to 66%). In the TBE, predation rates on aphids increased by 20 pp from monocultures to eight-species mixtures in both seasons (from 37% to 57% in May and from 70% to 90% in August). In contrast, predation rates on mealworms were positively affected in August but negatively affected by plant species richness in May but in both diversity gradients. Specifically, in May predation rates on mealworms declined along the plant species richness gradient by 51 pp (from 74% to 23%) in the vegetation and 12 pp (from 92% to 80%) on the ground in the Main

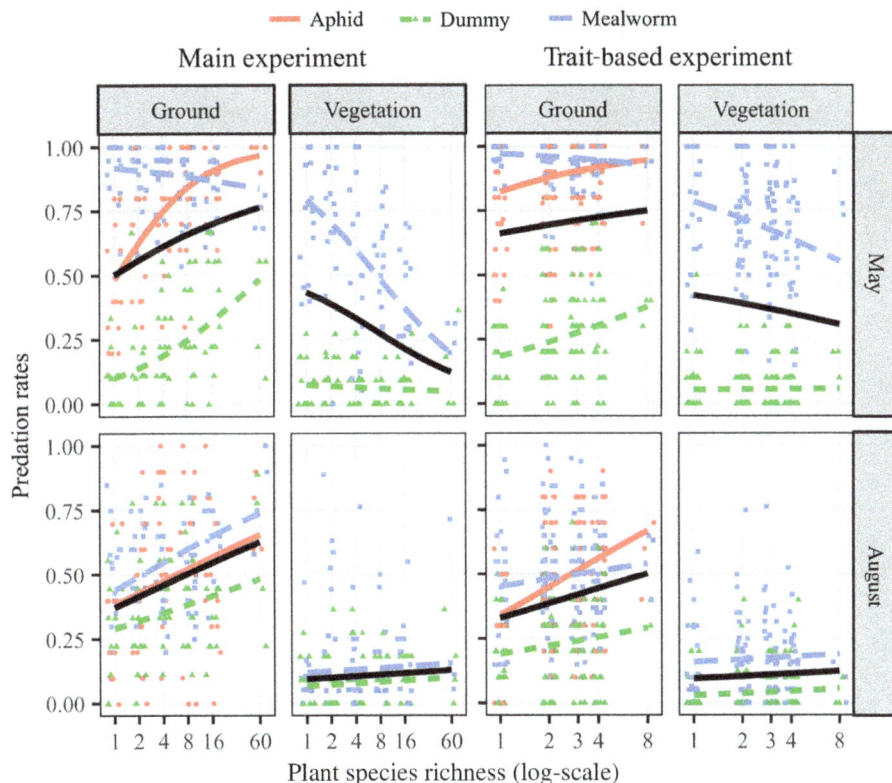

Fig. 1. Predation rates across the two plant species richness gradients: Main Experiment and Trait-Based Experiment, in two different strata (ground and vegetation), and in two different seasons (August and May). The colors indicate the different types of baits used in the experiment. Colored lines represent fitted regression curves from full generalized linear mixed models (see Table 1), and the thick black lines represent the average effect across the different types of baits.

Experiment and by 19 pp (from 75% to 56%) in the vegetation and 10 pp (from 92% to 82%) on the ground in the TBE.

Effects of excluding vertebrate predators on predation rates on mealworms

Because we observed large numbers of birds on the field site in May, the impact of vertebrate predation was tested in a separate experiment that excluded vertebrate predators using cages and compared patterns to uncaged controls. Plant species richness increased the predation rates on mealworms inside of cages, while it decreased predation rates on mealworms outside of cages (Fig. 2, Table 2).

Functional diversity effects on predation rates

Functional diversity had no significant main effects on predation rates across all three types of

baits, in both diversity experiments and for both baits on the ground and in the vegetation (Table 1). Also, no significant interactions between functional diversity and other explanatory variables were found. Plots of the fitted regression between predation rates and functional diversity are given in Appendix S1: Figs. S6 and S7.

Sensitivity analysis for the inclusion of scavenging slugs

Excluding gastropod marks from the estimates of predation rates on dummies weakened the relationship between plant species richness and predation rates on the ground. In the Main Experiment, the increase in predation rates with plant species richness became a nonsignificant trend (Appendix S1: Fig. S5, Table S3). In the TBE, the increase in predation rates with plant species richness remained significant, also when

Table 1. Generalized linear mixed model results for the relationships between predation rates and the type of baits, plant species richness, functional diversity (FunDiv), pools (only for the Trait-Based Experiment [TBE]), and season separately for each stratum (ground and vegetation) in the Main Experiment and in the TBE.

Explanatory variables	df	Main experiment		Trait-Based experiment	
		Ground	Vegetation	Ground	Vegetation
Bait type	2	189.3***	98.9***	438.4***	265.1***
Season	1	54.7***	51.1***	340.0***	207.4***
Pool	2	–	–	13.2**	3.0
Plant species richness (PSR)	1	33.4***	1.5	14.0***	0.1
Functional diversity (FunDiv)	1	0.0	2.4	0.0	0.0
Season:Pool	2	–	–	5.8	3.1
Season:PSR	1	0.3	9.1**	0.1	2.3
Season:FunDiv	1	1.5	0.6	0.0	1.5
Pool:FunDiv	2	–	–	5.8	0.5
Season:Type	2	119.9***	61.7***	257.8***	132.6***
Type:Pool	2	–	–	21.6***	5.3
Type:PSR	2	34.6***	5.5*	13.7**	1.3
Type:FunDiv	2	5.9	0.1	1.0	0.2
Season:Pool:FunDiv	2	–	–	0.8	2.4
Type:Season:Pool	4	–	–	9.5*	16.4***
Type:Season:PSR	2	26.7***	3.1	10.5**	0.0
Type:Season:FunDiv	2	0.2	0.0	1.6	0.5
Type:Pool:FDjena	5	–	–	3.5	0.6
Type:Season:Pool:FDjena	3	–	–	6.7	0.8

Notes: The models were sequentially simplified starting from the interactions up to a null model (intercept-only) one term at a time following the inverse row order. The significance of the terms was assessed using a likelihood ratio test comparing between two nested models with or without the focal term. The P-values were based on a chi-square test. Asterisks represent the significance level of the terms with $^*P < 0.05$, $^{**}P < 0.01$, and $^{***}P < 0.001$.

excluding marks caused by gastropods (Appendix S1: Fig. S5, Table S4).

DISCUSSION

Our experiment has demonstrated an increase in invertebrate predation with higher plant species richness, an effect that was most pronounced on the ground. This increase emerged from both experiments and all three bait types, despite variation between seasons. The strongest seasonal effect was shown by predation rates on mealworms in May which declined with plant species richness. An additional experiment excluding vertebrate predators revealed that this deviation from the general pattern was due to vertebrate predators (mainly birds) which were present in high abundance at the field site only at this time of the year. We speculate that these predators were impaired by the increasing plant cover with plant richness leading to the observed decline in predation rates with plant richness. Inside cages, also predation rates on mealworms increased consistently with plant species richness. As a significant proportion

of dummies showed marks from gastropods, we tested for the potential effect of these opportunistic scavengers on our analysis by excluding gastropods. We found that the positive effect of plant species richness on predation rates in the Main Experiment was weakened, while results from the TBE stayed significant. Therefore, results should be interpreted carefully as different groups of predators (and scavengers) might show different responses to plant species richness. Another caveat inherent with the use of sentinel prey is the fact that some predators use movement cues to detect their prey (Howe et al. 2009). Because of these limitations, estimates of predation rates using sentinel prey may not accurately reflect absolute levels of predation in natural communities but can estimate adequate proxies to contrast predation rates across environmental gradients (Gonzalez-Gomez et al. 2006, Lövei and Ferrante 2017).

Season and stratum effects

Predation rates on mealworms and aphids varied strongly between seasons, with higher predation rates in spring compared to summer. Such

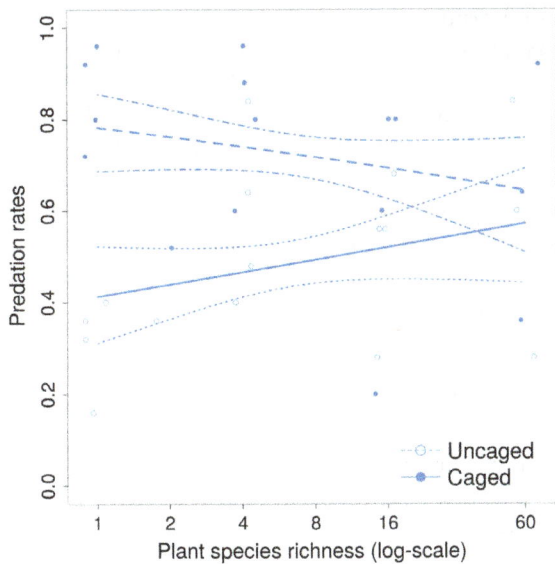

Fig. 2. Effect of removing vertebrate predation using cages on predation rates on mealworms across the plant species richness gradient in the Main Experiment in May. Points are the observed values and the solid lines the fitted regression curves from a generalized linear model (see Table 2) together with their 95% confidence intervals (dotted lines).

Table 2. Generalized linear model results for the relationship difference between predation rates on mealworm baits inside and outside of cages depending on plant functional group richness and plant species richness.

Explanatory variable	df	Deviance
Caged	1	45.6***
Plant species richness (PSR)	1	0.0
Functional group richness (FunRich)	1	0.0
Caged:FunRich	1	22.1***
Caged:PSR	1	4.1*

Notes: The P-values for the deviance were computed using a chi-square test. Asterisks represent the significance level of the terms with $*P< 0.05$ and $***P< 0.001$.

changes can be caused by differences in the populations of predators between seasons and also by changes in the height of the vegetation (Solovyeva 2015). In our study, predation rates on mealworms were probably higher in May than in August due to the high activity of birds (as documented with the additional cage experiment), which feed their offspring during spring (Martin 1987). Higher predation rates on aphids in May than in August have already been reported in the literature and are thought to have a strong impact on aphid population dynamics reducing the risks of outbreaks (Winder et al. 1994).

In general, we found very low predation rates for both mealworm and dummy baits in the vegetation compared to baits on the ground. Likely explanations for these differences between the two strata are variations in the abundance of predators between the two strata, in the cues used by predators to detect their prey, and also in the hunting modes of occurring predators. For example, actively hunting carabids occur on the ground, while web-building spiders are important predators in the vegetation (Miller et al. 2014). The potentially complex interactions between bait types, season, and stratum require future studies of the predator communities and their changes in abundance, diversity, and behavior along plant diversity gradients to mechanistically understand these context effects.

Effects of plant species richness and functional diversity

We found that plant species richness had larger and more consistent effects on predation rates than plant functional diversity. This contradicts the hypothesis that functional diversity is a better predictor of ecosystem functioning than taxonomic diversity (McGill et al. 2006, Gagic et al. 2015). This might be due to two reasons. First, we used the opportunity of available experimentally created gradients in plant functional diversity to test their effects on predation in our study. However, the traits used to define these gradients such as spatial and temporal plant resource acquisition traits (Roscher et al. 2004, Ebeling et al. 2014b) may not be strongly linked to predation. We would expect that predation rates depend more on plant traits that relate to vegetation structure (Schmitz 2008). However, we decided against screening a number of plant traits for their explanatory potential for observed predation rates by creating measures of functional diversity a posteriori. Second, in our dataset, plant productivity increased with plant species richness but not with plant functional diversity. If plant diversity effects on predation were mediated by changes in aboveground productivity, predation rates would consequently respond more strongly to plant species richness than to plant functional diversity (Hertzog et al. 2016).

Potential mechanisms of plant species richness effect on predation

Multiple mechanisms can potentially cause effects of plant species richness on predation rates. These are potentially counteracting, and our experiment does not allow us to test individual mechanisms explicitly, but based on our results, some mechanisms appear more relevant. We will briefly describe these in the following. First, the observed positive effect of plant diversity on predation may be caused by the increase in predator density with increasing plant diversity (Haddad et al. 2009, Hertzog et al. 2016) leading to higher predation through mass effects as documented in Ebeling et al. (2014a). Second, as predator diversity also increases with plant diversity (Hertzog et al. 2016), complementarity between predator species (Snyder et al. 2006, 2008), positive selection effects (Straub and Snyder 2006), or facilitation (Losey and Denno 1999) may lead to the observed patterns. Third, predator species may also show different levels of voracity along the richness gradient both through selection effects (Finke and Snyder 2010) and through compensatory feeding due to changes in prey quality (Abbas et al. 2014). Finally, plant diversity effects on predation may also be mediated by the increase in vegetation structural complexity which can cause changes in predation directly and indirectly by affecting prey and predator behavior, movement, and hunting efficiency (Brose 2003, Diehl et al. 2013). Higher predation rates at higher plant species richness could be caused by reduced intraguild predation (Finke and Denno 2002), changes in temperature and humidity conditions affecting predator activity (Heck and Crowder 1991), or reduced foraging efficiencies of top predator, such as birds, reducing predation pressure on intermediate predators (Thompson et al. 2016).

Further experimental work using moving prey and predator individuals in combination with behavioral observations or using more sophisticated methods such as gut-content analysis (Roubinet et al. 2015, Tiede et al. 2016) would enable further insights into the mechanisms underlying the positive effects of plant diversity on predation rates reported here. Another interesting result from our field site is that omnivorous species tend to become more predatory along the plant diversity gradient (Ebeling et al. 2017). Resasco et al. (2012) reported similar patterns, namely that the trophic position of omnivorous ants increased with plant richness. This implies that shifts in omnivore feeding behavior can contribute to higher predation rates with increasing plant diversity.

Ecological and applied implications

Elevated predation rates may have various effects on prey and plant communities as well as on ecosystem functioning. Higher predation rates could reduce prey population sizes and affect interspecific competition between prey species (Chase et al. 2002, Chesson and Kuang 2008) leading to changes in prey community composition. An increased predation may also affect prey foraging behavior (Preisser et al. 2007) because with increasing predation pressure prey organisms might avoid foraging in potentially risky habitats (Schmitz 2008). For example, recent evidence showed that predators can affect decomposition rates of plant litter by increasing stress levels in their prey (Hawlena et al. 2012) which would in turn affect nutrient cycling. These effects induced by predators may cascade down to lower trophic levels affecting plant community biomass (Schmitz et al. 2000, Halaj and Wise 2001), plant community composition (Schmitz 2003), and plant fitness (Romero and Koricheva 2011). Via trophic cascades, an increase in predation suppressing herbivores could cause an increase in plants biomass (Borer et al. 2005). However, at our field site community-level herbivory increased with plant species richness (Loranger et al. 2014, Meyer et al. 2017). This increase in herbivory occurred despite the increased predation pressure documented here. Taken together, these results imply that without the increase in topdown control exerted on herbivores at higher plant diversity, consumed plant biomass would be even larger in diverse plant communities.

Effects of plant diversity on predation also have implications in an applied context. In agricultural systems, biological control of herbivorous pest species by natural enemies has been of long-standing interest (Stern et al. 1959). Research on biological control has emphasized ways to enhance natural enemy populations (Landis et al. 2000, Tscharntke et al. 2007) to reduce the reliance on insecticides by increasing biological pest control. Our work shows that predation by invertebrate natural enemies increases with higher plant species richness in line with a meta-analysis showing higher herbivore suppression with a higher diversity of crop species (Letourneau et al. 2011). This evidence calls for management schemes that sustain natural enemy populations in agricultural landscapes by increasing plant and habitat diversity (Tscharntke et al. 2007) to sustain provisioning of the important ecosystem service of pest control.

Acknowledgments

Preparing, exposing, and collecting a large number of baits employed in this study would not have been possible without the help of Sylvia Creutzburg, Ilka Wolff, Gerlinde Kratzsch, Silke Schroeckh, Silke Hengelhaupt, Ute Köber, Heike Scheffler, and Steffen Eismann. Diana von Unruh helped to prepare, to collect, and to digitize the data during the spring campaign. Itziar Candeal and Javier Jorge helped in the analysis of the dummy bite marks. Jan-Hendrick Düdenhoffer kindly provided the cages used in the caging experiment. Jes Hines, Paolo Casula, and the TEREC journal club provided helpful comments to improve the manuscript. We also wish to thank Cameron Wagg and numerous student helpers for maintaining the diversity gradients. This study was funded by the Deutsche Forschungsgemeinschaft (FOR 1451). This work was supported by the German Research Foundation (DFG) and the Technical University of Munich (TUM) in the framework of the Open Access Publishing Program.

Literature Cited

Abbas, M., A.-M. Klein, A. Ebeling, Y. Oelmann, R. Ptacnik, W. W. Weisser, and H. Hillebrand. 2014. Plant diversity effects on pollinating and herbivorous insects can be linked to plant stoichiometry. Basic and Applied Ecology 15:169–178.

Bates, D., M. Mächler, B. Bolker, and S. Walker. 2015. Fitting linear mixed-effects models using lme4. Journal of Statistical Software 67:1–48.

Borer, E., E. Seabloom, J. Shurin, K. Anderson, C. Blanchette, B. Broitman, S. Cooper, and B. Halpern. 2005. What determines the strength of a trophic cascade? Ecology 86:528–537.

Brose, U. 2003. Bottom-up control of carabid beetle communities in early successional wetlands: Mediated by vegetation structure or plant diversity? Oecologia 135:407–413.

Cardinale, B. J., C. T. Harvey, K. Gross, and A. R. Ives. 2003. Biodiversity and biocontrol: emergent impacts of a multi-enemy assemblage on pest suppression and crop yield in an agroecosystem. Ecology Letters 6:857–865.

Cardinale, B. J., et al. 2012. Biodiversity loss and its impact on humanity. Nature 486:59–67.

Chase, J. M., P. A. Abrams, J. P. Grover, S. Diehl, P. Chesson, R. D. Holt, S. A. Richards, R. M. Nisbet, and T. J. Case. 2002. The interaction between predation and competition: a review and synthesis. Ecology Letters 5:302–315.

Chesson, P., and J. J. Kuang. 2008. The interaction between predation and competition. Nature 456:235–238.

Diehl, E., V. L. Mader, V. Wolters, and K. Birkhofer. 2013. Management intensity and vegetation complexity affect web-building spiders and their prey. Oecologia 173:579–589.

Douglass, J. G., J. E. Duffy, and J. F. Bruno. 2008. Herbivore and predator diversity interactively affect ecosystem properties in an experimental marine community. Ecology Letters 11:598–608.

Duffy, J. E. 2002. Biodiversity and ecosystem function: the consumer connection. Oikos 99:201–219.

Ebeling, A., A.-M. Klein, J. Schumacher, W. W. Weisser, and T. Tscharntke. 2008. How does plant richness affect pollinator richness and temporal stability of flower visits? Oikos 117:1808–1815.

Ebeling, A., S. T. Meyer, M. Abbas, N. Eisenhauer, H. Hillebrand, M. Lange, C. Scherber, A. Vogel, A. Weigelt, and W. W. Weisser. 2014a. Plant diversity impacts decomposition and herbivory via changes in aboveground arthropods. PLoS One 9:e106529.

Ebeling, A., S. Pompe, J. Baade, N. Eisenhauer, H. Hillebrand, R. Proulx, C. Roscher, B. Schmid, C. Wirth, and W. W. Weisser. 2014b. A trait-based experimental approach to understand the mechanisms underlying biodiversity–ecosystem functioning relationships. Basic and Applied Ecology 15:229–240.

Ebeling, A., M. Rzanny, M. Lange, N. Eisenhauer, L. R. Hertzog, S. T. Meyer, and W. W. Weisser. 2017. Plant diversity induces shifts in the functional structure and diversity across trophic levels. Oikos. https://doi.org/10.1111/oik.04210

Finke, D. L., and R. F. Denno. 2002. Intraguild predation diminished in complex-structured vegetation: implications for prey suppression. Ecology 83:643–652.

Finke, D. L., and W. E. Snyder. 2010. Conserving the benefits of predator biodiversity. Biological Conservation 143:2260–2269.

Gagic, V., et al. 2015. Functional identity and diversity of animals predict ecosystem functioning better than species-based indices. Proceedings of the Royal Society B 282:20142620.

Gonzalez-Gomez, P. L., C. F. Estades, and J. A. Simonetti. 2006. Strengthened insectivory in a temperate fragmented forest. Oecologia 148:137–143.

Griffin, J. N., J. E. Byrnes, and B. J. Cardinale. 2013. Effects of predator richness on prey suppression: a meta-analysis. Ecology 94:2180–2187.

Haddad, N. M., G. M. Crutsinger, K. Gross, J. Haarstad, and D. Tilman. 2011. Plant diversity and the stability of foodwebs. Ecology Letters 14:42–46.

Haddad, N. M., K. Gross, and D. Tilman. 2009. Plant species loss decreases arthropod diversity and shifts trophic structure. Ecology Letters 10:1029–1039.

Halaj, J., and D. H. Wise. 2001. Terrestrial trophic cascades: How much do they trickle? American Naturalist 157:262–281.

Harrison, X. A. 2014. Using observation-level random effects to model overdispersion in count data in ecology and evolution. PeerJ 2:e616.

Hawlena, D., M. S. Strickland, M. A. Bradford, and O. J. Schmitz. 2012. Fear of predation slows plant-litter decomposition. Science 336:1434–1438.

Heck Jr., K., and L. Crowder. 1991. Habitat structure and predator-prey interactions in vegetated aquatic systems. Pages 281–299 in S. S. Bell, E. D. McCoy, and H. R. Mushinsky, editors. Habitat structure. Springer, Dordrecht, The Netherlands.

Hertzog, L. R., A. Ebeling, S. T. Meyer, and W. W. Weisser. 2016. Experimental manipulation of grassland plant diversity induces complex shifts in aboveground arthropod diversity. PLoS One 11:e0148768.

Hines, J., et al. 2015. Chapter four-towards an integration of biodiversity–ecosystem functioning and food web theory to evaluate relationships between multiple ecosystem services. Advances in Ecological Research 53:161–199.

Hoffmann, K., W. Bivour, B. Frueh, M. Kossmann, and P. Voss. 2014. Klimauntersuchungen in Jena für die Anpassung an den Klimawandel und seine erwarteten Folgen. Selbstverlag des Deutschen Wetterdienstes, Offenbach am Main, Germany.

Hooper, D. U., et al. 2005. Effects of biodiversity on ecosystem functioning: a consensus of current knowledge. Ecological Monographs 75:3–35.

Howe, A., G. L. Lövei, and G. Nachman. 2009. Dummy caterpillars as a simple method to assess predation rates on invertebrates in a tropical agroecosystem. Entomologia Experimentalis et Applicata 131:325–329.

Ives, A. R., B. J. Cardinale, and W. E. Snyder. 2005. A synthesis of subdisciplines: predator–prey interactions, and biodiversity and ecosystem functioning. Ecology Letters 8:102–116.

Janssen, A., M. Montserrat, R. HilleRisLambers, A. M. de Roos, A. Pallini, and M. W. Sabelis. 2006. Intraguild predation usually does not disrupt biological control. Pages 21–44 in J. Brodeur and G. Boivin, editors. Trophic and guild in biological interactions control. Springer, Dordrecht, The Netherlands.

Katano, I., et al. 2015. A cross-system meta-analysis reveals coupled predation effects on prey biomass and diversity. Oikos 124:1427–1435.

Landis, D. A., S. D. Wratten, and G. M. Gurr. 2000. Habitat management to conserve natural enemies of arthropod pests in agriculture. Annual Review of Entomology 45:175–201.

Letourneau, D. K., J. A. Jedlicka, S. G. Bothwell, and C. R. Moreno. 2009. Effects of natural enemy biodiversity on the suppression of arthropod herbivores in terrestrial ecosystems. Annual Review of Ecology, Evolution, and Systematics 40:573–592.

Letourneau, D. K., et al. 2011. Does plant diversity benefit agroecosystems? A synthetic review. Ecological Applications 21:9–21.

Loranger, H., W. W. Weisser, A. Ebeling, T. Eggers, E. De Luca, J. Loranger, C. Roscher, and S. T. Meyer. 2014. Invertebrate herbivory increases along an experimental gradient of grassland plant diversity. Oecologia 174:183–193.

Losey, J. E., and R. F. Denno. 1999. Factors facilitating synergistic predation: the central role of synchrony. Ecological Applications 9:378–386.

Lövei, G. L., and M. Ferrante. 2017. A review of the sentinel prey method as a way of quantifying invertebrate predation under field conditions. Insect Science 24:528–542.

Low, P. A., K. Sam, C. McArthur, M. R. C. Posa, and D. F. Hochuli. 2014. Determining predator identity from attack marks left in model caterpillars: guidelines for best practice. Entomologia Experimentalis et Applicata 152:120–126.

Marquard, E., A. Weigelt, V. M. Temperton, C. Roscher, J. Schumacher, N. Buchmann, M. Fischer, W. W. Weisser, and B. Schmid. 2009. Plant species richness and functional composition drive overyielding in a six-year grassland experiment. Ecology 90:3290–3302.

Martin, T. E. 1987. Food as a limit on breeding birds: a life-history perspective. Annual Review of Ecology and Systematics 18:453–487.

McGill, B. J., B. J. Enquist, E. Weiher, and M. Westoby. 2006. Rebuilding community ecology from functional traits. Trends in Ecology and Evolution 21:178–185.

Meyer, S. T., C. Koch, and W. W. Weisser. 2015. Towards a standardized rapid ecosystem function assessment. Trends in Ecology and Evolution 30:390–397.

Meyer, S. T., L. Scheithe, L. Hertzog, A. Ebeling, C. Wagg, C. Roscher, and W. W. Weisser. 2017. Consistent increase in herbivory along two experimental plant diversity gradients over multiple years. Ecosphere 8:e01876.

Miller, J. R. B., J. M. Ament, and O. J. Schmitz. 2014. Fear on the move: Predator hunting mode predicts variation in prey mortality and plasticity in prey spatial response. Journal of Animal Ecology 83: 214–222.

Nakagawa, S., and H. Schielzeth. 2013. A general and simple method for obtaining R2 from generalized linear mixed-effects models. Methods in Ecology and Evolution 4:133–142.

Nelder, J. A., and R. J. Baker. 1972. Generalized linear models. In Encyclopedia of statistical sciences. Wiley, Hoboken, New Jersey, USA.

Östman, Ö., B. Ekbom, and J. Bengtsson. 2003. Yield increase attributable to aphid predation by ground-living polyphagous natural enemies in spring barley in Sweden. Ecological Economics 45: 149–158.

Preisser, E. L., J. L. Orrock, and O. J. Schmitz. 2007. Predator hunting mode and habitat domain alter nonconsumptive effects in predator-prey interactions. Ecology 88:2744–2751.

R Core Team. 2015. R: a language and environment for statistical computing. R Foundation for Statistical Computing, Vienna, Austria.

Resasco, J., D. J. Levey, and E. I. Damschen. 2012. Habitat corridors alter relative trophic position of fire ants. Ecosphere 3:1–9.

Romero, G. Q., and J. Koricheva. 2011. Contrasting cascade effects of carnivores on plant fitness: a meta-analysis. Journal of Animal Ecology 80:696–704.

Roscher, C., J. Schumacher, J. Baade, W. Wilcke, G. Gleixner, W. W. Weisser, B. Schmid, and E.-D. Schulze. 2004. The role of biodiversity for element cycling and trophic interactions: an experimental approach in a grassland community. Basic and Applied Ecology 5:107–121.

Roslin, T., et al. 2017. Higher predation risk for insect prey at low latitudes and elevations. Science 356: 742–744.

Rouabah, A., F. Lasserre-Joulin, B. Amiaud, and S. Plantureux. 2014. Emergent effects of ground beetles size diversity on the strength of prey suppression. Ecological Entomology 39:47–57.

Roubinet, E., C. Straub, T. Jonsson, K. Staudacher, M. Traugott, B. Ekbom, and M. Jonsson. 2015. Additive effects of predator diversity on pest control caused by few interactions among predator species. Ecological Entomology 40:362–371.

Sam, K., B. Koane, and V. Novotny. 2015. Herbivore damage increases avian and ant predation of caterpillars on trees along a complete elevational forest gradient in Papua New Guinea. Ecography 38:293–300.

Scherber, C., et al. 2010. Bottom-up effects of plant diversity on multitrophic interactions in a biodiversity experiment. Nature 468:553–556.

Schmitz, O. J. 2003. Top predator control of plant biodiversity and productivity in an old-field ecosystem. Ecology Letters 6:156–163.

Schmitz, O. J. 2008. Effects of predator hunting mode on grassland ecosystem function. Science 319:952–954.

Schmitz, O. J., P. A. Hambäck, and A. P. Beckerman. 2000. Trophic cascades in terrestrial systems: a review of the effects of carnivore removals on plants. American Naturalist 155:141–153.

Snyder, W. E., G. C. Chang, and R. P. Prasad. 2005. Conservation biological control: Biodiversity influences the effectiveness of predators. Pages 324–343 in P. Barbosa and I. Castellanos, editors. Ecology of predator-prey interactions. Oxford University Press, Oxford, UK.

Snyder, G. B., D. L. Finke, and W. E. Snyder. 2008. Predator biodiversity strengthens aphid suppression across single-and multiple-species prey communities. Biological Control 44:52–60.

Snyder, W. E., G. B. Snyder, D. L. Finke, and C. S. Straub. 2006. Predator biodiversity strengthens herbivore suppression. Ecology Letters 9:789–796.

Solovyeva, E. 2015. Seasonal and diel dynamics of predation in grassland. Thesis. Technische Universität München, München, Germany.

Stern, V. M., R. F. Smith, R. Van den Bosch, and K. S. Hagen. 1959. The integration of chemical and biological control of the spotted alfalfa aphid: the integrated control concept. Hilgardia 29:81–101.

Straub, C. S., and W. E. Snyder. 2006. Species identity dominates the relationship between predator biodiversity and herbivore suppression. Ecology 87: 277–282.

Thompson, S. J., C. M. Handel, R. M. Richardson, and L. B. McNew. 2016. When winners become losers: predicted nonlinear responses of arctic birds to increasing woody vegetation. PLoS One 11: e0164755.

Tiede, J., B. Wemheuer, M. Traugott, R. Daniel, T. Tscharntke, A. Ebeling, and C. Scherber. 2016. Trophic and non-trophic interactions in a biodiversity experiment assessed by next-generation sequencing. PLoS One 11:e0148781.

Tscharntke, T., R. Bommarco, Y. Clough, T. O. Crist, D. Kleijn, T. A. Rand, J. M. Tylianakis, S. van Nouhuys, and S. Vidal. 2007. Conservation biological control and enemy diversity on a landscape scale. Biological Control 43:294–309.

Tylianakis, J. M., and C. M. Romo. 2010. Natural enemy diversity and biological control: making sense of the context-dependency. Basic and Applied Ecology 11:657–668.

Wilby, A., and M. B. Thomas. 2002. Natural enemy diversity and pest control: patterns of pest emergence with agricultural intensification. Ecology Letters 5: 353–360.

Winder, L., D. Hirst, N. Carter, S. Wratten, and P. Sopp. 1994. Estimating predation of the grain aphid *Sitobion avenae* by polyphagous predators. Journal of Applied Ecology 31:1–12.

9

Habitat selection differs across hierarchical behaviors: selection of patches and intensity of patch use

Laura A. McMahon,[1] Janet L. Rachlow,[1],† Lisa A. Shipley,[2] Jennifer S. Forbey,[3]
and Timothy R. Johnson[4]

[1]*Department of Fish and Wildlife Sciences, University of Idaho, Moscow, Idaho, USA*
[2]*School of the Environment, Washington State University, Pullman, Washington, USA*
[3]*Department of Biological Sciences, Boise State University, Boise, Idaho, USA*
[4]*Department of Statistical Science, University of Idaho, Moscow, Idaho, USA*

Abstract. When animals select habitats, they integrate a suite of behaviors that are influenced by multiple competing resource requirements. Resources that influence decisions about habitat use are likely to differ across spatial scales and hierarchical behaviors. At coarser scales, animals are expected to select resources that are critical to fitness, and at finer scales, to intensively use resources that enhance fitness. Our goal was to contrast habitat selection at two hierarchical behavioral levels (patch selection and intensity of patch use) to test hypotheses about how resources shape habitat use. We applied a two-stage hurdle model to quantify both initial selection and intensity of use of resource patches by a burrowing herbivore, the pygmy rabbit (*Brachylagus idahoensis*). We expected security from predation to influence initial selection of patches, and forage availability to influence intensity of use of selected patches. We monitored locations of adults fitted with radio-collars during winter and summer. We measured vegetation, burrow characteristics, and concealment cover within patches that were used and unused by rabbits, and we quantified burrow densities surrounding patches in a GIS. Selection of used patches from available patches was largely influenced by security resources (presence and proximity of burrows, shrub height, and woody ground cover) during both seasons. Intensity of patch use was also influenced by forage availability and, consequently, differed between seasons. During winter, sagebrush is the primary forage, and greater sagebrush canopy within a patch was associated with increased intensity of use. During summer, rabbits more intensively used patches with greater availability of herbaceous forage. Elucidating how animals make choices about habitats across a diversity of spatial and temporal scales will continue to increase our understanding of the factors that govern distribution of populations and movements of individuals. By extending these concepts to hierarchical behaviors, we can enhance insights into the processes that shape the patterns of habitat use we observe.

Key words: *Brachylagus idahoensis;* habitat selection; hurdle model; intensity of use; predation risk; pygmy rabbit; resource selection; sagebrush.

† **E-mail:** jrachlow@uidaho.edu

Introduction

The habitat selection process represents an integrated suite of behaviors that are influenced by many stimuli, constraints, and tradeoffs (Johnson 1980, Orians and Wittenberger 1991, Mayor et al.

2009). Animals select habitat across spatial and temporal scales (DeCesare et al. 2012, McGarigal et al. 2016), but their choices about resources also can be evaluated at multiple behavioral resolutions. Hierarchical behaviors can be identified by distinguishing among different steps in the selection process. For example, initial selection, defined as the process of choosing resources for use, can be distinguished as a first-step behavior, whereas variation in intensity of use of selected resources (based on frequency or duration of use) represents a secondary step in the selection process. Such a behavioral process is akin to Johnson's (1980) orders of habitat selection in two important ways. First, the scales or levels identified for spatial analyses represent human constructs in what is an inherently continuous process. Second, selection for resources across spatial scales represents a nested hierarchy with selection at lower orders constraining selection at higher ones (Senft et al. 1987, Rettie and Messier 2000, Frye et al. 2013). Both of these properties characterize selection of habitat across behavioral levels.

Resources that influence decisions about habitat use differ across spatial scales and are likely to vary in a similar manner across hierarchical behaviors. At coarse spatial scales (and behavioral resolutions), individuals should select resources that address the greatest threats to fitness, for example, avoiding mortality from predation (Rettie and Messier 2000, Spencer 2002) or agonistic interactions with conspecifics (Van Horne 1983, Moorcroft et al. 2006). At finer scales, individuals are expected to select resources that enhance components of fitness (e.g., accelerating growth or increasing fecundity; Rettie and Messier 2000, Payer and Harrison 2003, Gaillard et al. 2010). Applying hierarchy theory to the habitat selection process, Lesmerises et al. (2013) documented that woodland caribou (*Rangifer tarandus*) displayed differing responses to patch size; occurrence within patches increased with patch area, but intensity of patch use (assessed as counts of animal locations within a patch) decreased with patch area. Likewise, blue gills (*Lepomis macrochirus*) selected nesting habitat by first selecting for a homogeneous substrate and once this resource was attained, prioritized other resource needs including vegetation coverage and shoreline protection (Stahr et al. 2013). Such examples provide empirical evidence that resources that shape animal choices about habitat differ across behavioral resolutions.

Separating the resource selection process into two steps, initial selection of habitat and intensity of habitat use, requires a combination of quantitative approaches. A hurdle model is a two-stage

modeling process that provides flexibility to assess factors that contribute to both coarse- and fine-scale processes (Wenger and Freeman 2008, Santini et al. 2015). The hurdle component of the model separates zero and non-zero data by first estimating the probability of a non-zero count (e.g., probability of selection or occurrence). If the hurdle is crossed (meaning a non-zero count obtained), a truncated count model is used to model only the data without zero counts (Zurr et al. 2009). Previous applications of hurdle models in animal ecology include evaluation of variables that influence presence–absence and abundance (Heinänen et al. 2008, Eskelson et al. 2009), and assessment of occurrence within patches and intensity of patch use (Lesmerises et al. 2013). To evaluate habitat characteristics influencing both probability of selection and intensity of resource use in a hurdle model framework, binary data for selection and count data representing intensity of use can be modeled separately (Martin et al. 2005).

We applied a kind of two-stage hurdle model to test hypotheses about hierarchical behaviors associated with resource selection (i.e., selection of resource patches and intensity of use of patches across a season) by a specialist herbivore in a highly heterogeneous environment. The pygmy rabbit (*Brachylagus idahoensis*) is a sagebrush (*Artemisia* spp.) habitat specialist that occurs only in the sagebrush-steppe of the western USA. Predation is a primary cause of mortality for the species throughout the year (Estes-Zumpf and Rachlow 2009, Crawford et al. 2010, Price et al. 2010), and like other lagomorphs, predation likely represents a strong evolutionary force that has shaped their morphology and behavior (Lima and Dill 1990, Smith and Litvaitis 2000). Pygmy rabbits are obligate burrowers that excavate and use burrow systems that serve as effective refuges from all but a few predators. Additionally, sagebrush shrubs create habitat structure that provides security from aerial and terrestrial predators, and other vegetation and woody material on the ground also provide concealment and reduce perceptions of predation risk by this species (Camp et al. 2012). During winter, the height and structure of shrubs above the snow surface is reduced, which likely diminishes concealment (Olsoy et al. 2015). However, snow also can provide additional cover because pygmy rabbits readily create and use subnivean tunnels (Katzner and Parker 1997).

Because predation is the overriding proximate cause of mortality for pygmy rabbits, we hypothesized that security resources that decrease risk of predation would strongly influence selection of habitat patches during both summer and winter. We expected that individuals would select to use patches with excavated burrow systems, especially those that were close to other burrow systems. Additionally, we predicted that selection of habitat patches would be positively influenced by shrub structure (i.e., shrub canopy and height) and presence of woody debris at the ground level that provide concealment from predators.

At the next behavioral resolution, we hypothesized that forage availability would influence intensity of use of selected habitat patches across a season. Our ability to test this relationship rigorously, however, is restricted during winter. Because sagebrush comprises most of the winter diet of pygmy rabbits (Thines et al. 2004, Shipley et al. 2006), sagebrush shrubs provide both forage and security resources. Therefore, although we predicted that intensity of patch use during winter would be influenced by the availability of sagebrush, we cannot distinguish between the dual functions of forage availability and security from predators. In contrast, during summer, we predicted that intensity of patch use would be positively related to abundance of herbaceous forage (i.e., grasses and forbs, which comprise about half of the summer diet) as well as sagebrush. Finally, if habitat parameters associated with security (e.g., presence of burrows and features that provide concealment from predators) strongly influence selection of patches, then we expected that these factors might be less influential in shaping intensity of use of selected patches because individuals addressed security requirements at the coarser behavioral level. Data for both behavioral levels were analyzed at the same spatial scale (patch) and temporal scale (season). An understanding of how resources shape the occurrence and distribution of animals across a landscape requires integration across behaviors as well as spatial and temporal scales (Bélisle 2005).

METHODS

Study area

We conducted research within the Lemhi Valley in east-central Idaho, USA. The climate of the Lemhi Valley was typical of high-elevation sagebrush-steppe habitats. Winters are characterized by freezing temperatures (daily average = −7.1°C; Western Regional Climate Center 1965–2005), and summer is dominated by warm (daily average = 26°C) and dry periods (Western Regional Climate Center 1965–2005). The region is comprised of a mix of private and public lands, which support spring cattle grazing and alfalfa (*Medicago sativa*) production. Our study site, Cedar Gulch (44°41′ N, 113°17′ W), encompassed approximately 120 ha of continuous sagebrush-steppe habitat characterized by mima mounds, distinct dome-like mounds of sediments. At Cedar Gulch, the mean diameter of mima mounds was 10.6 m (Parsons et al. 2016), and Wyoming big sagebrush (*Artemisia tridentate* spp. *wyomingensis*) shrubs occurred predominantly clumped on these mounds (Fig. 1). Sagebrush shrubs are typically taller and denser on mima mounds compared to the off-mound matrix (Parsons et al. 2016), and vegetation between mounds was relatively sparse and short, creating a highly heterogeneous landscape. Black sagebrush (*Artemisia nova*) and three-tip sagebrush (*Artemisia tripartite*) were distributed less commonly throughout Cedar Gulch. Grasses and forbs occurred seasonally throughout the study area at relatively low densities.

Sagebrush shrubs created overstory and vertical habitat structure that provide both cover and forage for pygmy rabbits and a suite of other species. Three other lagomorphs occurred at relatively low densities at Cedar Gulch: mountain cottontails (*Sylvilagus nuttallii*), white-tailed jackrabbits (*Lepus townsendii*), and black-tailed jackrabbits (*Lepus californicus*). Additionally, pygmy rabbits experience high rates of predation (Estes-Zumpf and Rachlow 2009, Crawford et al. 2010, Price et al. 2010), and several species of terrestrial and avian predators occurred at the study site, including American badgers (*Taxidea taxus*), coyotes (*Canis latrans*), long-tailed weasels (*Mustela frenata*), red foxes (*Vulpes vulpes*), northern harriers (*Circus cyaneus*), short-eared owls (*Asio flammeus*), and great horned owls (*Bubo virginianus*).

Capture and radiotelemetry

We radiotagged adult pygmy rabbits during winter (January–March) and summer (June–August) of 2015 and 2016. Animals were trapped in wire box traps (Tomahawk Live

Fig. 1. Aerial photograph of the Cedar Gulch study site in Idaho, USA. Darker areas on the imagery acquired from the National Agriculture Imagery Program represent areas of relatively dense shrub cover typically associated with microtopographic features known as mima mounds (inset photograph).

Traps, Hazelhurst, Wisconsin, USA) set at burrow entrances. We handled rabbits in a mesh bag, recorded weight, identified sex, and attached a very high frequency (VHF) radio-collar weighing ~6 g to adults (>400 g). We collected location data on rabbits for 4–6 weeks, after which the rabbits were trapped to remove the collar. To identify individuals that might be recaptured in subsequent seasons, we implanted all study animals with passive integrated transponder tags following collar removal.

Individuals were radiotracked daily during daylight hours, and we approximated their location to minimize animal disturbance. To avoid disturbing individuals, we used VHF homing techniques to find the location of the animal from a distance of >40 m. To approximate the coordinates of the animal location, the observer recorded his or her location using a handheld Global Positioning System (GPS) unit and then estimated the distance and orientation to the animal using a range finder and compass. Because of the highly heterogeneous distribution of vegetation, animal locations could be estimated within 10 m, which was consistent with the scale at which we

assessed habitat use. All methods used in this study were approved by the University of Idaho Animal Care and Use Committee (Protocol Number 2015-12) and are in compliance with the guidelines for use of wild mammals in research published by the American Society of Mammalogists (Sikes et al. 2016).

Habitat sampling

To evaluate factors shaping both initial selection of habitat patches and intensity of patch use, we sampled habitat features at used and available patches. We defined habitat patches as areas of relatively dense sagebrush where rabbits tend to cluster activity surrounded by a matrix of relatively sparse vegetation. Although pygmy rabbits will cross open areas among shrub patches, such areas are rarely used for burrowing, foraging, or resting (Estes-Zumpf and Rachlow 2009), and consequently, we did not include the open areas in our sampling frame. Most animal locations were within the relatively dense vegetation on mima mounds, although rabbits occasionally exploited sagebrush patches that were not associated with mounded microtopography. We identified patches used by individuals via radiotelemetry, and we sampled a minimum of four used patches per individual (unless the animal was located at fewer patches, in which case we sampled all patches used by that individual). For each rabbit, we randomly selected the same number of available patches for habitat sampling. Available patches had no documented use by the individual to which they were assigned based on radiotelemetry locations. If use was documented after sampling, a new patch was selected as a replacement available patch. We identified activity areas for each individual by buffering used patches by 75 m in a GIS (ArcView 10.3; ESRI, Redlands, California, USA), and we randomly selected available patches from those areas identified in aerial imagery obtained by the National Agriculture Imagery Program.

At each used and available patch, we quantified habitat features within the patch and in the surrounding area. Within each patch, we established two perpendicular transect lines that intersected at the center of the patch, with the first line set in a random direction. Transect lengths varied based on the width and length of the patch, and because most habitat sampling occurred on mima mounds, we used the edge of the mound to establish the patch boundary. At non-mound patches without clear boundaries, transect length was determined using the average diameter (11 m) for mima mounds at this study site (Parsons et al. 2016). Along these transects, we measured shrub canopy of live and dead sagebrush and rabbitbrush using the line-intercept method (Canfield 1941). We identified the Wyoming big sagebrush (>15 cm) rooted closest to the quadrat center and recorded height of the tallest branch excluding inflorescences. On each of the four resulting transect segments, we randomly placed a 0.5 × 0.5 m quadrat for vegetation sampling. We recorded cover of herbaceous plants (i.e., grasses and forbs) and woody debris within quadrats by visually estimating cover into standard cover classes (0 [0%], 1 [0–5%], 2 [5–25%], 3 [25–50%], 4 [50–75%], 5 [75–95%], and 6 [95–100%]; Bonham 1989). Finally, in each of the four quadrats, we estimated concealment of a rabbit-sized animal from terrestrial and aerial predators by viewing a 15 × 15 cm cube placed at the center of the quadrat from a height of 1 m at a distance of 4 m in the four cardinal directions and from a height of 1.5 m

directly above the cube (Camp et al. 2013). If a burrow system was present in a patch, we counted the number of open burrow entrances and estimated terrestrial and aerial concealment at up to three entrances, using the methods described previously. We used a GIS data layer of burrow locations generated during annual burrow surveys at Cedar Gulch (Sanchez et al. 2009, Parsons et al. 2016) to calculate the distance (m) to the nearest neighboring burrow system, and burrow system density within a 50 m radius surrounding the center of each sampled patch.

During both seasons and years, we distributed sampling effort for each animal temporally over the course of the season to capture conditions as the animals experienced them. However, during winter, significant snow accumulation and creation of subnivean tunnels by rabbits made it impossible to sample habitat characteristics without substantial disturbance that could influence habitat use by rabbits and potentially bias our results. Consequently, we measured snow depth over the course of the season coincident with use of the patches, but shrub and burrow characteristics were measured in March when much of the snow had melted. We estimated available shrub height during winter by subtracting the average snow depth of the patch from the average shrub height.

Data analysis

We evaluated the influence of habitat characteristics on both initial selection of habitat patches and variation in intensity of patch use. We first grouped habitat variables into three categories for variable reduction (burrow characteristics, vegetation, and concealment cover; Table 1). Within each category, we ran models for both initial selection and intensity of patch use with all possible combinations of variables. The variable or multiple variables from the top model in each single-category model set were carried forward into the final set of candidate models to represent that category. The final candidate set of models included the three most plausible single-category models, variables from combinations of two and three categories (i.e., burrow–vegetation, burrow–concealment, vegetation–concealment, and burrow–vegetation–concealment), and an intercept-only model. We checked for multicollinearity using Pearson's correlation coefficients, and all values were <0.06. To assess strength of

Table 1. Habitat variables potentially influencing selection of habitat patches and intensity of patch use by pygmy rabbits in Idaho, USA, were grouped into three categories (burrow characteristics, vegetation, and concealment cover) for variable selection.

Burrow characteristics	Vegetation	Concealment cover
Distance to neighboring burrow	Sagebrush canopy	Terrestrial concealment (S)
Number of burrow entrances	Shrub height	Aerial concealment (S)
Surrounding burrow density	Herbaceous cover (S)	Woody debris cover (S)
Aerial concealment at burrow (S)		Snow depth (W)
Terrestrial concealment at burrow (S)		

Note: Some variables were included only in summer (S) or winter (W) models.

support, we evaluated all models using Akaike information criterion corrected for small sample sizes (Burnham and Anderson 2002).

Because we expected that seasonal differences in resource distribution and animal behavior would influence habitat selection, we modeled data separately for winter and summer. We did not include concealment variables (i.e., terrestrial and aerial concealment, percent woody debris) in the winter models because we assumed that these variables would change over the course of the season as snow depth fluctuated. Additionally, we could not sample these properties without destroying snow characteristics of the patch. Instead, we used shrub height above snow to represent potential concealment during winter.

To evaluate the influence of security and forage resources on initial selection of patches and intensity of patch use, we employed a hurdle model approach using conditional logistic regression to test hypotheses about selection (i.e., use vs. availability) and zero-truncated Poisson (ZTP) regression to evaluate variation in intensity of use (Eskelson et al. 2009). This particular hurdle model is a variation on the more commonly used hurdle models in that a conditional logistic regression model was used for the hurdle portion of the model rather than a traditional logistic regression. But because we sampled a fixed number of used and available patches per animal, we employed conditional logistic regression because it is

appropriate for matched use-availability designs (Compton et al. 2002, Boyce 2006). To account for individual variation, our models incorporated individual as a stratifying variable (Lendrum et al. 2012). Therefore, the models described the difference between available and used locations for each animal. To assess the relative importance of habitat on intensity of patch use, we removed locations with no use (i.e., available patches) and employed a ZTP regression model (Zurr et al. 2009). The response variable for our ZTP model was the total count of VHF locations per individual at each sampled patch, which, by definition, had recorded counts of at least one. In addition, we included a random effect for individual in the ZTP regression models to account for individual differences in factors influencing intensity of use (Duchesne et al. 2010). An offset term was included in the models to account for the number of days that each individual was monitored. For ease of interpretation, we exponentiated all parameter estimates to generate odds ratios for the conditional logistic regression output and incident rate ratios for the ZTP output. We inferred variable significance if the 85% confidence intervals for the odds or incident rate ratios did not overlap one (Arnold 2010). All statistical analyses were conducted using R 3.2.3 (R Development Core Team 2015). Conditional logistic regressions were run using the survival package (Therneau 2015), and mixed-effect ZTP regression models were conducted in the glmmADMB package (Skaug et al. 2016). Mean values are presented with standard errors.

Results

Individual pygmy rabbits used multiple habitat patches and burrow systems during both summer and winter. We monitored daily movements of 29 individuals during winter (14 females and 15 males) and 13 (9 females and 4 males) during summer. Most ($n = 39$) were tracked for a single season, but three individuals were recaptured and contributed data to two consecutive seasons. Rabbits were tracked for an average of 39 ± 0.8 d, and we collected 37 ± 1.0 locations per animal, for a total of 1580 telemetry locations. During winter, $89\% \pm 2.3\%$ of locations were within patches that contained burrow systems, and rabbits exploited an average of six different habitat

Table 2. Habitat characteristics used to evaluate selection of habitat patches by pygmy rabbits in Idaho, USA.

| | Winter | | | | Summer | | | |
| | Used | | Available | | Used | | Available | |
Habitat parameter	Mean	SE	Mean	SE	Mean	SE	Mean	SE
Number of burrow entrances	2.7	0.15	0.3	0.07	1.5	0.21	0.4	0.12
Surrounding burrow density†	2.8	0.12	1.9	0.12	2.7	0.16	2.5	0.14
Distance to neighboring burrow (m)	37.9	1.59	38.7	1.06	32.7	1.05	36.4	1.23
Snow depth (cm)	19.9	1.52	21.0	1.60	–	–	–	–
Sagebrush canopy (%)	45.9	0.94	50.2	0.90	48.3	1.14	51.5	1.02
Shrub height (cm)	30.5	1.87	23.0	1.91	52.2	1.48	46.9	1.26
Herbaceous cover (%)	–	–	–	–	12.6	1.3	11.7	1.3
Terrestrial concealment (%)	–	–	–	–	76.4	1.31	74.7	1.43
Aerial concealment (%)	–	–	–	–	20.7	1.42	19.8	1.47
Woody debris (%)	–	–	–	–	4.0	0.78	3.0	0.70

Notes: SE, standard error. Values for each season represent averages across all used (winter: $n = 144$; summer: $n = 85$) and available (winter: $n = 144$; summer: $n = 85$) patches sampled during that season.
    † Variable recorded in a 50 m radius surrounding the center of the habitat patch.

patches (range = 2–11). During summer, use of patches with burrow systems declined (only 62% ± 6.1% of locations were within patches that included burrows), and although individuals exploited an average of nine different habitat patches (range = 2–14), only four of those typically had burrow systems. We sampled habitat variables at 458 patches during winter ($n = 288$) and summer ($n = 170$), half of which were used and half of which were available (Table 2).

Initial selection of habitat patches from available patches was largely driven by security resources during both winter and summer. As expected, strong selection for burrows within the patches and in the surrounding area was evident during both seasons (Table 3). Model results suggested that for each additional burrow entrance, the odds of selection increased by a factor of 4.3 during winter and 1.6 during summer (Fig. 2A). Additionally, security resources that might enhance concealment at a patch were positively associated with selection. Although there were four competing models in the candidate set describing initial patch selection during winter, the significant positive influence of the number of burrow entrances within a patch and density of burrow systems in the surrounding area was evident in all models (Table 3). The best-supported model indicated a significant

Table 3. Model selection results for the 95% confidence set of models describing selection of habitat patches by pygmy rabbits in Idaho, USA, during winter and summer (2015–2016).

Variables	K	AIC_c	ΔAIC_c	Wt	Rank
Winter					
BurrowDensity* + #BurrowEntr* + ShrubHt* − ShrubCanopy	4	147.78	0.00	0.56	1
BurrowDensity* + #BurrowEntr*	2	149.83	2.05	0.20	2
BurrowDensity* + #BurrowEntr* − SnowDepth	3	150.12	2.34	0.17	3
BurrowDensity* + #BurrowEntr* − SnowDepth + ShrubCanopy	4	152.12	4.34	0.06	4
Summer					
BurrowEntr* − NearestBurr* + WD* + ShrubHt* − ShrubCanopy	5	168.73	0.00	0.45	1
BurrowEntr* − NearestBurr* + WD*	3	169.20	0.47	0.36	2
BurrowEntr* − NearestBurr* + ShrubHt* − ShrubCanopy*	4	170.94	2.22	0.15	3

Notes: AIC_c, Akaike information criterion. Selection models were analyzed with conditional logistic regression using individual as the stratifying variable. Parameters are as follows: K, the number of parameters estimated; ΔAIC_c, the change in AIC_c; and Wt, model weight. BurrowDensity, density of burrow systems within 50 m; #BurrowEntr, number of burrow entrances at patch (0 = no burrow system); NearestBurr, distance (m) to the nearest neighboring burrow system; ShrubHt, shrub height (cm) on the patch; ShrubCanopy, total percent canopy cover on the patch, includes living and dead canopy cover; SnowDepth, depth (cm) of snow at patch; WD, percentage of woody debris cover.
    * Significant at the 85% confidence level.

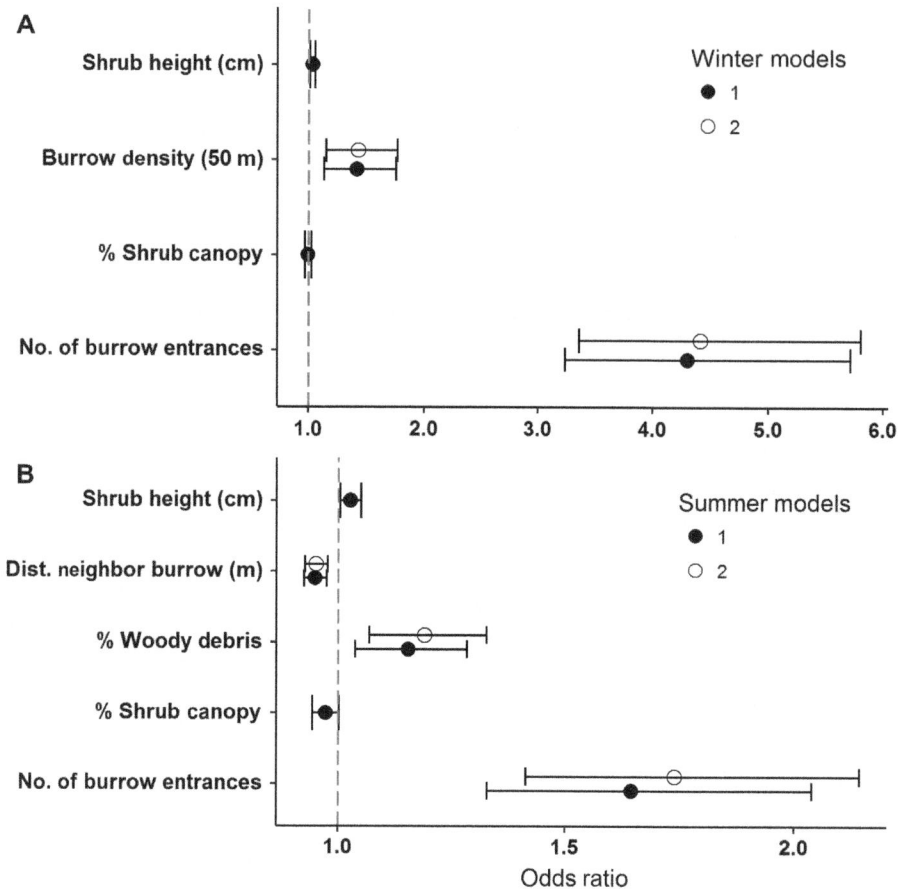

Fig. 2. Odds ratios (+85% CI) for parameter estimates generated from the first- and second-ranked top models (ΔAIC < 2) describing intensity of habitat patch use by pygmy rabbits during winter ($n = 29$; A) and summer ($n = 13$; B) of 2015–2016 in Idaho, USA. AIC, Akaike information criterion; CI, confidence interval.

association between height of sagebrush shrubs and patch selection; however, because of the complementary properties of sagebrush, selection for taller shrubs might be associated with concealment cover, forage availability, or both. Like winter, initial selection of habitat patches during summer was consistently associated with variables only providing security, such as burrows (within patches and in surrounding areas) and woody debris on the ground, and variables that potentially provide security and forage resources, such as shrub height and canopy (Fig. 2B). Summer patch selection was best described by two similar, top-ranking models that collectively received 81% of the model weight. Both models included burrow characteristics and security concealment provided by woody debris, but not herbaceous forage (Table 3).

Intensity of use of selected patches during winter was influenced by resources providing security and by resources that potentially provide both forage and security. Contrary to our expectations, presence of burrow systems significantly and positively influenced intensity of patch use during winter. In addition, individuals more intensively used patches with greater sagebrush canopy (Fig. 3A). A 5% increase in sagebrush canopy was associated with an 11% increase in intensity of patch use. Intensity of use during winter was best described by two similar models that collectively received 100% of the model weight (Table 4). Although sagebrush is the primary winter food source for this species, we cannot decouple the influence of forage and security on intensive use of patches with greater sagebrush canopy during winter.

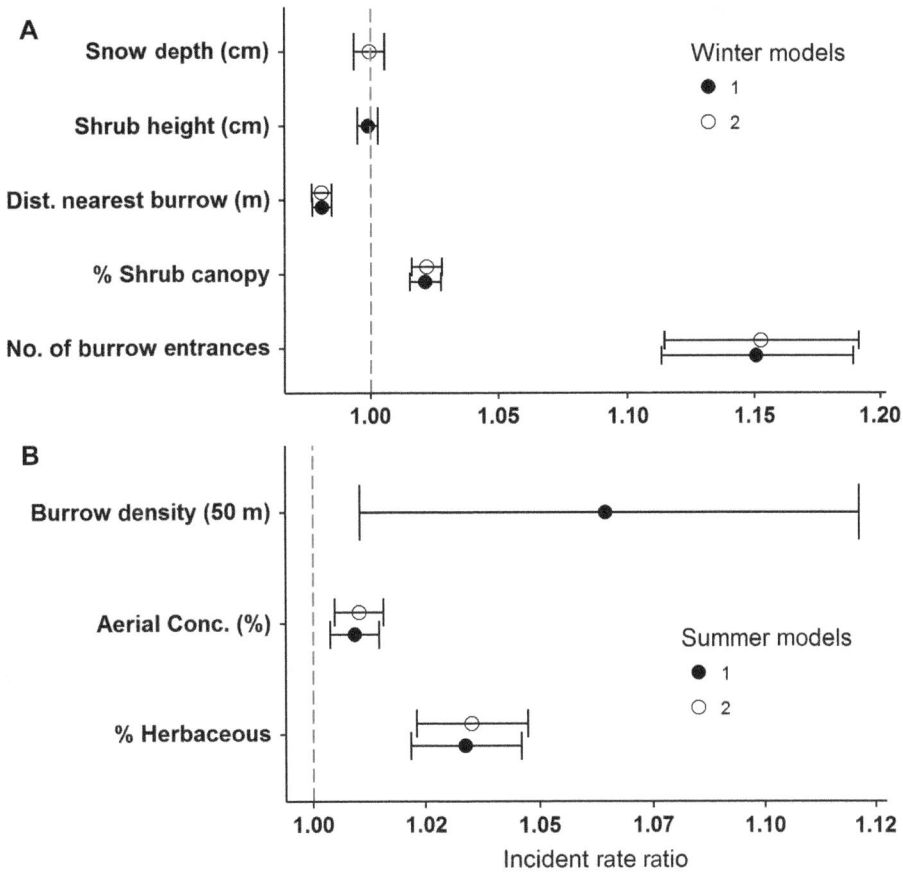

Fig. 3. Incident rate ratios (+85% CI) for parameter estimates generated from the first- and second-ranked top models (ΔAIC < 2) describing intensity of habitat patch use by pygmy rabbits during winter ($n = 29$; A) and summer ($n = 13$; B) of 2015–2016 in Idaho, USA. AIC, Akaike information criterion; CI, confidence interval.

In contrast to winter, intensity of patch use was clearly influenced by forage availability as well as security during summer. In this season, rabbits more intensively used patches with greater availability of herbaceous forage, which comprises about half of their summer diet. A 5% increase in cover of grasses and forbs resulted in an 18% increase in patch use (Fig. 3B). In addition to herbaceous forage, security resources also influenced intensity of patch use during summer. Individuals more intensively used habitat patches with greater aerial concealment that were surrounded by more burrow systems (Table 4, Fig. 3B). Intensity of use during summer was described best by two models that held 89% of the model weight, which was almost evenly distributed between both models (Table 4).

DISCUSSION

Our work suggested that animals met diverse fundamental resource needs by choosing habitat differently across hierarchical behaviors. At a relatively coarse behavioral level (i.e., initial selection of patches for use), individuals chose patches that provided security from predation. In contrast, availability of forage influenced intensity of patch use during summer, but not initial selection of patches. Contrary to our expectations, however, security resources (e.g., presence and proximity of burrows and concealment cover) also strongly influenced intensity of patch use, even though those resources were selected at the coarser behavioral level. Given the high fitness consequences of predation, security likely represents a strong selective factor driving

Table 4. Model selection results for the 95% confidence set of models describing intensity of use of habitat patches by pygmy rabbits in Idaho, USA, during winter and summer (2015–2016).

Variables	K	AIC$_c$	ΔAIC$_c$	Wt	Rank
Winter					
#Entrances* − NearestBurrow* − ShrubHt + ShrubCanopy*	6	1213.77	0.00	0.52	1
#Entrances* − NearestBurrow* + ShrubCanopy* − SnowDepth	6	1213.89	0.12	0.48	2
Summer					
BurrowDensity + Herb* + AerialConc*	5	541.86	0.00	0.52	1
Herb* + AerialConc*	4	542.55	0.70	0.37	2
BurrowDensity* + Herb*	4	545.58	3.72	0.08	3

Notes: AIC$_c$, Akaike information criterion. Intensity of use was analyzed with zero-truncated Poisson regression model with an offset to account for the number of days the animal was tracked and a random variable to account for individual-level variation. Parameters are as follows: K, the number of parameters estimated; ΔAIC$_c$, the change in AIC$_c$; and Wt, model weight. #Entrances, number of burrow entrances; NearestBurrow, distance (m) to the nearest neighboring burrow system; BurrowDensity, burrow density within 50 m of mound; ShrubHt, height (cm) of sagebrush shrubs at patch; AerialConc, aerial concealment on the patch; ShrubCanopy, total percent canopy cover on the patch, includes living and dead canopy cover; Herb, percentage of herbaceous cover sums grass and forb cover.
   * Significant at the 85% confidence level.

fundamental resource requirements, especially for species that experience high rates of predation. Quantifying selection and use of resource across hierarchical behaviors facilitated testing hypotheses about the mechanisms influencing the habitat selection process.

During both seasons, pygmy rabbits demonstrated strong selection for habitat patches that reduced risk of predation. Choice of patches on the landscape was predominately shaped by the presence of burrow systems, and selected patches tended to have taller shrubs than available patches. Similar summer habitat associations have been documented for this species (Gabler et al. 2001, Heady and Laundré 2005, Schmalz et al. 2014). Few studies of winter habitat selection have been published (but see Wilson et al. 2010); however, Katzner and Parker (1997) reported that pygmy rabbits in a heavy snow year restricted use to the tallest patches of sagebrush. During winter, tall shrubs provide the only structure that can conceal rabbits above the snow surface. Shrub height above the snow is likely to be less associated with forage availability, however, because pygmy rabbits also can access sagebrush forage beneath the snow in subnivean tunnels. Nonetheless, we cannot exclude forage availability as a contributing factor for initial selection of patches during winter. Species that experience high rates of predation are expected to select habitats at coarse spatial scales to minimize predation risk (Rettie and Messier 2000, Apps et al. 2001). Our work suggests that selection at coarse hierarchical

behaviors by pygmy rabbits was influenced by similar factors.

In contrast to initial selection of patches, intensity of patch use was influenced by availability of season-specific forages. During winter, rabbits intensively exploited habitat patches with greater shrub canopy cover (Fig. 3A). The diet of pygmy rabbits is comprised predominately of sagebrush during winter (Thines et al. 2004), and shrub canopy cover likely reflects the biomass of available forage more than other sagebrush characteristics that we evaluated such as shrub height. However, the multiple functions provided by sagebrush make it impossible to completely disentangle the complementary properties of forage and security during winter. Intensity of patch use during summer, however, more clearly reflected the influence of forage on behavior. The summer diet of pygmy rabbits includes a large proportion of grasses and forbs (Thines et al. 2004), and as expected, intensity of use increased at patches with higher cover of herbaceous plants. At our study site, these plants were low growing and relatively sparsely distributed and, consequently, provided little concealment from predators.

Despite differences between hierarchical behaviors and seasons, pygmy rabbits consistently demonstrated a strong association with burrows. This species is an obligate burrower unlike other North American lagomorphs, and burrows facilitate heat dumping during summer (Long et al. 2005) and reduce metabolic costs associated with low winter temperatures (Kinlaw 1999, Milling

2017). In addition, burrows provide protection from most predators (Camp et al. 2012), and proximity to burrows was a key factor in selection of resting locations by rabbits at our study site (Milling et al. 2017). Although patch-level burrow characteristics did not significantly influence intensity of patch use during summer, patches that were surrounded by elevated burrow densities were used more intensively (Fig. 3). Availability of multiple burrow systems likely enhances opportunities for forging while maintaining the relative safety of nearby refuges (Wilson et al. 2012, Crowell et al. 2016).

Our work aimed to identify fundamental resource needs that shape patterns of habitat selection by pygmy rabbits (i.e., security and forage) at the patch scale, but we acknowledge that other factors also can influence selection. We did not include the effect of sex in our models because of sample size limitations. Although behavioral patterns differ between males and females, especially during the summer reproductive season (Heady and Laundré 2005, Burak 2006, Sanchez and Rachlow 2008), security and forage are essential for all individuals, especially for species that are subject to high rates of predation and have relatively high metabolic requirements (Shipley et al. 2006). Nonetheless, examination of more complex models that include sex and other habitat resources and potential constraints on selection (e.g., population density or interspecific interactions) would enhance understanding of habitat relationships for this species. Finally, in our study area, vegetation patches used by pygmy rabbits were typically associated with mima mounds and, consequently, delineated from the surrounding matrix. When sampling patches in non-mound areas, we used the average size of a mima mound (11 m diameter) to define patches, which is the area that typically encompasses a single burrow system. The size of appropriate patches might differ in other portions of the species range depending on typical burrow size and distribution of burrow entrances.

We examined habitat selection at two hierarchical behaviors across a season, but selection also likely occurs at other behavioral levels. For example, intensity of use could be characterized at a finer resolution by distinguishing frequency of repeated use from longer vs. shorter durations of use during individual visits to patches (i.e.,

patch residency time; Bastille-Rousseau et al. 2010). Availability of GPS locations collected at short time intervals (McMahon et al. 2017) could provide the data needed to evaluate such fine-scale behavioral processes.

The hierarchical behaviors we examined share some conceptual properties with the spatial or temporal scales commonly used to frame habitat selection studies. Decisions made by animals across hierarchical behaviors likely occur simultaneously, and the separation of behavioral levels for analyses parallels delineation of spatiotemporal scales associated with analyses of habitat selection (Johnson 1980, Turner 1989, Boyce 2006). By modeling the behavioral process as two steps, however, we detected the influence of differing resources on initial selection of habitat patches and intensity of patch use. Dissecting behavioral strategies is a critical step in understanding the complex behavioral processes that result in habitat use and distribution of animals across landscapes. Elucidating how animals make choices about habitats across a diversity of spatial and temporal scales has increased understanding of the factors that govern distribution of populations and movements of individuals (Johnson et al. 2002, Mayor et al. 2009). By extending these concepts to hierarchical behaviors, we can enhance insights into the processes that shape the patterns we observe.

Acknowledgments

Funding and other support were provided by the National Science Foundation (DEB-1146166 to JLR, DEB-1146368 to LAS, DEB-1146194 to JSF), Berklund Graduate Assistantship, University of Idaho, Bureau of Land Management, and U.S. Forest Service. We thank M. Crowell, M. Whetzel, C. Milling, M. Hernandez, N. Carter, R. Hohbein, and B. Shipley for field work assistance. We appreciate the comments of R. Long and S. DeMay who helped to improve the manuscript.

Literature Cited

Apps, C. D., B. N. McLellan, T. A. Kinley, and J. P. Flaa. 2001. Scale-dependent habitat selection by mountain caribou, Columbia Mountains, British Columbia. Journal of Wildlife Management 65:65.

Arnold, T. W. 2010. Uninformative parameters and model selection using Akaike's information criterion. Journal of Wildlife Management 74:1175–1178.

Bastille-Rousseau, G., D. Fortin, and C. Dussault. 2010. Inference from habitat-selection analysis depends on foraging strategies. Journal of Animal Ecology 79:1157–1163.

Bélisle, M. 2005. Measuring landscape connectivity: the challenge of behavioral landscape ecology. Ecology 86:1988–1995.

Bonham, C. D. 1989. Measurements for terrestrial vegetation. Wiley, New York, New York, USA.

Boyce, M. S. 2006. Scale for resource selection functions. Diversity and Distributions 12:269–276.

Burak, G. S. 2006. Home ranges, movements, and multi-scale habitat use of pygmy rabbits (*Brachylagus idahoensis*) in southwestern Idaho. Dissertation. Boise State University, Boise, Idaho, USA.

Burnham, K., and D. Anderson. 2002. Model selection and multimodel inference: a practical information-theoretic approach. Springer-Verlag, New York, New York, USA.

Camp, M. J., J. L. Rachlow, B. A. Woods, T. R. Johnson, and L. A. Shipley. 2012. When to run and when to hide: the influence of concealment, visibility, and proximity to refugia on perceptions of risk. Ethology 118:1010–1017.

Camp, M. J., J. L. Rachlow, B. A. Woods, T. R. Johnson, and L. A. Shipley. 2013. Examining functional components of cover: the relationship between concealment and visibility in shrub-steppe habitat. Ecosphere 4:1–19.

Canfield, R. 1941. Application of the line interception method in sampling range vegetation. Journal of Forestry 39:388–394.

Compton, B. W., J. M. Rhymer, and M. McCollough. 2002. Habitat selection by wood turtles (*Clemmys insculpta*): an application of paired logistic regression. Ecology 83:833.

Crawford, J. A., R. G. Anthony, J. T. Forbes, and G. A. Lorton. 2010. Survival and causes of mortality for pygmy rabbits (*Brachylagus idahoensis*) in Oregon and Nevada. Journal of Mammalogy 91:838–847.

Crowell, M. M., L. A. Shipley, M. J. Camp, J. L. Rachlow, J. S. Forbey, and T. R. Johnson. 2016. Selection of food patches by sympatric herbivores in response to concealment and distance from a refuge. Ecology and Evolution 6:2865–2876.

DeCesare, N. J., M. Hebblewhite, F. Schmiegelow, D. Hervieux, G. J. McDermid, L. Neufeld, M. Bradley, J. Whittington, K. G. Smith, and L. E. Morganti. 2012. Transcending scale dependence in identifying habitat with resource selection functions. Ecological Applications 22:1068–1083.

Duchesne, T., D. Fortin, and N. Courbin. 2010. Mixed conditional logistic regression for habitat selection studies. Journal of Animal Ecology 79:548–555.

Eskelson, B. N. I., H. Temesgen, and T. M. Barrett. 2009. Estimating cavity tree and snag abundance using negative binomial regression models and nearest neighbor imputation methods. Canadian Journal of Forest Research 39:1749–1765.

Estes-Zumpf, W. A., and J. L. Rachlow. 2009. Natal dispersal by pygmy rabbits (*Brachylagus idahoensis*). Journal of Mammalogy 90:363–372.

Frye, G. G., J. W. Connelly, D. D. Musil, and J. S. Forbey. 2013. Phytochemistry predicts habitat selection by an avian herbivore at multiple spatial scales. Ecology 94:308–314.

Gabler, K. I., L. T. Heady, and J. W. Laundré. 2001. A habitat suitability model for pygmy rabbits (*Brachylagus idahoensis*) in southeastern Idaho. Western North American Naturalist 61:480–489.

Gaillard, J.-M., M. Hebblewhite, A. Loison, M. Fuller, R. Powell, M. Basille, and B. Van Moorter. 2010. Habitat-performance relationships: finding the right metric at a given spatial scale. Philosophical Transactions of the Royal Society of London B: Biological Sciences 365:2255–2265.

Heady, L. T., and J. W. Laundré. 2005. Habitat use patterns within the home range of pygmy rabbits (*Brachylagus idahoensis*) in southeastern Idaho. Western North American Naturalist 65:490–500.

Heinänen, S., M. Rönkä, and M. von Numers. 2008. Modelling the occurrence and abundance of a colonial species, the arctic tern *Sterna paradisaea* in the archipelago of SW Finland. Ecography 31:601–611.

Johnson, D. H. 1980. The comparison of usage and availability measurements for evaluating resource preference. Ecology 61:65–71.

Johnson, C. J., K. L. Parker, D. C. Heard, and M. P. Gillingham. 2002. Movement parameters of ungulates and scale-specific responses to the environment. Journal of Animal Ecology 71:225–235.

Katzner, T. E., and K. L. Parker. 1997. Vegetative characteristics and size of home ranges used by pygmy rabbits (*Brachylagus idahoensis*) during winter. Journal of Mammalogy 78:1063–1072.

Kinlaw, A. 1999. A review of burrowing by semi-fossorial vertebrates in arid environments. Journal of Arid Environments 41:127–145.

Lendrum, P. E., C. R. Anderson, R. A. Long, J. G. Kie, and R. T. Bowyer. 2012. Habitat selection by mule deer during migration: effects of landscape structure and natural-gas development. Ecosphere 3:1–19.

Lesmerises, R., J.-P. Ouellet, C. Dussault, and M.-H. St-Laurent. 2013. The influence of landscape matrix on isolated patch use by wide-ranging animals: conservation lessons for woodland caribou. Ecology and Evolution 3:2880–2891.

Lima, S. L., and L. M. Dill. 1990. Behavioral decisions made under the risk of predation: a review and prospectus. Canadian Journal of Zoology 68:619–640.

Long, R. A., T. J. Martin, and B. M. Barnes. 2005. Body temperature and activity patterns in free-living arctic ground squirrels. Journal of Mammalogy 86:314–322.

Martin, T. G., B. A. Wintle, J. R. Rhodes, P. M. Kuhnert, S. A. Field, S. J. Low-Choy, A. J. Tyre, and H. P. Possingham. 2005. Zero tolerance ecology: improving ecological inference by modelling the source of zero observations: modelling excess zeros in ecology. Ecology Letters 8:1235–1246.

Mayor, S. J., D. C. Schneider, J. A. Schaefer, and S. P. Mahoney. 2009. Habitat selection at multiple scales. Écoscience 16:238–247.

McGarigal, K., H. Y. Wan, K. A. Zeller, B. C. Timm, and S. A. Cushman. 2016. Multi-scale habitat selection modeling: a review and outlook. Landscape Ecology 31:1161–1175.

McMahon, L. A., J. L. Rachlow, L. A. Shipley, J. S. Forbey, T. R. Johnson, and P. J. Olsoy. 2017. Evaluation of micro-GPS receivers for tracking small-bodied mammals. PLoS ONE 12:e0173185.

Milling, C. R. 2017. The role of physiology, behavior, and habitat in seasonal thermoregulation by pygmy rabbits. Dissertation. University of Idaho, Moscow, Idaho, USA.

Milling, C. R., J. L. Rachlow, T. R. Johnson, J. S. Forbey, and L. A. Shipley. 2017. Seasonal patterns of behavior reveal variation in strategies for thermoregulation and predator avoidance by a small-bodied endotherm. Behavioral Ecology arx084. https://doi.org/10.1093/beheco/arx084

Moorcroft, P. R., M. A. Lewis, and R. L. Crabtree. 2006. Mechanistic home range models capture spatial patterns and dynamics of coyote territories in Yellowstone. Proceedings of the Royal Society of London B: Biological Sciences 273:1651–1659.

Olsoy, P. J., J. S. Forbey, J. L. Rachlow, J. D. Nobler, N. F. Glenn, and L. A. Shipley. 2015. Fearscapes: mapping functional properties of cover for prey with terrestrial LiDAR. BioScience 65:74–80.

Orians, G. H., and J. F. Wittenberger. 1991. Spatial and temporal scales in habitat selection. American Naturalist 137:S29–S49.

Parsons, M. A., T. C. Barkley, J. L. Rachlow, J. L. Johnson-Maynard, T. R. Johnson, C. R. Milling, J. E. Hammel, and I. Leslie. 2016. Cumulative effects of an herbivorous ecosystem engineer in a heterogeneous landscape. Ecosphere 7:1–17.

Payer, D. C., and D. J. Harrison. 2003. Influence of forest structure on habitat use by American marten in an industrial forest. Forest Ecology and Management 179:145–156.

Price, A. J., W. Estes-Zumpf, and J. Rachlow. 2010. Survival of juvenile pygmy rabbits. Journal of Wildlife Management 74:43–47.

R Development Core Team. 2015. R: a language and environment for statistical computing. R Foundation for Statistical Computing, Vienna, Austria.

Rettie, W. J., and F. Messier. 2000. Hierarchical habitat selection by woodland caribou: its relationship to limiting factors. Ecography 23:466–478.

Sanchez, D. M., and J. L. Rachlow. 2008. Spatiotemporal factors shaping diurnal space use by pygmy rabbits. Journal of Wildlife Management 72:1304–1310.

Sanchez, D. M., J. L. Rachlow, A. P. Robinson, and T. R. Johnson. 2009. Survey indicators for pygmy rabbits: temporal trends of burrow systems and pellets. Western North American Naturalist 69:426–436.

Santini, M. S., M. E. Utgés, P. Berrozpe, M. Manteca Acosta, N. Casas, P. Heuer, and O. D. Salomón. 2015. *Lutzomyia longipalpis* presence and abundance distribution at different micro-spatial scales in an urban scenario. PLOS Neglected Tropical Diseases 9:e0003951.

Schmalz, J. M., B. Wachocki, M. Wright, S. I. Zeveloff, and M. M. Skopec. 2014. Habitat selection by the pygmy rabbit (*Brachylagus idahoensis*) in Northeastern Utah. Western North American Naturalist 74:456–466.

Senft, R. L., M. B. Coughenour, D. W. Bailey, L. R. Rittenhouse, O. E. Sala, and D. M. Swift. 1987. Large herbivore foraging and ecological hierarchies. BioScience 37:789–799.

Shipley, L. A., T. B. Davila, N. J. Thines, and B. A. Elias. 2006. Nutritional requirements and diet choices of the pygmy rabbit (*Brachylagus idahoensis*): a sagebrush specialist. Journal of Chemical Ecology 32:2455–2474.

Sikes, R. S., and The Animal Care and Use Committee of the American Society of Mammalogists. 2016. 2016 Guidelines of the American Society of Mammalogists for the use of wild mammals in research and education. Journal of Mammalogy 97:663–688.

Skaug, H., D. Fournier, A. Nielsen, A. Magnusson, and B. Bolker. 2016. Generalized linear mixed models using "AD Model Builder". https://rdrr.io/rforge/glmmADMB/

Smith, D. F., and J. A. Litvaitis. 2000. Foraging strategies of sympatric lagomorphs: implications for differential success in fragmented landscapes. Canadian Journal of Zoology 78:2134–2141.

Spencer, R.-J. 2002. Experimentally testing nest site selection: fitness trade-offs and predation risk in turtles. Ecology 83:2136–2144.

Stahr, K. J., M. A. Kaemingk, and D. W. Willis. 2013. Factors associated with bluegill nest site selection within a shallow, natural lake. Journal of Freshwater Ecology 28:283–292.

Therneau, T. M. 2015. A package for survival analysis in S. https://cran.r-project.org/web/packages/survival/index.html

Thines, N. J., L. A. Shipley, and R. D. Sayler. 2004. Effects of cattle grazing on ecology and habitat of Columbia Basin pygmy rabbits (*Brachylagus idahoensis*). Biological Conservation 119:525–534.

Turner, M. G. 1989. Landscape ecology: the effect of pattern on process. Annual Review of Ecology and Systematics 20:171–197.

Van Horne, B. 1983. Density as a misleading indicator of habitat quality. Journal of Wildlife Management 47:893.

Wenger, S. J., and M. C. Freeman. 2008. Estimating species occurrence, abundance, and detection probability using zero-inflated distributions. Ecology 89:2953–2959.

Western Regional Climate Center. 1965–2005. Period of record monthly climate summary. Western Regional Climate Center, Leadore, Idaho, USA. http://www.wrcc.dri.edu/cgi-bin/cliMAIN.pl?idlead

Wilson, T. L., J. B. Odei, M. B. Hooten, and T. C. Edwards Jr. 2010. Hierarchical spatial models for predicting pygmy rabbit distribution and relative abundance. Journal of Applied Ecology 47:401–409.

Wilson, T. L., A. P. Rayburn, and T. C. Edwards. 2012. Spatial ecology of refuge selection by an herbivore under risk of predation. Ecosphere 3:6.

Zurr, A. F., E. N. Ieno, N. J. Walker, A. A. Saveliev, and G. M. Smith. 2009. Zero-truncated and zero-inflated models for count data. Pages 261–293 *in* Mixed effects models and extensions in ecology with R. Springer, New York, New York, USA.

10

Vegetation dynamics during last 35,000 years at a cold desert locale: preferential loss of forbs with increased aridity

ROBERT S. NOWAK,[1],† CHERYL L. NOWAK,[2] AND ROBIN J. TAUSCH[2]

[1]*Department of Natural Resources & Environmental Sciences, University of Nevada Reno,
MS 186, 1664 North Virginia Street, Reno, Nevada 89557 USA*
[2]*U.S. Forest Service Great Basin Research Laboratory, 920 Valley Road, Reno, Nevada 89521 USA*

Abstract. Paleoecological records are an important source of data to better understand ecological responses to climate. To help understand vegetation–climate relationships in the Great Basin cold desert of North America, we analyzed 154 plant taxa from 52 fossil woodrat middens that spanned both an 800-m elevation gradient and the last 35,000 yr (35 ka) within a single study area. Vegetation was analyzed by community assemblage (CA), by plant functional type, and by individual species. Concordant with a predominant trend of increased aridity since glacial maximum, CA change was largely unidirectional despite centennial-scale cycles of climate variation. Richness of forb and other herbaceous plant functional types peaked from 15 to 26 ka during glacial maximum and then gradually decreased from ~15 ka to present. Analysis of individual taxa indicates that once a herbaceous species was lost from the study area, that species did not re-occur. In contrast, woody (shrubs and trees) species richness peaked from 8 to 15 ka during and following the Bolling-Allerod rapid warming. However, most tree taxa that established during this period of warming were subsequently lost as climate became more arid after the beginning of Mid-Holocene Temperature Maximum warming ~8 ka. These shifts in plant functional types decreased relative richness of forbs compared to shrubs (i.e., decreased forb/shrub ratio), and these shifts continue to the present despite intervening cycles of climate cooling and warming. We conclude that the relative importance of herbaceous species in current CAs of the Great Basin cold desert has been decreasing for the last 15 ka. We also speculate that decreased richness of herbaceous species has, in part, provided opportunities for exotic species to establish and proliferate in the Great Basin during the last 100 yr. Thus, observations of past vegetation change suggest that increased aridity with future climate warming will continue to favor woody vegetation over herbaceous species at our cold desert locale, at least until invasive species displace native shrub species.

Key words: Bolling-Allerod warming; climate change; cluster analysis; forbs; Great Basin shrub steppe; invasive species; late Pleistocene; Mid-Holocene Temperature Maximum; plant functional type; shrubs; vegetation assemblages; woodrat middens.

† **E-mail:** nowak@cabnr.unr.edu

INTRODUCTION

Terrestrial ecosystems have experienced rapid climate change over the last century, experiencing a rate of warming that far exceeds that of any previous period of the Holocene (Blois et al. 2013) as well as the Pleistocene (Diffenbaugh and Field 2013). This rapid climate change will impact ecosystems that are already stressed, and the extent that individual plant species and associated communities will be affected remains uncertain (Moritz and Agudo 2013). Nonetheless, fragmented habitats, species attrition, ecosystem simplification, and invasion of exotic species likely will result, with local to regional species extinctions occurring (Diffenbaugh and Field 2013).

Because climate-driven dispersal and migration are common in paleoecological records, paleo-records provide data to better understand the patterns and processes governing ecological responses to climate change (Alley et al. 2003, Williams et al. 2011). Both late Pleistocene and Holocene climate changes (century to millennial timescales) in the western United States affected ecological processes and drove vegetation changes in plant community assemblages (CAs) at local and regional scales (e.g., Thompson 1990, Swetnam et al. 1999, Wigand and Rhode 2002, Tausch et al. 2004, Gray et al. 2006). These documented shifts in CAs during past climate change serve as model systems to understand vegetation response to ongoing and future climate change (e.g., Swetnam et al. 1999, Lyford et al. 2003).

The Great Basin of western North America is an ideal location for paleoecological studies of plant response to climate change because its inherent arid climate helps preserve plant materials (Nowak et al. 1994a, Tausch et al. 2004). Paleobotanical data from pollen records (e.g., Mensing et al. 2004, 2013) and fossil woodrat (*Neotoma* Say and Ord, 1825) middens (e.g., Betancourt et al. 1990, Tausch et al. 2004) significantly aid the reconstruction of Great Basin vegetation response to past climate change. However, the western portion of the Great Basin has little midden research for the late Pleistocene and Holocene (Betancourt et al. 1990, Nowak et al. 1994a, b). Furthermore, data from woodrat middens generally cannot discriminate shorter-term vegetation change because middens are generally discontinuous in their temporal and spatial distributions (Coats et al. 2008).

Here, we report on patterns of vegetation change across the late Pleistocene and Holocene at a cold desert locale in the northwestern Great Basin of North America. Vegetation data were obtained from a series of woodrat middens that were located in the Painted Hills region of the Virginia Mountains in northwestern Nevada and provided a relatively continuous record over the last 35,000 yr. Our goal was to assess vegetation responses to past climate changes. Vegetation

responses were quantified as changes in CAs, changes in plant functional types, and changes in individual species occurrences. We were particularly interested in how vegetation differed among four major climate periods over the last 35,000 yr (35 ka): (1) the relatively warm climate period of Pleistocene interstadials between 26 and 35 ka that occurred before the Last Glacial Maximum (LGM); (2) the cold climate period before, during, and after LGM (15–26 ka); (3) the post-LGM warming climate period 8–15 ka that began with the Bolling-Allerod rapid warming; and (4) the most recent warming climate period after the onset of the Mid-Holocene Temperature Maximum (<8 ka). This long time-series of vegetation data shows that directional changes in CAs, preferential loss of herbaceous taxa, and local extinctions of individual species occurred with the predominant trend of increasing aridity over the last 15,000 yr. This deep historical perspective of past vegetation and climate change not only provides an understanding of likely vegetation response to future climate change, but also provides insight into modern vegetation dynamics, including those associated with invasive species.

Summary of climate change over past 35,000 yr

In this section, we briefly describe major climate periods that occurred over the last ~35 ka, that is, the period of time that corresponds with our paleovegetation record. We reference local proxy climate data when available, especially sedimentology, isotopic composition, and pollen from cores of Pyramid Lake (Benson et al. 2002, 2013, Mensing et al. 2004), which is 8 km northeast of our study area (Fig. 1) and is a major remnant of Pluvial Lake Lahontan. We primarily focus on major climate periods given the temporal resolution of our midden samples and lag time response of vegetation to climate change, and therefore, we do not discuss all the climate periods that are thought to have occurred during the last 35 ka. We also use widely accepted terms for climate periods when local climate periods are generally concordant with regional or global climate cycles.

Pre-LGM interstadials (26–35 ka).—Overall, this climate period is characterized by the last series of warm interstadials that precede the LGM near the end of the Pleistocene. A gradual cooling

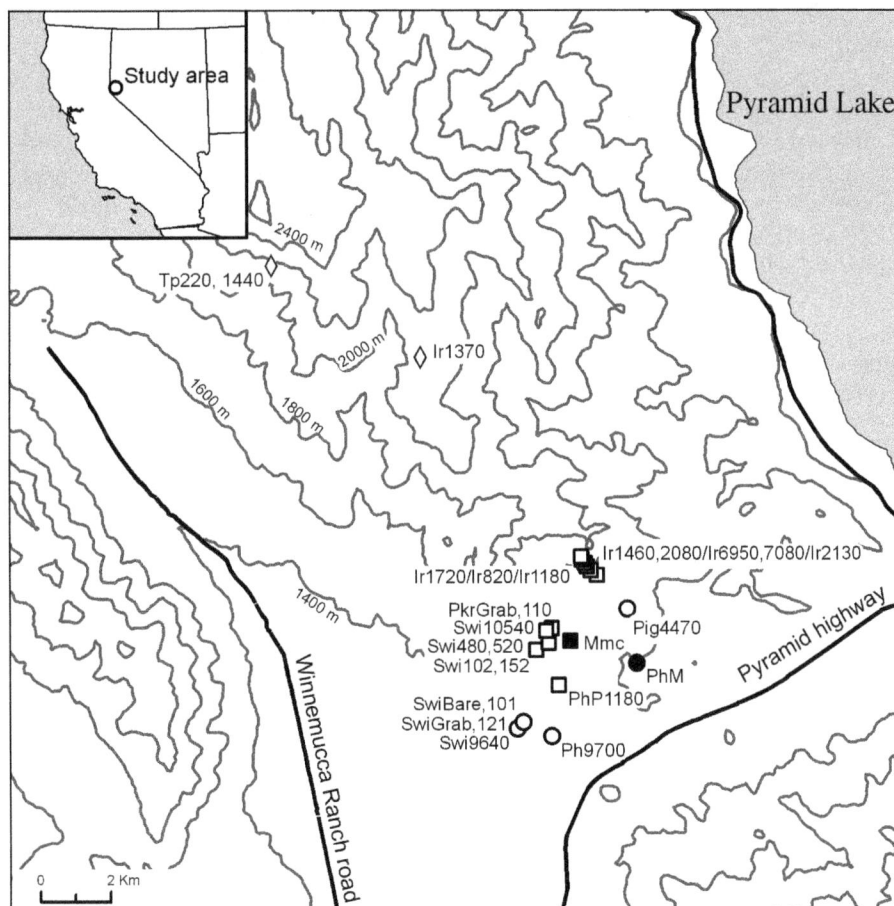

Fig. 1. Map of the Painted Hills study area in the Virginia Mountains of northwestern Nevada, USA, showing midden locales, major highways, and Pyramid Lake, which is a major remnant lake of Pluvial Lake Lahontan. Geometric symbols indicate midden locales. Circles indicate low-elevation locales, squares mid-elevation, and diamonds high elevation; filled symbols are locales with >2 middens and open symbols with one to two middens. Note that because some locales cannot be differentiated at the scale of this map, labels for some symbols include locales that are too close together to differentiate; specific locations for each midden are in Appendix S1. Midden codes associated with each symbol correspond with those listed in Appendix S1.

between 36 and 38 ka (Kindler et al. 2014) was followed by a rapid warming interstadial that reached its driest conditions locally at ~35.5 ka, as evidenced by a shallow lake level of Pluvial Lake Lahontan (Benson et al. 2013). More mesic conditions followed, which rapidly increased lake level. Climate then cycled between cool and warm periods, with gradually increasing cool periods before climate cooled to the very cold Heinrich Stadial 2 (HS 2), which began ~26 ka (Cooper et al. 2015, Rhodes et al. 2015). Locally, this period of gradual cooling that was punctuated by warm events is evident between 25.5 and

31.8 ka, with lake level initially shallow and then beginning to rise at 29.5 ka (Benson et al. 2013).

Glacial maximum (15–26 ka).—This climate period is characterized by the cold periods before, during, and after LGM. Lake level of Pluvial Lake Lahontan was very high over much of the time period from 15.0 to 24.2 ka (Benson et al. 2013), indicating a favorable water balance. The glacial maximum time period began with the very cold HS 2 (23–26 ka), a period when lake level of Pluvial Lake Lahontan increased and then largely stabilized (Benson et al. 2013). Globally, LGM occurred 20–23 ka (Schaefer et al.

2006, Liu et al. 2009, Denton et al. 2010), with global temperatures reaching a minimum during glacial maximum (Lyle et al. 2012). Gradual warming and gradual glacial melting began by ~20 ka and continued until ~18 ka (Becklin et al. 2014), which corresponds with a transitory decrease in lake level for Pluvial Lake Lahontan (Benson et al. 2013). After 18 ka, a period of cold dry winters, called both the HS 1 (Cooper et al. 2015, Rhodes et al. 2015) and the Older Dryas (Buizert et al. 2014), occurred across the Northern Hemisphere until ~14.7–15.0 ka (Liu et al. 2009, Denton et al. 2010, Buizert et al. 2014). Highstand for Pluvial Lake Lahontan occurred ~15.5 ka (Benson et al. 2013).

Post-LGM warming (8–15 ka).—This climate period is characterized by a series of sustained warm periods that are interrupted by cold periods and ends ~8 ka when major continental ice sheets are completely melted. This climate period began with the Bolling-Allerod rapid warming ~15 ka, which resulted in a warmer climate that included the warmest temperatures to this point in time since ~35 ka (Liu et al. 2009, Lyle et al. 2012, Benson et al. 2013, Becklin et al. 2014, Buizert et al. 2014, Kindler et al. 2014). The relative aridity of the Bolling-Allerod is evident from decreasing lake level of Pluvial Lake Lahontan between 14.5 and 14.8 ka, and Pluvial Lake Lahontan was relatively shallow after 14.5 ka until 13.0 ka (Benson et al. 2013). Lake level rose, stabilized, then rose again between 11.8 and 13.0 ka, indicating a predominantly cool/mesic climate that was contemporaneous with the Younger Dryas. Locally, warming returned after 11.8 ka, further decreasing lake level, but unfortunately the available lake cores end at 11.7 ka (Benson et al. 2013). For the Northern Hemisphere, warming returned ~11.7 ka and melted the remains of the Laurentide Ice Sheet such that melting was completed by ~8.0 ka (Williams et al. 2011).

Mid- and late Holocene warming (<8 ka).—This most recent climate period is characterized by the warmest and driest climate conditions since LGM and persistent droughts (Mensing et al. 2004). A major shift in atmospheric circulation patterns after ~8.0 ka (Williams et al. 2011) resulted in rapid warming and drying (Shuman et al. 2002) that continued to the Mid-Holocene Temperature Maximum, which was the warmest and driest climate since LGM. Maximum temperatures in the

Great Basin during the mid- and late Holocene occurred 5.5–7.5 ka (Lindstrom 1990, Grayson 2000, 2011, Benson et al. 2002, Wigand and Rhode 2002, Mensing et al. 2004, Tausch et al. 2004, Wigand 2006). Cooler/wetter conditions occurred 2.8–3.2 ka (often called the Neoglacial; Williams et al. 2011), but increased aridity returned 1.7–2.8 ka across the Great Basin (often called the Late Holocene Dry Period or Megadrought; Miller et al. 2001, 2004, Wigand and Rhode 2002, Mensing et al. 2004, 2013). The Megadrought in the western Great Basin was especially severe 1.9–2.5 ka (Mensing et al. 2008, 2013). Other major climate periods during the last 2000 yr included the Medieval Climate Anomaly (0.65–1.15 ka; Stine 1994, Kleppe et al. 2011) and the Little Ice Age (0.1–0.6 ka; Cook et al. 2004, Gray et al. 2006). Pollen records in the Pyramid Lake region of the northwestern Great Basin indicated especially severe drought periods before, during, and after the Medieval Climate Anomaly at 1.25–1.5, 0.725–0.8, and 0.45–0.6 ka, respectively (Mensing et al. 2004, 2008).

METHODS

Study area

Our field study area consists of approximately 80 km² of mountainous terrain in the Virginia Mountains of northwestern Nevada in an area called the Painted Hills that is located near Pyramid Lake (Fig. 1). This area has many Tertiary age, moderately welded volcanic tuff, andesite, and basalt rock outcrops (Fig. 2a). These rock materials, especially the tuff, have small cavities that make suitable homes for woodrats to construct middens or nests. Because these cavities protect middens from weathering, we found a large number of fossil woodrat middens in the study area.

Modern vegetation for our field study area includes species from the Great Basin cold desert shrub steppe and from pinyon–juniper woodland (Fig. 2a) and predictably changes with elevation (West 1988). The most common contemporary plant community members are trees: *Juniperus osteosperma*; shrubs: *Artemisia tridentata, Atriplex canescens, Atriplex confertifolia, Brickellia microphylla, Chrysothamnus nauseousus, Chrysothamnus viscidiflorus, Ephedra viridis, Ericameria nana, Grayia spinosa, Purshia tridentata, Salvia dorrii, Tetradymia*

Fig. 2. (a) General view of mid-elevation vegetation within the study area, showing present vegetation of Great Basin shrub steppe in the foreground, grading to shrub steppe with scattered *Juniperus osteosperma* trees in the middle of the photograph, and then to *Juniperus* woodland at higher elevations. In the upper left corner of the photograph (arrow) is one of the midden locales, the *Mmc* mid-elevation multi-temporal locale. (b) Freshly collected, indurated midden from the *Mmc* mid-elevation multi-temporal locale. *Pinus albicaulis* needles (arrow) are evident in the top right of the midden, which was dated at 9.6 ka.

glabrata, and *Tetradymia canescens*; and grasses: *Achnatherum hymenoides*, *Elymus elymoides*, *Pappostipa speciosa*, *Poa secunda*, and *Pseudoroegneria spicata*. (Nomenclature follows the Integrated Taxonomic Information System, http://www.itis.gov/) Common forbs found throughout the area are in the genera: *Chaenactis*, *Crepis*, *Cryptantha*, *Eriogonum*, *Galium*, *Lupinus*, *Malacothrix*, *Microseris*, and *Phacelia*. Within this region, three isolated *Pinus monophylla* trees occur near the northwestern regional geographic limit for this species. More mesic pines such as *Pinus albicaulis* currently are on Peavine Mountain, approximately 40 km

to the southwest, and in the Sierra Nevada range, approximately 50 km to the west and south.

Midden processing and grouping

Woodrat middens were collected from various locales within the study area ranging in elevation from approximately 1350 to 2150 m (Fig. 1; Appendix S1). As with most woodrat midden researchers (Betancourt et al. 1990), we define "midden" as a homogeneous matrix of plant parts, woodrat fecal pellets, and other materials (usually encased by woodrat urine) that represents a single temporal sample. We use the term "locale" to indicate an individual location where one or more middens were obtained. Fifty-two middens from 22 locales were used (Appendix S1).

We collected both unconsolidated and indurated middens. Unconsolidated middens consisted of plant parts recently collected and incorporated into the woodrats midden, but plant parts were not encased by woodrat urine. These unconsolidated middens represent modern vegetation samples that reflect woodrat foraging behavior and helped connect the dateline of past plant history closer to modern time. In the laboratory, unconsolidated middens were washed in a geologic sieve stack to clean the midden contents and to size-sort the contents. Contents were then air-dried prior to taxonomic identification.

Solid or indurated middens (Fig. 2b) were collected from the field and processed as described previously (Nowak et al. 1994a, b). In brief, middens were removed from rock cavities, using hammer and chisel if needed when middens were anchored to rock by crystallized woodrat urine. Most indurated middens were small (400–600 g) and isolated. However, some larger middens had obvious individual layers, and these latter middens were divided along naturally occurring breaks into individual middens. To avoid cross-contamination of middens, the interface was gently rinsed with water to remove debris and loose material, and then, the interface was carefully inspected to insure that each midden was self-contained and was a homogeneous matrix without discontinuities. Either the entire midden (if 400–500 g) or a 400–500 g subsample was processed by soaking in water for 7–14 d to disaggregate the contents. Contents were then cleaned and size-sorted by wet sieving through a geologic sieve stack, air-dried, and taxonomically identified.

Retrieved plant parts were identified and assigned taxonomic labels by comparison to a reference collection of seeds, leaves, and twigs at the University of Nevada Reno (RENO) herbarium (Nowak et al. 1994*a*, *b*). Macrofossil specimens were identified to the most specific taxonomic level possible. Although many taxa could be assigned to species, those more problematic were identified to genus or family level. Plant macrofossils that could be conclusively identified only to the family level typically were plant parts indicative of non-woody taxa; thus, we inferred that these taxa were forbs. A total of 154 different plant taxa were identified across all 52 middens.

Radiocarbon analysis was used to assign an age for each indurated midden (Appendix S1). For approximately half of the indurated middens, sufficient juniper twigs were available for conventional radiometric techniques. For most of the remaining indurated middens, woodrat fecal pellets were dated by conventional radiometric techniques. For the oldest *Mmc* midden, needles from *P. albicaulis* were dated by accelerator mass spectrometry. Radiocarbon age was then converted to thousands of calendar years (ka) before present (i.e., 1950) using the web converter maintained by R.G. Fairbanks, Lamont-Doherty Earth Observatory, Columbia University (http://radioca rbon.ldeo.columbia.edu/research/radcarbcal.htm). All radiocarbon analyses were performed by Beta Analytic International (Miami, Florida, USA) and included their conventional one sigma standard error.

Most locales consisted of a single midden, but two locales yielded 6 and 20 middens (midden codes "*Mmc*" and "*PhM*"). Middens spanned >20,000 calendar years for each of those two locales. In addition, six other locales yielded two middens each, but each of these locales did not cover as long a timeline, with the difference in calendar ages at any one locale averaging 360 yr (range: 7–1113). To facilitate interpretation of data, we grouped the 52 individual middens based on two factors: (1) whether the midden was either from the *Mmc* or *PhM* locales with the longest timelines (hereafter referred to as "multi-temporal locale") or from any of the other locales (hereafter, "uni-temporal locale") and (2) whether the midden was from a locale at low (<1500 m), mid (1500–1700 m)-, or high (>2100 m) elevation, which were elevation groups consistent with elevation groups based on modern vegetation assemblages. The two "multi-temporal" locales represent individual locations with multiple temporal samples that span a long timeline (>20,000 yr), whereas the "uni-temporal" locales represent individual locations with one to two middens that represent single temporal samples. Because topographic factors such as slope, elevation, and aspect modify climate, the two multitemporal locales are important controls for topographic effects on climate, and thus, any vegetation change observed through time is due to long-term climate shifts. In contrast, single temporal locales vary by slope, elevation, aspect, rock type, etc., and thus, long-term effects of climate change on vegetation assemblage may be confounded.

Midden vegetation analyses

Although some paleoecologists measure abundance of macrofossils in middens, we used presence/absence data in our analyses for three reasons. First, our midden sampling techniques yield reliable information both for presence and for absence of taxa in the local plant community (Nowak et al. 2000). Second, species richness (Coats et al. 2008) but not necessarily abundance of those species (Thompson 1985, 1990, Dial and Czaplewski 1990, Spaulding 1990) in the local plant community (i.e., within 100 m radius) is well represented. Third, presence/absence vegetation data have strong and stable correlations with environment (Wilson 2012). Thus, plant macrofossils extracted from a woodrat midden contain a relatively complete list of plants growing near the midden locale, yielding a snapshot of vegetation composition in space and time.

To analyze plant taxa present in each midden, we first used vegetation analysis techniques to group middens into CAs based upon plant taxa composition. We used two multivariate methods to group middens (hierarchical agglomerative cluster analysis and two-way indicator species analysis) and one multivariate method to ordinate middens (non-metric multi-dimensional scaling, NMDS). These analyses, as well as the indicator species analyses described below, used the software package PC-Ord V4 (MjM Software

Design, Gleneden Beach, Oregon, USA). Because infrequent species influence the reliability of inferences made during grouping or ordination analyses (McCune and Grace 2002, Azaele et al. 2010), we removed 44 infrequent taxa (i.e., taxa that occurred only once among all middens) before multivariate vegetation analysis. For cluster analysis, we followed the recommendations of McCune and Grace (2002) and used Ward's linkage method, which minimizes the increase in the error sum of squares during clustering, with Euclidean distance measures. The cluster analysis dendrogram was pruned to achieve the best compromise between within-group homogeneity and the number of ecologically meaningful groups (i.e., paleovegetation CAs in our case), taking into account criteria such as number of groups that is appropriate for the study, ecological interpretability of individual groups, and results from indicator species analyses at levels of pruning from 2 to 15 groups (McCune and Grace 2002). Two-way indicator species analysis, NMDS, and cluster analysis yielded similar groupings, and thus for brevity, we report only cluster analysis results in this paper.

To determine which individual taxa were associated with different CAs, we used indicator species analysis to determine whether the occurrence of individual taxa significantly differed among the CAs. Indicator species analysis detects the ability of different species to indicate a particular group (McCune and Grace 2002). A perfect indicator species should be present in all samples from a particular group and found exclusively in that group; thus, indicator values for each species range from zero (no indication) to 100 (perfect indicator). To determine whether a species has a significant indicator value ($P \leq 0.05$), species are randomly assigned to groups in 1000 Monte Carlo simulations, and the null hypothesis that the indicator value from the actual data does not exceed the indicator value by chance (i.e., from the 1000 Monte Carlo simulations) is tested.

Next, we examined vegetation composition based on plant functional types and used one-way analysis of variance to determine whether taxa richness for each functional type as well as total taxa richness differed among the four major climate periods defined in this paper. Plant functional types are a measure of community structure that is relatively independent of individual

species and thus provide a robust framework for estimating community and ecosystem responses to climate changes (Eronen et al. 2010, Polly et al. 2011, Blois et al. 2013). The functional structure of communities tends to respond consistently to disturbance (Mouillot et al. 2013), and functional types are recognizable at all scales of study, providing a dynamic understanding of community processes (Woodward et al. 2011). In semi-arid regions, plant functional types most often used are trees, shrubs, grasses, forbs, and annuals (Sala et al. 2011, Westoby and Leishman 2011). (Note: In this paper, the grass functional type includes grasses, sedges, and other graminoids.) Thus, our complete list of 154 taxa was used to compute total taxa richness as well as taxa richness for the major plant functional types of tree, shrub, grass, and forb. Because forbs and shrubs had the most number of taxa and because they are important ecological indicators in arid ecosystems, we also computed a forb/shrub taxa ratio for each midden to use as an index of relative change in abundance of herbaceous to woody taxa over time. Other combinations of plant functional types that we analyzed included total herbaceous taxa (forbs + grasses), total woody taxa (trees + shrubs), and total taxa (herbaceous + woody).

To determine whether the occurrence of individual taxa changed through time, we used indicator species analysis to determine whether the occurrence of individual taxa significantly differed among the four climate periods.

RESULTS

Paleovegetation community assemblages

Hierarchical agglomerative cluster analysis of 52 middens based on plant taxa composition of each midden identified six paleovegetation community assemblages (CA1–CA6; Fig. 3).

CA1 combined most of the youngest middens from uni-temporal locales located at both low and mid-elevations. Three of these 10 locales were unconsolidated middens and represented modern vegetation. CA1 contained taxa that today are common in the Great Basin cold desert scrub along with a high occurrence of saltbushes. For example, shrubs that are common in modern cold desert vegetation, such as *Artemisia tridentata*, *Brickellia*, *Chrysothamnus*, and *Tetradymia glabrata*, occurred in >60% of the middens in this

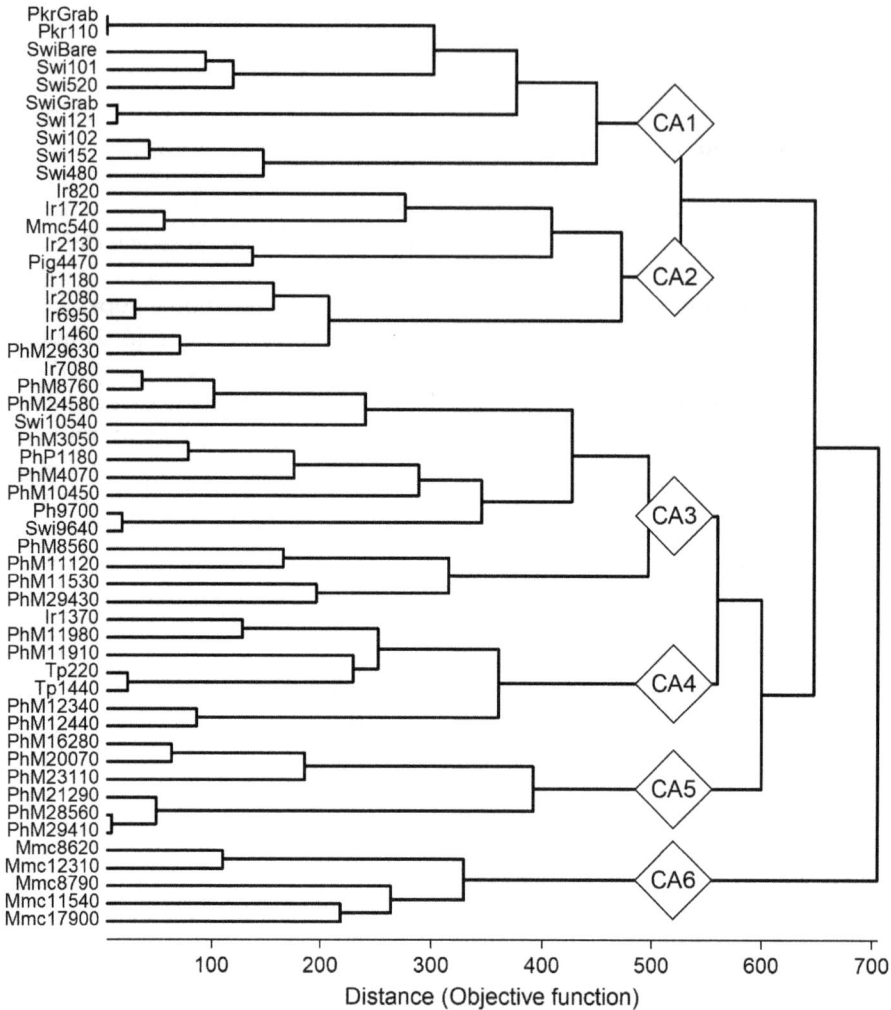

Fig. 3. Dendrogram from hierarchical agglomerative cluster analysis of 52 middens based on taxa composition of each midden. Six paleovegetation community assemblages (CA1–CA6) were identified based on 110 taxa that were present in two or more middens; 44 taxa that were present in only one midden were excluded from this analysis. Middens codes are shown in Appendix S1.

CA, and the saltbush *Atriplex confertifolia* was present in all CA1 middens (Table 1). One significant indicator species for CAs, *Grayia spinosa*, was present in 90% of CA1 middens. Two taxa recently introduced to North America, the grass *Bromus tectorum* and the forb *Erodium cicutarium*, were also significant indicator species that were found in this CA and no other. Another significant indicator species, the common cold desert forb *Chaenactis douglasii*, occurred in half of the CA1 middens. However, many forbs were absent from CA1 middens, including the forb indicator

species *Cryptantha watsonii*, *Galium*, and *Galium aparine*. Forb, herbaceous, and total taxa richness of CA1 was the second lowest of all six CAs (Table 2), and CA1 also had the second lowest forb/shrub and herb/woody ratios. The high occurrences of cold desert shrubs and saltbushes coupled with the low occurrences of many forbs suggest that CA1 is analogous to the cold desert–saltbush scrub CA that today occupies lower-elevation areas in the Great Basin.

Most of the CA2 middens (8 of 10) were from uni-temporal locales at mid-elevations (Fig. 3;

Table 1. Percentage of middens with an individual taxon within each of six paleovegetation CAs and results from indicator species analysis for taxa found in six CAs.

Plant functional type†	Taxon	CA1 (N = 10)‡	CA2 (N = 10)	CA3 (N = 14)	CA4 (N = 7)	CA5 (N = 6)	CA6 (N = 5)	P value
		Percentage of middens with taxon within each CA						
Tree	*Cercocarpus ledifolius*	0	0	0	14	0	20	0.386
Tree	*Juniperus osteosperma*	80	100	100	100	100	100	0.717
Tree	***Pinus albicaulis***	0	0	7	43	17	100	**<0.001**
Shrub	*Artemisia* sec. Tridentatae	100	100	93	100	83	100	0.505
Shrub	***Artemisia tridentata***	70	40	14	29	0	100	**0.004**
Shrub	*Atriplex canescens*	0	10	14	14	0	0	0.965
Shrub	*Atriplex confertifolia*	100	40	93	29	83	100	0.102
Shrub	*Brickellia*	80	70	57	57	50	80	0.858
Shrub	***Chrysothamnus***	80	50	29	86	0	100	**0.039**
Shrub	***Ephedra***	60	90	57	86	100	0	**0.020**
Shrub	*Ericameria nana*	70	40	29	57	0	60	0.292
Shrub	*Eriogonum heermannii*	10	0	7	14	0	0	0.909
Shrub	*Eriogonum microthecum*	0	0	0	29	0	20	0.117
Shrub	***Glossopetalon spinescens* var. *aridum***	0	10	43	0	0	20	**0.053**
Shrub	***Grayia spinosa***	90	60	14	0	0	80	**0.013**
Shrub	*Holodiscus discolor*	0	10	7	29	0	40	0.130
Shrub	*Kochia americana*	10	0	0	14	0	0	0.720
Shrub	*Linanthus pungens*	30	20	7	0	0	0	0.227
Shrub	*Picrothamnus desertorum*	20	0	0	0	0	0	0.105
Shrub	*Prunus andersonii*	0	10	14	0	0	0	0.615
Shrub	*Prunus emarginata*	0	0	7	14	0	0	0.577
Shrub	*Purshia tridentata*	10	50	29	0	0	40	0.186
Shrub	***Ribes***	0	0	36	100	0	60	**0.001**
Shrub	***Ribes velutinum***	0	0	29	86	0	60	**0.002**
Shrub	*Rosa woodsii*	10	0	7	14	0	0	0.902
Shrub	***Salvia dorrii***	50	100	64	29	67	20	**0.001**
Shrub	***Symphoricarpos***	20	10	29	57	33	80	**0.049**
Shrub	***Tetradymia canescens***	0	0	7	43	0	0	**0.009**
Shrub	*Tetradymia glabrata*	60	40	7	29	0	60	0.260
Grass	*Achnatherum hymenoides*	50	60	57	86	100	80	0.251
Grass	*Agropyron*	0	30	21	14	50	0	0.112
Grass	***Bromus tectorum***	50	0	0	0	0	0	**0.001**
Grass	***Deschampsia***	0	0	0	0	50	0	**0.001**
Grass	***Eleocharis quinqueflora***	0	10	14	0	100	0	**<0.001**
Grass	***Elymus elymoides***	40	20	43	29	0	100	**0.001**
Grass	*Hesperostipa comata*	0	30	36	0	0	20	0.317
Grass	*Pappostipa speciosa*	30	0	0	0	0	20	0.154
Grass	***Poa***	0	10	0	14	17	60	**0.010**
Grass	***Poa secunda* ssp. *secunda***	0	10	0	0	0	60	**0.001**
Grass	*Poaceae*	100	90	93	100	100	100	0.827
Grass	*Pseudoroegneria spicata*	0	10	7	14	0	0	0.911
Grass	*Stipa*	60	60	64	43	33	40	0.882
Forb	*Allium*	10	0	0	0	0	20	0.306
Forb	*Amaranthaceae*	30	10	93	86	100	60	0.060
Forb	*Amaranthus*	0	10	14	0	0	0	0.622
Forb	***Amsinckia tessellata***	80	70	100	43	50	100	**0.031**
Forb	*Arabis*	0	0	7	0	17	20	0.443
Forb	*Arenaria*	30	10	0	0	17	0	0.224
Forb	*Aster*	0	20	0	0	0	20	0.403
Forb	*Astragalus*	20	20	7	43	33	20	0.474

(Table 1. *Continued*)

Plant functional type†	Taxon	Percentage of middens with taxon within each CA						P value
		CA1 (N = 10)‡	CA2 (N = 10)	CA3 (N = 14)	CA4 (N = 7)	CA5 (N = 6)	CA6 (N = 5)	
Forb	Boraginaceae	90	100	100	86	100	100	0.434
Forb	Cactaceae	10	0	29	14	17	0	0.488
Forb	***Castilleja***	0	80	21	0	17	100	**0.001**
Forb	*Chaenactis*	60	10	43	14	0	60	0.208
Forb	***Chaenactis douglasii***	50	10	36	0	0	0	**0.046**
Forb	***Chaenactis stevoides***	0	0	7	14	0	60	**0.003**
Forb	*Chenopodium*	30	10	86	86	83	60	0.227
Forb	***Cirsium***	20	0	0	0	0	80	**<0.001**
Forb	*Claytonia*	10	10	0	0	0	0	1.000
Forb	*Cordylanthus*	0	10	14	14	0	0	0.960
Forb	***Crepis***	10	10	29	0	67	20	**0.014**
Forb	***Cryptantha ambigua***	0	0	0	0	0	40	**0.007**
Forb	*Cryptantha circumscissa*	10	50	7	0	50	20	0.192
Forb	*Cryptantha gracilis*	0	20	36	29	17	0	0.503
Forb	***Cryptantha muricata***	0	0	0	29	0	0	**0.033**
Forb	*Cryptantha nevadensis*	20	10	0	0	0	0	0.217
Forb	*Cryptantha pterocarya*	30	50	36	0	0	0	0.115
Forb	*Cryptantha torreyana*	10	20	0	0	0	0	0.229
Forb	***Cryptantha watsonii***	0	50	14	57	83	100	**0.013**
Forb	***Descurainia***	30	50	21	29	100	100	**0.031**
Forb	***Eatonella nivea***	10	50	29	14	83	100	**0.011**
Forb	*Eriogonum*	30	30	29	14	50	40	0.662
Forb	*Eriogonum cernuum*	0	0	21	0	17	0	0.273
Forb	*Eriophyllum lanatum*	0	0	14	14	0	0	0.765
Forb	***Erodium cicutarium***	40	0	0	0	0	0	**0.007**
Forb	*Euphorbia*	0	0	14	0	0	0	0.223
Forb	Euphorbiaceae	0	0	21	0	0	0	0.093
Forb	Fabaceae	70	50	14	86	67	80	0.337
Forb	***Galium***	0	20	79	14	17	80	**0.019**
Forb	***Galium aparine***	0	10	57	14	17	80	**0.005**
Forb	*Gilia*	20	10	0	0	0	0	0.221
Forb	*Grindelia*	0	0	14	0	0	0	0.222
Forb	*Helianthus*	0	0	21	0	0	0	0.101
Forb	*Hypericum*	0	0	7	0	17	0	0.364
Forb	Lamiaceae	0	0	0	14	0	20	0.385
Forb	***Lappula occidentalis var. occidentalis***	0	0	21	0	50	0	**0.013**
Forb	*Lepidium*	0	0	14	0	0	0	0.221
Forb	*Lupinus*	40	30	7	29	0	60	0.121
Forb	***Mentzelia albicaulis***	30	70	86	0	83	100	**0.022**
Forb	***Montia***	0	0	0	29	0	0	**0.036**
Forb	*Oenothera*	30	0	7	0	17	20	0.457
Forb	Onagraceae	30	10	7	0	33	20	0.533
Forb	***Orobanche corymbosa***	0	30	7	0	0	80	**0.001**
Forb	***Phacelia***	30	50	71	0	100	20	**0.001**
Forb	*Phacelia crenulata*	30	0	21	0	17	0	0.319
Forb	***Phacelia glandlifera***	0	0	7	0	83	0	**<0.001**
Forb	***Phacelia humilis***	0	20	36	0	83	0	**0.001**
Forb	***Phlox***	0	0	0	0	0	40	**0.007**
Forb	*Plagiobothrys kingii*	0	30	43	43	83	80	0.092
Forb	***Polygonom minimum***	0	0	0	0	33	0	**0.018**
Forb	***Polygonum***	0	10	0	14	83	0	**<0.001**

(Table 1. *Continued*)

Plant functional type†	Taxon	Percentage of middens with taxon within each CA						P value
		CA1 (N = 10)‡	CA2 (N = 10)	CA3 (N = 14)	CA4 (N = 7)	CA5 (N = 6)	CA6 (N = 5)	
Forb	*Potentilla*	0	30	0	0	0	60	0.007
Forb	Saxifragaceae	0	10	0	0	17	0	0.509
Forb	*Scrophularia desertorum*	0	10	0	0	0	80	<0.001
Forb	**Scrophulariaceae**	0	80	43	29	50	100	0.013
Forb	*Senecio*	0	0	0	0	0	100	<0.001
Forb	*Stephanomeria*	30	0	7	0	0	0	0.063
Forb	*Thelypodium*	10	0	0	0	0	40	0.017
Forb	*Trifolium*	0	10	0	0	67	0	<0.001
Forb	**Bryophyte**	0	0	0	0	0	40	0.007

Notes: The six community assemblages (CAs) were delineated by cluster analysis of 52 middens (Fig. 3) from the Painted Hills area near Pyramid Lake, northwestern Nevada, USA. Results are only for the 110 taxa that occurred in more than one midden. Significant indicator taxa from the Monte Carlo indicator species test ($P \leq 0.05$) are in bold text.

† For ease of locating a particular taxon, taxa are sorted by plant functional type and then alphabetically by name within each functional type.

‡ Number of middens (N) for each CA.

Appendix S1). The two remaining middens in CA2 were the most recent midden from the *Mmc* mid-elevation multi-temporal locale and the oldest midden from the *PhM* low-elevation multi-temporal locale. The percentages of CA2 middens that contained common cold desert shrubs were generally smaller than those of CA1 middens except for the significant shrub indicator species *Salvia dorrii*. All CA2 middens contained

S. dorrii (Table 1). The only other significant indicator species for CAs that had high occurrence in CA2 was the forb *Castilleja*. Taxa richness of each plant functional type for CA2 as well as for forb/shrub and herb/woody ratios was not significantly different between CA1 and CA2 (Table 2). We also note that the dendrogram from the cluster analysis (Fig. 3) has CA1 and CA2 on the same branch. Thus, CA2 is similar to CA1 except

Table 2. Comparison of taxa richness within six paleovegetation CAs.

Functional type	CAs†						P value
	CA1 (cold desert–saltbush scrub)	CA2 (cold desert–*Salvia* scrub)	CA3 (mixed shrub steppe)	CA4 (mixed montane shrubland)	CA5 (herbaceous meadow)	CA6 (*Pinus albicaulis*–mixed shrub woodland)	
Forb	$10.2^A \pm 1.2$	$12.5^{A,B} \pm 1.2$	$14.4^B \pm 1.0$	$9.1^A \pm 1.4$	$18.7^C \pm 1.6$	$24.2^D \pm 1.7$	<0.001
Shrub	$8.8^{C,D} \pm 0.6$	$7.5^{B,C} \pm 0.6$	$6.9^B \pm 0.5$	$9.0^{C,D} \pm 0.7$	$4.2^A \pm 0.8$	$10.2^D \pm 0.8$	<0.001
Forb/shrub ratio	$1.2^A \pm 0.2$	$1.7^{A,B} \pm 0.2$	$2.3^B \pm 0.2$	$1.0^A \pm 0.3$	$4.7^C \pm 0.3$	$2.4^B \pm 0.3$	<0.001
Grass	3.5 ± 0.4	3.5 ± 0.4	3.4 ± 0.4	3.0 ± 0.5	4.5 ± 0.5	5.0 ± 0.6	0.070
Tree	$0.8^A \pm 0.1$	$1.0^A \pm 0.1$	$1.1^A \pm 0.1$	$1.9^B \pm 0.2$	$1.2^A \pm 0.2$	$2.2^B \pm 0.2$	<0.001
Total taxa	$23.3^A \pm 1.8$	$24.3^A \pm 1.8$	$25.9^A \pm 1.5$	$23.0^A \pm 2.1$	$28.5^A \pm 2.3$	$41.6^B \pm 2.5$	<0.001
Herbaceous‡	$13.7^A \pm 1.4$	$15.8^{A,B} \pm 1.4$	$17.8^B \pm 1.2$	$12.1^A \pm 1.7$	$23.2^C \pm 1.8$	$29.2^D \pm 2.0$	<0.001
Woody§	$9.0^{B,C} \pm 0.7$	$8.5^B \pm 0.7$	$8.1^B \pm 0.6$	$10.9^{C,D} \pm 0.8$	$5.3^A \pm 0.8$	$12.4^D \pm 0.9$	<0.001
Herb/woody ratio¶	$1.5^{A,B} \pm 0.2$	$1.9^{B,C} \pm 0.2$	$2.4^C \pm 0.2$	$1.1^A \pm 0.3$	$4.9^D \pm 0.3$	$2.4^C \pm 0.3$	<0.001

Notes: Mean values (with standard error) are for all middens within each community assemblage (CA). The complete list of 154 taxa was used to compute taxa richness. P values from one-way analysis of variance are shown. Means with different uppercase letters within each row are significantly different.

† CAs were delineated by cluster analysis of 52 middens (Fig. 3) from the Painted Hills area near Pyramid Lake, northwestern Nevada, USA.

‡ Herbaceous richness is forb + grass richness.

§ Woody richness is shrub + tree richness.

¶ Ratio of herbaceous to woody taxa.

that high *Salvia* occurrence replaces high saltbush occurrence, suggesting that CA2 is analogous to a cold desert–*Salvia* scrub CA that today occupies slightly higher elevations than the CA1 cold desert–saltbush communities in the Great Basin.

CA3 had the most middens (14) and included middens from both low- and mid-elevation locales and from both uni-temporal and multi-temporal locales (Fig. 3; Appendix S1). Nine middens in CA3 were from the *PhM* low-elevation multi-temporal locale. The other five middens were from uni-temporal locales: two from low elevation and three from mid-elevation. One significant shrub indicator species for CAs, *Glossopetalon spinescens* var. *aridum*, and two significant indicator forbs, *Amsinckia tessellata* and *Galium*, had their highest occurrence in CA3 middens (Table 1). Common cold desert shrubs typically occurred with similar or smaller frequency in CA3 than in CA1 and CA2, except for the more mesic shrub taxa *Ribes* and *Ribes velutinum*. However, these two more mesic shrub taxa along with another mesic shrub *Symphoricarpos* occurred less frequently in CA3 than in CA4. CA3 had significantly greater forb richness than both CA1 and CA4 but significantly smaller shrub richness (Table 2). Consequently, CA3 had a significantly greater forb/shrub ratio than both CA1 and CA4, even though total taxa richness was not significantly different among CA1, CA3, and CA4. Thus, CA3 is intermediate in taxa composition between CA1 and CA4, indicating a mixed montane–desert shrub steppe, that is, a mix of forbs and shrubs from both montane and cold desert modern plant communities.

CA4 grouped the three middens from high-elevation uni-temporal locales with four middens from the *PhM* low-elevation multi-temporal locale (Fig. 3; Appendix S1). *Pinus albicaulis*, a significant tree indicator species for CAs, was present in almost half of these middens (Table 1). Three significant shrub indicator species had their greatest occurrence in CA4: *Ribes*, *R. velutinum*, and *Tetradymia canescens*. Most other shrubs occurred in CA4 at a similar or slightly greater frequency than CA3. Although two significant forb indicator species (*Cryptantha muricata* and *Montia*) only occurred in CA4, most other forbs occurred in CA4 middens at a similar or smaller frequency than in CA3 middens. Thus, CA4 had significantly greater tree, shrub, and

woody richness than CA3 (Table 2), but significantly smaller forb and herbaceous richness as well as significantly smaller forb/shrub ratio. In fact, the forb/shrub (and herb/woody) ratio was the smallest of all CAs. Thus, the high occurrence of montane shrub taxa in CA4 indicates a mixed montane shrubland.

CA5 consisted of six middens that were only from the *PhM* low-elevation multi-temporal locale (Fig. 3; Appendix S1). Eleven significant herbaceous indicator species for CAs had their highest occurrence or only occurrence in CA5: two grass (*Deschampsia* and *Eleocharis quinqueflora*) and nine forbs (*Crepis*, *Descurainia*, *Lappula occidentalis* var. *occidentalis*, *Phacelia*, *Phacelia glandulifera*, *Phacelia humilis*, *Polygonom*, *Polygonom minimum*, and *Trifolium*; Table 1). Three significant indicator species were conspicuously absent from CA5: the two cold desert shrubs *A. tridentata* and *Chrysothamnus* and the common cold desert bunchgrass *Elymus elymoides*. In addition, five other significant shrub indicator species were also absent from all CA5 middens. Overall, CA5 had significantly smaller shrub richness than all the other five CAs (Table 2), significantly smaller woody richness than all other CAs, significantly greater forb richness than all CAs except CA6, and significantly greater forb/shrub (and herb/woody) ratio than all other CAs. The abundance of forb taxa and sparse occurrence of woody taxa coupled with predominantly wet meadow taxa such as *Deschampsia* and *E. quinqueflora* indicate predominantly herbaceous meadow plant community that today are found in some of the most mesic areas of the Great Basin.

The last cluster group, CA6, had the fewest middens (5), and all middens were from the *Mmc* mid-elevation multi-temporal locale (Fig. 3; Appendix S1). A consistent component of CA6 was the presence of *P. albicaulis*, which is a significant tree indicator species for CAs (Table 1) and today is a high-elevation conifer. Many shrub taxa had high occurrences in CA6, including significant indicator shrubs such as the cold desert *A. tridentata* and *Chrysothamnus*, saltbushes such as *A. confertifolia* and *G. spinosa*, and more mesic montane shrubs such as *Ribes*, *R. velutinum*, and *Symphoricarpos*. Twenty-two significant herbaceous indicator species had their highest occurrence in CA6, including the grasses *Poa*, *Poa secunda* ssp. *secunda*, and *E. elymoides* and the forbs *Amsinckia tessellata, Castilleja, Chaenactis stevioides, Cirsium, Cryptantha ambigua, C. watsonii, Descurainia, Eatonella nivea, Galium, G. aparine, Mentzelia albicaulis, Orobanche corymbosa, Phlox, Potentilla, Scrophularia desertorum*, Scrophulariaceae, *Senecio, Thelypodium*, and Bryophyte. Not surprisingly given these high occurrences of many taxa, CA6 had significantly greater forb, herbaceous, and total taxa richness than all other CAs (Table 2). CA6 also had the greatest shrub, tree, and woody richness, although richness of these plant functional types was not always significantly greater than other CAs. The consistent occurrence of *P. albicaulis*, high occurrences of both montane and cold desert shrubs, and high occurrences of forbs suggest a *P. albicaulis*–mixed shrub woodland.

Change in community assemblage through time

To place trends in species and vegetation assemblages in a climate context, we plotted timelines of well-established climatic periods from local climate proxy data (see *Summary of climate change* subsection at end of *Introduction*) in Fig. 4. When CA results from cluster analysis are overlaid with the climate timeline (Fig. 4), changes in CA are largely concordant with major changes in climate but do not always respond to each climate cycle. Our longest vegetation record is from the *PhM* low-elevation multi-temporal locale (Fig. 4: filled circles connected by solid line). Fossil CA at *PhM* starts ~35 ka as CA2 (cold desert–*Salvia* scrub), which was just after a time period when Pluvial Lake Lahontan was relatively shallow (~35.5 ka). As time moves toward the present, CA rapidly changes to CA3 (mixed shrub steppe) and then to CA5 (herbaceous meadow) in ~250 yr. These CA changes follow a similarly rapid increase in lake level, indicating a climate shift to greater water availability. Community assemblage changes back to CA3 coincident with a shallow lake level by ~29 ka, and then back to CA5 within 1500 yr and stays as CA5 as lake level increased up to LGM. Unfortunately, we have not been able to find middens at *PhM* dated to the period of maximum glaciation, but middens before (24.0 ka) and after (19.4 ka) LGM had CA5 (herbaceous meadow). By 14.3 ka, which was after the Bolling-Allerod rapid warming began and Pluvial Lake Lahontan had reached a relatively shallow lake level, CA had changed to CA4

(mixed montane shrubland; Fig. 4). CA4 persisted for at least 600 yr at this low-elevation locale, but by 1300 yr after the Bolling-Allerod rapid warming began (i.e., about two-thirds through the Bolling-Allerod), CA changed to CA3 (mixed shrub steppe). CA3 persisted at that locale through the end of the Bolling-Allerod warming, through the cold Younger Dryas, and through the Mid-Holocene Temperature Maximum until the end of its fossil record at 3.3 ka.

Community assemblages from the uni-temporal locales at low and mid-elevations generally changed almost unidirectionally since ~12 ka from CA3 to CA2 to CA1 (Fig. 4: open circles and squares), even though climate cycled between cold/wet and warm/dry during the same time period. Prior to the Mid-Holocene Temperature Maximum (7.5–5.5 ka), most CAs at low- and mid-elevation locales were CA3 (mixed shrub steppe). After the Mid-Holocene Temperature Maximum, most CAs at both low and mid-elevations changed to CA2 (cold desert–*Salvia* scrub) and persisted as CA2 through the Medieval Climate Anomaly except for one mid-elevation locale that had a CA of CA3 near the beginning of the Medieval Climate Anomaly (Fig. 4 inset). The final change in CA at low- and mid-elevation locales occurred following the Medieval Climate Anomaly (~1.0 ka), and all low- and mid-elevation middens had CA1 (cold desert–saltbush scrub) through the Little Ice Age and into the present. Interestingly, CAs for middens at high-elevation locales (Fig. 4 inset: open diamonds), which are only dated between 0.2 and 1.3 ka, are CA4 (mixed montane shrubland), which also occurred ~14 ka at the lowest elevation.

Our *Mmc* mid-elevation multi-temporal locale had a different assemblage of plants than the low-elevation multi-temporal locale as well as other locales in this study area (Fig. 4: filled squares connected by solid line). This *Mmc* mid-elevation multi-temporal locale differs from the *PhM* low-elevation multi-temporal locale in that the mid-elevation *Mmc* locale maintains the same CA for a longer period of time during the Pleistocene–Holocene transition, that is, CA6 (*P. albicaulis*–mixed shrub woodland) from its oldest midden during the middle of LGM through the Bolling-Allerod rapid warming until the beginning of the Holocene. Although CA appeared to persist at this mid-elevation multi-temporal

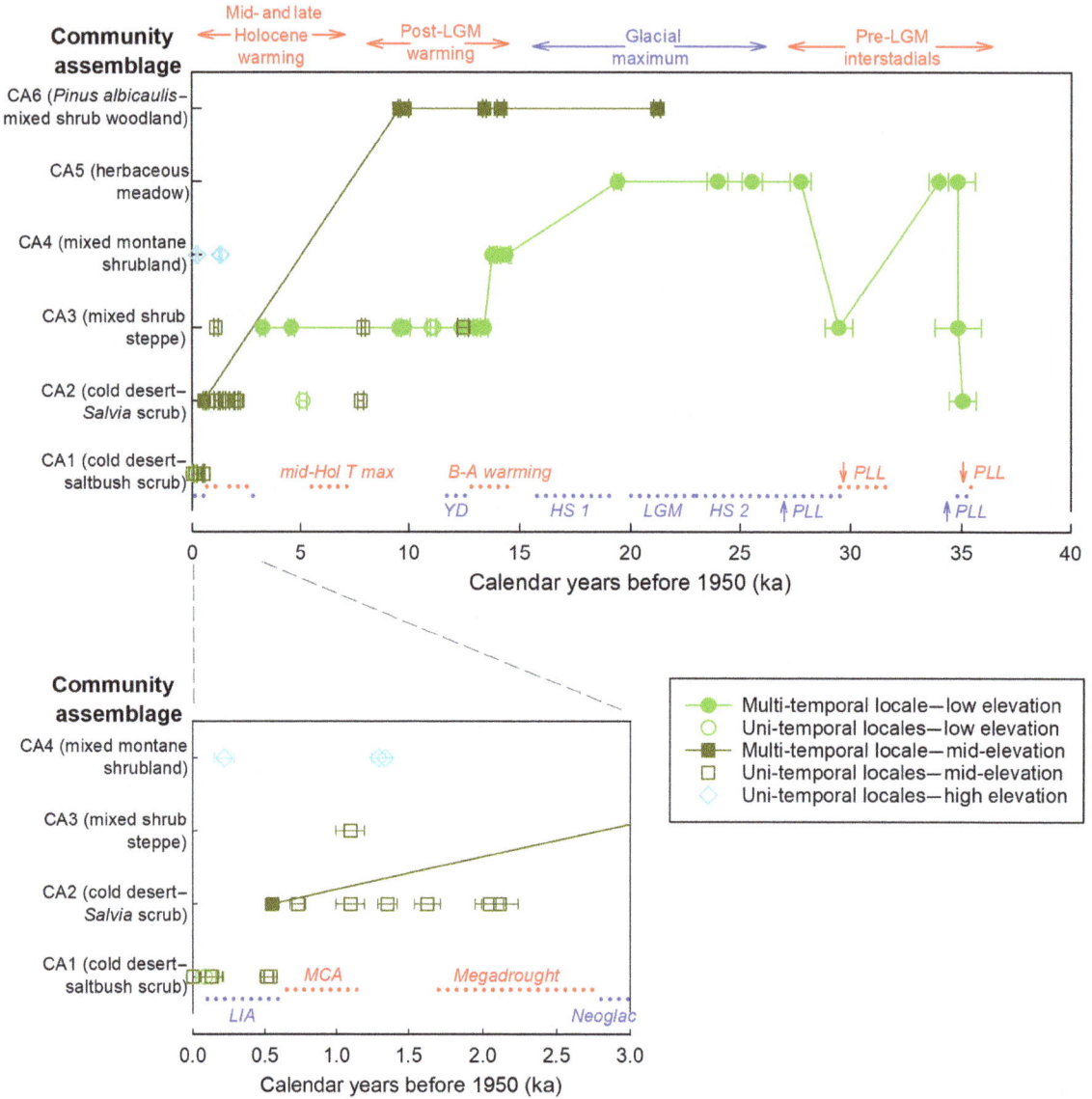

Fig. 4. Change in community assemblages over the last 35,000 yr (upper panel) and over the last 3000 yr (lower inset panel), along with information on climate change. Different symbol shapes and colors indicate elevation groups: low (green circles), mid (dark-yellow squares)-, or high (cyan diamonds) elevations. Different symbol fills indicate locale group: multi-temporal locales with >2 middens that spanned >20,000 yr (filled symbols connected by line) or uni-temporal locales with one to two middens (open symbols). Error bars are 1-sigma range in age from radiocarbon analyses. Climate information is shown in two ways. First along the top of the upper panel are shown four major climate periods (red and blue text and associated horizontal arrows; red indicating predominantly warm/dry climate and blue indicating predominantly cold/wet climate) that are used in this paper. Second, select climate periods that occur locally are shown near the x-axis as horizontal dotted lines (red indicating warm/dry climate; blue indicating cold/wet climate). These climate periods include (ordered from oldest to youngest): in the upper panel, shallow lake level of Pluvial Lake Lahontan (↓PLL), rising lake level (↑PLL), shallow lake level (↓PLL), rising lake level (↑PLL), Heinrich Stadial 2 (HS 2), Last Glacial Maximum (LGM), HS 1, Bolling-Allerod rapid warming (B-A warming), Younger Dryas (YD), and Mid-Holocene Temperature Maximum (mid-Hol T max); and in the inset panel, Neoglacial (Neoglac), late Holocene Megadrought (Megadrought), Medieval Climate Anomaly (MCA), and Little Ice Age (LIA).

locale through major climate change that occurred during the post-LGM rapid warming climate period, ultimately the CA succumbed to Holocene warming and increased aridity: The last midden from this locale, dated near the beginning of the Little Ice Age ~600 yr ago, was CA2 (cold desert–*Salvia* scrub).

Change in plant functional types through time

For all plant functional types, forbs and shrubs had the most number of taxa in the data set, and the number of taxa within the forb and shrub functional types was significantly different among the four major climate periods during the last 35,000 yr (Table 3). Forb taxa richness was greatest during the glacial maximum climate period (15–26 ka) and was smaller both in more recent and in older climate periods (Table 3; Appendix S2). Forb richness at its peak during glacial maximum was almost twice that (and significantly greater than) in the mid- and late Holocene warming climate period (<8 ka) and was ~50% (and significantly) greater than the oldest climate period, the pre-LGM interstadials climate period (26–35 ka). In contrast for shrubs, shrub richness was greatest during the post-LGM

warming climate period (8–15 ka), with only a small (and not significant) decline during mid- and late Holocene warming (Table 3; Appendix S2). Peak shrub taxa richness during post-LGM warming was significantly greater than the two previous climate periods (glacial maximum and pre-LGM interstadials).

Similar to shrubs, tree richness also was greatest during post-LGM warming (Table 3), but unlike shrubs, tree richness significantly declined during mid- and late Holocene warming. Total taxa richness was greatest during glacial maximum (Table 3; Appendix S2), but total richness during glacial maximum was not significantly different than that during post-LGM warming. However, total richness during mid- and late Holocene warming was significantly smaller than that of the previous two climate periods. Grass richness was not significantly different among the four climate periods.

The ratio of forb richness to shrub richness during glacial maximum was significantly greater than that of any of the other three climate periods (Table 3). These differences in forb/shrub ratio among climate periods occur because forb and shrub richness change relatively independent of

Table 3. Comparison of taxa richness within each plant functional type among four climate periods during the last 35,000 yr.

| Functional type | Climate period† | | | | P value |
	Mid- and late Holocene warming (<8 ka)	Post-LGM warming (8–15 ka)	Glacial maximum (15–26 ka)	Pre-LGM interstadials (26–35 ka)	
Forb	$11.5^A \pm 0.9$	$16.1^{B,C} \pm 1.3$	$21.2^C \pm 2.8$	$14.3^{A,B} \pm 2.6$	0.002
Shrub	$7.8^{B,C} \pm 0.4$	$8.9^C \pm 0.5$	$5.8^{A,B} \pm 2.4$	$5.3^A \pm 0.8$	0.005
Forb/shrub ratio	$1.6^A \pm 0.2$	$1.9^{A,B} \pm 0.2$	$4.8^C \pm 0.9$	$2.8^B \pm 0.4$	<0.001
Grass	3.2 ± 0.3	3.9 ± 0.3	4.0 ± 0.6	4.5 ± 0.4	0.124
Tree	$1.0^A \pm 0.1$	$1.7^B \pm 0.2$	$1.2^{A,B} \pm 0.2$	$1.3^{A,B} \pm 0.2$	0.001
Total taxa	$23.4^A \pm 1.3$	$30.5^B \pm 1.7$	$32.2^B \pm 3.4$	$25.5^{A,B} \pm 2.8$	0.006
Herbaceous‡	$14.7^A \pm 1.1$	$19.9^B \pm 1.5$	$25.2^B \pm 2.9$	$18.8^{A,B} \pm 2.4$	0.002
Woody§	$8.7^A \pm 0.5$	$10.6^B \pm 0.6$	$7.0^A \pm 1.2$	$6.7^A \pm 1.0$	0.004
Herb/woody ratio¶	$1.8^A \pm 0.2$	$2.0^A \pm 0.2$	$4.5^C \pm 0.4$	$2.9^B \pm 0.4$	<0.001

Notes: Mean values (with standard error) are for all middens within each climate period. The complete list of 154 taxa was used to compute taxa richness. *P* values from one-way analysis of variance are shown. Means with different uppercase letters within each row are significantly different. LGM, Last Glacial Maximum.

† Four climate periods are as follows: (1) mid- and late Holocene warming, which begins ~8 ka with rapid increase in temperature up to the Mid-Holocene Temperature Maximum and ends at present; (2) post-LGM warming, which begins with the Bolling-Allerod rapid warming ~15 ka and ends after continental ice sheets have completely melted ~8 ka; (3) cold period just before, during, and after glacial maximum, which begins ~26 ka with the Heinrich Stadial 2 (HS 2) cooling and ends before the Bolling-Allerod rapid warming ~15 ka; and (4) the last series of warm interstadials before the LGM, which includes the beginning of our vegetation record at ~35 ka that coincides with a very warm interstadial to before the HS 2 rapid cooling ~26 ka.

‡ Herbaceous richness is forb + grass richness.

§ Woody richness is shrub + tree richness.

¶ Ratio of herbaceous to woody taxa.

each other (Table 3; Appendix S2). Thus, the forb/shrub ratio is a reliable indicator of the herbaceous composition of CAs. The use of forb/shrub ratio as a measure of relative herbaceous composition is supported by other results (Table 3). First, changes in herbaceous and woody richness among climate periods were similar to those for forbs and shrubs, respectively. Second, changes in the herb/woody ratio were similar to those of the forb/shrub ratio.

Change in individual species through time

We used indicator species analyses to determine whether the occurrence of individual taxa significantly differed among the four major climate periods. Below, we present results for those taxa that were significant indicator species; results for all taxa are given in Appendix S3. Results are organized first by plant functional type and then in time order from oldest to youngest (i.e., those taxa more frequent in the pre-LGM interstadials, then those more frequent during glacial maximum). We also include observations on taxa that commonly occurred among all four climate periods (i.e., taxon was present in 33% or more of the middens within all four climate periods).

Eleven of 68 forb taxa were significant indicator species for climate periods, and eight forb taxa commonly occurred among all four major climate periods (Appendix S3). Two significant indicator forbs (*Crepis* and *Hypericum*) had greatest occurrence during the pre-LGM interstadials, and *Hypericum* was present only during that climate period. Seven significant indicator forbs (*Arabis*, *C. watsonii*, *Descurainia*, *Phacelia humulis*, *Plagiobothrys kingii*, *Polygonom*, and *P. minimum*) had greatest occurrence during glacial maximum, with *P. minimum* present only during glacial maximum. One significant indicator forb (*Trifolium*) occurred with equal frequency in both pre-LGM interstadials and glacial maximum and did not occur in any middens younger than 15 ka. The last significant indicator forb (*Galium aparine*) had greatest occurrence during post-LGM warming. None of the significant indicator forbs had greatest occurrence during mid- and late Holocene warming. Eight forb taxa that commonly occurred among all four major climate periods were Amaranthaceae, *A. tessellata*, Boraginaceae, *Chenopodium*, Fabaceae, *Mentzelia albicaulis*, *Phacelia*, and Scrophulariaceae.

Although no shrubs were significant indicator species for climate periods at $P \leq 0.05$, five shrubs were significant indicator species at $P \leq 0.10$ (Appendix S3). Four of these shrubs (*G. spinescens* var. *aridum*, *Holodiscus discolor*, *Ribes*, and *Symphoricarpus*) had greatest occurrence during post-LGM warming, and the first two taxa were present only in middens <15 ka. The last significant indicator shrub (*Artemisia* sec. Tridentatae) occurred in 75% or more of middens among all four major climate periods but had it lowest occurrence during glacial maximum. Three shrub taxa (*A. confertifolia*, *Brickellia*, and *Ephedra*) were commonly found in middens from all four time periods. We also note that in addition to two significant indicator shrubs (*G. spinescens* var. *aridum* and *H. discolor*), eight shrub taxa (*Atriplex canescens*, *Eriogonum heermannii*, *Kochia americana*, *Linanthus pungens*, *Picrothamnus desertorum*, *Prunus andersonii*, *Rose woodsia*, and *Tetradymia canescens*) were only present in middens <15 ka.

Two of 13 grass taxa were significant indicator species for climate periods (Appendix S3). One significant indicator species (*E. quinqueflora*) had high occurrence during both pre-LGM interstadials and glacial maximum, but only occurred in one midden after glacial maximum. The other significant indicator grass (*E. elymoides*) had greatest occurrence during post-LGM warming. Two grass taxa (*Achnatherum hymenoides* and Poaceae) commonly occurred among all four time periods. We also note that *Deschampsia* occurred only in middens >15 ka, that *Pappostipa speciosa* and *Pseudoroegneria spicata* occurred only in middens <15 ka, and that *Stipa* was conspicuously absent from the glacial maximum climate period.

Although no tree taxa were significant indicator species for climate periods (Appendix S3), two trees (*Cercocarpus ledifolius* and *P. albicaulis*) did not occur in any middens < 8 ka. *Pinus albicaulis* had greatest occurrence during post-LGM warming, and *C. ledifolius* occurred only during that climate period. Finally, *Juniperus osteosperma* is notable in that it was present in all but two middens and that it was the only tree taxon present in middens during mid- and late Holocene warming.

Discussion

A distinctive strength of our paleoecological study is the high density of temporal and spatial samples within a relatively small geographic area: 52 middens that span the last 35,000 yr from 22 locales along an elevation gradient. This high sample density provides a high degree of temporal and spatial continuity to our results, which in turn facilitates inferences on vegetation responses to climate changes. Although paleoecological changes in individual species and in vegetation assemblage have been reported in the Great Basin (Thompson 1990, Nowak et al. 1994a, b), this paper is the first to report changes in CAs over an elevation gradient at one study area and to report changes in plant functional types in the Great Basin. Our use of vegetation analysis techniques to define CAs is especially revealing because it demonstrates the extent that plant communities re-occur through time and across space. Our results consistently indicate directional changes in CAs, reduced proportions of forbs in the vegetation, and more arid species in the vegetation since glacial maximum.

Past vegetation change: community assemblages

The overall pattern of paleovegetation CA change in our study area since glacial maximum generally represents a change to more arid CAs (Fig. 4). This pattern occurred for middens from both multi-temporal locales as well as for uni-temporal locales within each elevation group. CA5, which is analogous to a herbaceous meadow (i.e., one of the most mesic CAs in modern Great Basin vegetation), occurred at the *PhM* low-elevation multi-temporal locale during glacial maximum. A CA with montane shrubs (CA4) quickly followed the beginning of the Bolling-Allerod rapid warming at ~14.7 ka and in turn was quickly replaced by a slightly more arid CA with fewer montane but more cold desert shrubs (CA3). This mixed shrub steppe (CA3) occurred at most low- and mid-elevation locales until mid- and late Holocene warming. An even more arid community assemblage (CA2, cold desert–*Salvia* scrub) begins to occur at low and mid-elevations soon after the beginning of mid- and late Holocene warming, and only the most arid community assemblage (CA1, cold desert–saltbush scrub) was present at low- and mid-elevation locales (except for the *Mmc* mid-elevation multi-temporal locale) following the locally severe droughts that occurred before and during the Medieval Climate Anomaly (Mensing et al. 2004, 2008).

An additional pattern was apparent in paleovegetation CAs: Conspicuous recurrences in vegetation did occur with climate cycles prior to glacial maximum, but vegetation recurrences were rare after glacial maximum despite climate cycles. We speculate that the relatively rapid rate and extended longevity of warm/dry periods during the last 15 ka, especially during the Mid-Holocene Temperature Maximum, resulted in an ecological threshold where the capacity of species and communities to buffer the effects of environmental change was exceeded (Briske et al. 2006). Local extinction of mesic species during the last 15 ka is consistent with this concept that ecological aridity thresholds were exceeded in our study area. Although lag time in adjustment of taxa's geographic ranges to climate change may also explain the more recent lack of vegetation recurrences, we note that at ~35 ka, CAs changed from CA2 to CA3 to CA5 in ~250 yr following a similarly rapid change in local climate, suggesting that lag time is not sufficient to account for the lack of vegetation recurrences during the last 15 ka. Unfortunately, midden data do not provide sufficient evidence to account for another alternative factor: more rapid climate variations during the last 15 ka as compared to pre-LGM. Vegetation samples from middens have an inherent time sensitivity. Each midden represents the accumulation of plants over a time span of decades (Thompson 1985, 1990), and thus, changes in vegetation assemblages with decadal climate cycles cannot be distinguished. Nonetheless, midden samples would contain plants that were present during all extremes of decadal-level climate cycles and thus provide a reliable long-term (centennial- to millennial-scale) measure of vegetation as species migrate into the area, persist through time, and are locally extirpated.

Although our results are consistent with increased aridity as the factor driving long-term vegetation changes, we cannot conclusively rule out that other factors, such as changes in climate seasonality, herbivory patterns, or fire cycles, may also have influenced vegetation. However, these other factors typically affect plant abundance rather than cause permanent extirpation. Although these other factors likely caused changes in plant abundance in our study area over the last 35 ka, we cannot detect plant abundance changes from our midden vegetation record, only presence/absence (Thompson 1985, 1990, Dial and Czaplewski 1990, Spaulding 1990, Nowak et al. 2000, Coats et al. 2008). Nonetheless, plant distribution is ultimately tied to climate (Woodward 1987), and thus, our presence/absence vegetation record from middens reliably reflects over-riding climate trends.

Microclimate influence on plant assemblage was evident at the *Mmc* mid-elevation multitemporal locale. All middens from *Mmc* except for the most recent midden were the only members of CA6 (Figs. 3, 4). CA6 has the greatest total taxa diversity as well as among the greatest diversity for each major functional type (Table 2). These middens were in a small cave located at the base of a large rock face over 30 m high (Fig. 2a). Snow cornicing over this rock face likely provided significant water runoff throughout the midden record, which in turn resulted in higher available soil water around the cave than was present at other sites, thus supporting more mesic plants, such as *Pinus albicaulis*, and greater plant diversity. We have observed this snow cornicing and subsequent water harvesting in the Sierra Nevada Mountains and at other montane locations, even at lower-elevation mountains during heavy snow years when snow cornices and snowfields persist until mid-summer, allowing melt water to feed the local vegetation. Even though the microclimate influence affected the CA at *Mmc*, vegetation changes at *Mmc* followed the general pattern as other locales in the study area, such as preferential loss of forbs with increasing aridity, but with delayed timing.

Past vegetation change: plant functional types

The effects of climate change during the last 35,000 yr on forb richness differed from those on shrubs and indicate a preferential loss of forbs with increased aridity. Forbs were most prevalent and comprised the greatest proportion of the plant community during glacial maximum (Table 3; Appendix S2), but forb richness decreased 24% during post-LGM warming with an additional 29% decrease during mid- and late Holocene warming. In contrast, shrub richness was low during both the pre-LGM interstadials and glacial maximum, but then increased 61% during post-LGM warming, followed by a non-significant decline (12%) during mid- and late Holocene warming (Table 3). Consequently, the forb/shrub

ratio declined 61% from glacial maximum to post-LGM warming due to both a decrease in forb richness and an increase in shrub richness, then decreased an additional 17% from post-LGM warming to mid- and late Holocene warming because of a relatively greater decrease in forb than in shrub richness. Differences in forb richness and forb/shrub ratio between the two oldest climate periods (Table 3) are consistent with preferential loss of forbs with increasing aridity. Going back in time from glacial maximum to pre-LGM interstadials represents a climate change from cool/wet to warm/dry, and forb/shrub ratio decreased. Analyses of all herbaceous (forb + grass) and all woody (shrub + tree) taxa had similar results as those for forbs and shrubs (Table 3).

Overall, these results indicate that forbs responded positively during climate cooling that occurred before and during LGM, but shrub richness was not significantly affected. In contrast, the climate change from relatively cold during glacial maximum to relatively warm during post-LGM warming was large enough to allow establishment of additional woody taxa in the study area, including trees, but a loss in herbaceous taxa also occurred. However, as climate became more arid during mid- and late Holocene warming and was marked by extended drought periods up to the present, loss of herbaceous species continued at a similar rate, whereas shrub richness declined slightly. Our results are similar to those from a contrasting environment, the arctic, which also shows a large decrease in the relative abundance of forbs during the Holocene (Willerslev et al. 2014).

Past vegetation change: individual species

The preferential loss of the forb functional type during the last 15,000 yr of increasing aridity is also reflected in results for individual species. Ten of the 11 significant indicator species for climate periods >15 ka were forbs, and the last was a grass. Three of these significant indicator forbs became locally extinct <15 ka, and the significant indicator grass was locally extinct before the beginning of mid- and late Holocene warming. In addition, the grass *Deschampsia*, although not a significant indicator species, only occurred in middens >15 ka.

In contrast to forbs, four significant shrub indicator species had greatest occurrence during post-LGM warming. In addition, two tree taxa had their greatest occurrence during post-LGM warming, and 10 shrub taxa occurred only in middens <15 ka. This increased occurrence of woody shrubs and trees associated with post-LGM warming suggests that increased warmth, especially during winter, increased the ability of woody species to establish in our study area. However, the four significant indicator shrubs as well as all three tree taxa had lower occurrences or became locally extinct following mid- and late Holocene warming beginning ~8 ka. Thus although woody species initially increased in occurrence with warming after glacial maximum, some were subsequently lost as climate became even warmer and more arid during and after the Mid-Holocene Temperature Maximum.

Interestingly, none of the 18 significant indicator taxa for climate periods had greatest occurrence in middens <8 ka. This result along with the observations about individual species above suggests that taxa were lost during mid- and late Holocene warming but not gained. The forbs, grasses, and trees that were lost during post-LGM and mid- and late Holocene periods of warming climate are today usually associated with cool/wet environments, a result consistent with the pattern of increasingly more arid climate over the last 15,000 yr. Furthermore, 13 taxa were commonly found in at least one-third of the middens from each major time period and represent taxa that persisted through major climate shifts. Taxa that persisted are typically widely distributed today, which suggests an ability to acclimate or adapt to climate change (Nowak et al. 1994b). Based on these results, we speculate that Great Basin plant communities, like montane mammal communities (Hadley and Maurer 2001), follow a "nested subset" pattern of distribution in space and time. Those species that are more widely distributed in space also tend to be more widely distributed through time and less susceptible to local extinction.

What past vegetation change tells us about the future

Climate warming is projected for the rest of this century, but at a rate ~10 times faster than occurred at any previous time during the late Pleistocene and will result in temperatures at least as warm as any previous time during the Holocene (Diffenbaugh and Field 2013) and drought at least as severe (Cook et al. 2015). Our results indicate that preferential loss of forbs occurred with increased aridity during and after both the Bolling-Allerod rapid warming and Mid-Holocene Temperature Maximum and resulted in a greatly reduced richness for that plant functional type and a greater proportion of shrubs in plant CAs over the last 15,000 yr. This preferential loss of forbs occurred not only because the ecological tolerance thresholds of some forbs to climate change were exceeded (Briske et al. 2006), but also because a nested subset pattern of species–area relationships occurs in both space and time (Hadley and Maurer 2001). We expect that the preferential loss of forbs and other herbaceous species will continue as future climate change brings increased aridity to the Great Basin.

This preferential loss of forbs also may have left potential gaps in resource use, which then, in part, allowed invasive species such as *Bromus tectorum*, *Taeniatherum caput-medusae*, *Lepidium latifolium*, and *Tribulus terrestris* to establish and proliferate in the Great Basin (Klemmedson and Smith 1964, Young and Evans 1973, Davis et al. 2000, Anderson and Inouye 2001, Evans et al. 2001, Mata-Gonzalez et al. 2008, Chambers et al. 2014). Resources, such as early-season soil water (Harris and Wilson 1970, Melgoza et al. 1990), may be appropriated by exotic species, facilitating their establishment and expansion in the region. Furthermore, frequent and extensive wildfires associated with the invasive annual grass *B. tectorum* (Knapp 1998) adversely affect perennial species, especially woody species. Thus, we speculate that as future severe droughts emerge in the Great Basin with climate warming (Cook et al. 2015), native herbaceous plants will continue to be lost from plant communities in the Great Basin, and woody plant species will continue to increase their dominance of the Great Basin, at least until woody plants succumb to the ravages of fire, after which invasive species will rapidly increase.

CONCLUSIONS

Abrupt, rapid climate change appears to be a common phenomenon in semi-arid systems, but because of site-to-site heterogeneity in aridity, predicting the timing of local response is difficult

(Williams et al. 2011). However, by using a suite of vegetation analysis techniques to examine plant species composition of many middens that span 35,000 yr of climate change over an elevation gradient in one geographic area, we observed three distinct trends: (1) generally unidirectional change in CA since LGM; (2) preferential loss of forbs and increasing prevalence of shrubs with increasing aridity since LGM; and (3) once an herbaceous species was lost from the study area during the last 15 ka, that species did not re-occur. Our ability to analyze CAs, plant functional types, and individual species from our middens makes it possible for us to discern these important, over-arching vegetation trends, which in turn allow us to better estimate Great Basin vegetation response to future climate change. Because future warming is most likely to continue the climate trajectory of the most recent past, we expect the three trends to continue. If, as we speculate, loss of native forb richness with increasing aridity facilitated the establishment and expansion of invasive species in the Great Basin over the last century, then human actions are both accelerating vegetation change and altering the trajectory of vegetation change.

ACKNOWLEDGMENTS

We greatly appreciate reviews of an earlier draft of this manuscript by Stan Kitchen, Scott Mensing, and Connie Millar, and anonymous reviewers provided valuable feedback that greatly improved the manuscript. Jerry Tiehm checked taxonomic nomenclature for accuracy. Tom Dilts prepared Fig. 1. Many student research assistants helped collect and process middens over many years, and we especially thank Julie Allen, Craig Biggart, and Katie Tanner for their assistance. This research was supported in part by the USDA Forest Service Rocky Mountain Research Station and by the Nevada Agricultural Experiment Station (NEV52XC).

LITERATURE CITED

Alley, R. B., et al. 2003. Abrupt climate change. Science 299:2005–2010.

Anderson, J. E., and R. S. Inouye. 2001. Landscape-scale changes in plant species abundance and biodiversity of a sagebrush steppe over 45 years. Ecological Monographs 71:531–556.

Azaele, S., R. Muneepeerakul, A. Rinaldo, and I. Rodriguez-Iturbe. 2010. Inferring plant ecosystem organization from species occurrences. Journal of Theoretical Biology 262:323–329.

Becklin, K. M., J. S. Medeiros, K. R. Sale, and J. K. Ward. 2014. Evolutionary history underlies plant physiological responses to global change since the last glacial maximum. Ecology Letters 17:691–699.

Benson, L., M. Kashgarian, R. Rye, S. Lund, F. Paillet, J. Smoot, C. Kester, S. Mensing, D. Meko, and S. Lindstrom. 2002. Holocene multidecadal and multicentennial droughts affecting Northern California and Nevada. Quaternary Science Reviews 21:659–682.

Benson, L. V., J. P. Smoot, S. P. Lund, S. A. Mensing, F. F. Foit Jr., and R. O. Rye. 2013. Insights from a synthesis of old and new climate-proxy data from the Pyramid and Winnemucca lake basins for the period 48 to 11.5 cal ka. Quaternary International 310:62–82.

Betancourt, J. L., T. R. Van Devender, and P. S. Martin. 1990. Packrat middens: the last 40,000 years of biotic change. University of Arizona Press, Tucson, Arizona, USA.

Blois, J. L., P. L. Zarnetske, M. C. Fitzpatrick, and S. Finnegan. 2013. Climate change and the past, present, and future of biotic interactions. Science 341:499–504.

Briske, D. D., S. D. Fuhlendorf, and F. E. Smeins. 2006. A unified framework for assessment and application of ecological thresholds. Rangeland Ecology and Management 59:225–236.

Buizert, C., et al. 2014. Greenland temperature response to climate forcing during the last deglaciation. Science 345:1177–1180.

Chambers, J. C., R. F. Miller, D. I. Board, J. B. Grace, D. A. Pyke, B. A. Roundy, E. W. Schupp, and R. J. Tausch. 2014. Resilience and resistance of sagebrush ecosystems to management treatments: implications of state and transition models. Rangeland Ecology and Management 67:440–454.

Coats, L. L., K. L. Cole, and J. I. Mead. 2008. 50,000 years of vegetation and climate history on the Colorado Plateau, Utah and Arizona. Quaternary Research 70:322–338.

Cook, B. I., T. R. Ault, and J. E. Smerdon. 2015. Unprecedented 21st century drought risk in the American Southwest and Central Plains. Scientific Advances 1:e1400082.

Cook, E. R., C. A. Woodhouse, C. M. Eakin, D. M. Meko, and D. W. Stahle. 2004. Long-term aridity changes in the western United States. Science 306:1015–1018.

Cooper, A., C. Turney, K. A. Hughen, B. W. Brook, H. G. McDonald, and C. J. A. Bradshaw. 2015. Abrupt warming events drove late Pleistocene Holarctic megafaunal turnover. Science 349:602–606.

Davis, M. A., J. P. Grime, and K. Thompson. 2000. Fluctuating resources in plant communities: a general theory of invisibility. Journal of Ecology 88: 528–534.

Denton, G. H., R. F. Anderson, J. R. Toggweiler, R. L. Edwards, J. M. Schaefer, and A. E. Putnam. 2010. The last glacial termination. Science 328:1652–1656.

Dial, K. P., and N. J. Czaplewski. 1990. Do woodrat middens accurately represent the animals' environments and diets? The Woodhouse Mesa study. Pages 43–58 in J. L. Betancourt, T. R. Van Devender, and P. S. Martin, editors. Packrat middens: the last 40,000 years of biotic change. University of Arizona Press, Tucson, Arizona, USA.

Diffenbaugh, N. S., and C. B. Field. 2013. Changes in ecologically critical terrestrial climate conditions. Science 341:486–492.

Eronen, J. T., P. D. Polly, M. Fred, J. Damuth, D. C. Frank, V. Mosbrugger, C. Scheidegger, N. C. Stenseth, and M. Fortelius. 2010. Ecometrics: the traits that bind the past and present together. Integrative Zoology 5:88–101.

Evans, R. D., R. Rimer, L. Sperry, and J. Belnap. 2001. Exotic plant invasion alters nitrogen dynamics in an arid grassland. Ecological Applications 11:1301–1310.

Gray, S. T., J. L. Betancourt, S. T. Jackson, and R. G. Eddy. 2006. Role of multidecadal climate variability in a range expansion of pinyon pine. Ecology 87:1124–1130.

Grayson, D. K. 2000. Mammalian responses to Middle Holocene climatic change in the Great Basin of the western United States. Journal of Biogeography 27:181–192.

Grayson, D. K. 2011. The Great Basin: a natural prehistory. University of California Press, Berkeley, California, USA.

Hadley, E. A., and B. A. Maurer. 2001. Spatial and temporal patterns of species diversity in montane mammal communities of western North America. Evolutionary Ecology Research 3:477–486.

Harris, G. A., and A. M. Wilson. 1970. Competition for moisture among seedlings of annual and perennial grasses as influenced by root elongation at low temperature. Ecology 51:530–534.

Kindler, P., M. Guillevic, M. Baumgartner, J. Schwander, A. Landais, and M. Leuenberger. 2014. Temperature reconstructions from 10 to 120 kyr b2k from the NGRIP ice core. Climate of the Past 10: 887–902.

Klemmedson, J. O., and J. G. Smith. 1964. Cheatgrass (Bromus tectorum L.). Botanical Review 30:226–262.

Kleppe, J. A., D. S. Brothers, G. M. Kent, F. Biondi, S. Jensen, and N. W. Driscoll. 2011. Duration and severity of medieval drought in the Lake Tahoe Basin. Quaternary Science Reviews 30:3269–3279.

Knapp, P. A. 1998. Spatio-temporal patterns of large grassland fires in the Intermountain West, U.S.A. Global Ecology and Biogeography Letters 7:259–272.

Lindstrom, S. 1990. Submerged tree stumps as indicators of mid-Holocene aridity in the Lake Tahoe Basin. Journal of California and Great Basin Anthropology 12:146–157.

Liu, Z., et al. 2009. Transient simulation of last deglaciation with a new mechanism of Bolling-Allerod warming. Science 325:310–314.

Lyford, M. E., S. T. Jackson, J. L. Betancourt, and S. T. Gray. 2003. Influence of landscape structure and climate variability on a late Holocene plant migration. Ecological Monographs 73:567–583.

Lyle, M., L. Heusser, C. Ravelo, M. Yamamoto, J. Barron, N. S. Diffenbaugh, T. Herbert, and D. Andreasen. 2012. Out of the tropics: the Pacific, Great Basin lakes, and Late Pleistocene water cycle in the western United States. Science 337:1629–1633.

Mata-Gonzalez, R., R. G. Hunter, C. L. Coldren, T. McLendon, and M. W. Paschke. 2008. A comparison of modeled and measured impacts of resource manipulations for control of Bromus tectorum in sagebrush steppe. Journal of Arid Environments 72:836–846.

McCune, B., and J. B. Grace. 2002. Analysis of ecological communities. MjM Software Design, Gleneden Beach, Oregon, USA.

Melgoza, G., R. S. Nowak, and R. J. Tausch. 1990. Soil water exploitation after fire: competition between Bromus tectorum (cheatgrass) and two native species. Oecologia 83:7–13.

Mensing, S. A., L. V. Benson, M. Kashgarian, and S. Lind. 2004. A Holocene pollen record of persistent drought from Pyramid Lake, Nevada, USA. Quaternary Research 62:29–38.

Mensing, S. A., S. E. Sharpe, I. Tunno, D. W. Sada, J. M. Thomas, S. Starratt, and J. Smith. 2013. The Late-Holocene dry period: multiproxy evidence for an extended drought between 2800 and 1850 cal yr BP across the central Great Basin, USA. Quaternary Science Reviews 78:266–282.

Mensing, S. A., J. Smith, K. B. Norman, and M. Allan. 2008. Extended drought in the Great Basin of western North America in the last two millennia reconstructed from pollen records. Quaternary International 188:79–89.

Miller, J., D. Germanoski, K. Waltman, R. Tausch, and J. Chambers. 2001. Influence of late Holocene hillslope processes and landforms on modern channel dynamics in upland watersheds of central Nevada. Geomorphology 38:373–391.

Miller, J., K. House, D. Germanoski, R. Tausch, and J. Chambers. 2004. Fluvial geomorphic responses to

Holocene climate change. Pages 49–87 *in* J. C. Chambers and J. R. Miller, editors. Great Basin riparian ecosystems: ecology, management and restoration. Island Press, Covelo, California, USA.

Moritz, C., and R. Agudo. 2013. The future of species under climate change: Resilience or decline? Science 341:504–508.

Mouillot, D., N. A. J. Graham, S. Villeger, N. W. H. Mason, and D. R. Bellwood. 2013. A functional approach reveals community responses to disturbances. Trends in Ecology and Evolution 28:167–177.

Nowak, R. S., C. L. Nowak, and R. J. Tausch. 2000. Probability that a fossil absent from a sample is also absent from the paleolandscape. Quaternary Research 54:144–154.

Nowak, C. L., R. S. Nowak, R. J. Tausch, and P. E. Wigand. 1994*a*. Tree and shrub dynamics in northwestern Great Basin woodland and shrub steppe during the late Pleistocene and Holocene. American Journal of Botany 81:265–277.

Nowak, C. L., R. S. Nowak, R. J. Tausch, and P. E. Wigand. 1994*b*. A 30,000 year record of vegetation dynamics at a semi-arid locale in the Great Basin. Journal of Vegetation Science 5:579–590.

Polly, P. D., J. T. Eronen, M. Fred, G. P. Dietl, V. Mosbrugger, C. Scheidegger, D. C. Frank, J. Damuth, N. C. Stenseth, and M. Fortelius. 2011. History matters: ecometrics and integrative climate change biology. Proceedings of the Royal Society of London B: Biological Sciences 278:1131–1140.

Rhodes, R. H., E. J. Brook, J. C. H. Chiang, T. Blunier, O. J. Maselli, J. R. McConnell, D. Romanini, and J. P. Severinghaus. 2015. Enhanced tropical methane production in response to iceberg discharge in the North Atlantic. Science 348:1016–1019.

Sala, O. E., W. K. Lauenroth, and R. A. Golluscio. 2011. Plant functional types in temperate semi-arid regions. Pages 217–233 *in* T. M. Smith, H. H. Shugart, and F. I. Woodward, editors. Plant functional types. Cambridge University Press, New York, New York, USA.

Schaefer, J. M., G. H. Denton, D. J. A. Barrell, S. Ivy-Ochs, P. W. Kubik, B. G. Andersen, F. M. Phillips, T. V. Lowell, and C. Schluchter. 2006. Near-synchronous interhemispheric termination of the last glacial maximum in mid-latitudes. Science 312:1510–1513.

Shuman, B., P. Bartlein, N. Logar, P. Newby, and T. Webb III. 2002. Parallel climate and vegetation responses to the early Holocene collapse of the Laurentide Ice Sheet. Quaternary Science Reviews 21:1793–1805.

Spaulding, W. G. 1990. Vegetational and climatic development of the Mojave Desert: the last glacial maximum to present. Pages 166–199 *in* J. L. Betancourt,

T. R. Van Devender, and P. S. Martin, editors. Packrat middens: the last 40,000 years of biotic change. University of Arizona Press, Tucson, Arizona, USA.

Stine, S. 1994. Extreme and persistent drought in California and Patagonia during Mediaeval time. Nature 369:546–549.

Swetnam, T. W., C. D. Allen, and J. L. Betancourt. 1999. Applied historical ecology: using the past to manage for the future. Ecological Applications 9:1189–1206.

Tausch, R., C. Nowak, and S. Mensing. 2004. Climate change and associated vegetation dynamics during the Holocene: the paleoecological record. Pages 24–48 *in* J. C. Chambers and J. R. Miller, editors. Great Basin riparian ecosystems: ecology, management and restoration. Island Press, Covelo, California, USA.

Thompson, R. S. 1985. Palynology and Neotoma middens. Contribution Series 16, American Association Stratigraphic Palynologists Foundation, Dallas, Texas, USA. Pages 89–112.

Thompson, R. S. 1990. Late Quaternary vegetation and climate in the Great Basin. Pages 200–239 *in* J. L. Betancourt, T. R. Van Devender, and P. S. Martin, editors. Packrat middens: the last 40,000 years of biotic change. University of Arizona Press, Tucson, Arizona, USA.

West, N. E. 1988. Intermountain deserts, shrub steppes, and woodlands. Pages 209–230 *in* M. G. Barbour and W. D. Billings, editors. North American terrestrial vegetation. Cambridge University Press, New York, New York, USA.

Westoby, M., and M. Leishman. 2011. Categorizing plant species into functional types. Pages 104–121 *in* T. M. Smith, H. H. Shugart, and F. I. Woodward, editors. Plant functional types. Cambridge University Press, New York, New York, USA.

Wigand, P. E. 2006. Postglacial pollen records of southwestern North America. Pages 2773–2783 *in* S. A. Elias, editor. Encyclopedia of Quaternary science. Elsevier, London, UK.

Wigand, P. E., and D. Rhode. 2002. Great Basin vegetation history and aquatic systems: the last 150,000 years. Pages 309–367 *in* R. Hershler, D. B. Madsen, and D. R. Currey, editors. Great Basin aquatic ecosystems history, Smithsonian Contributions to Earth Sciences 33. Smithsonian Institution Press, Washington, District of Columbia, USA.

Willerslev, E., et al. 2014. Fifty thousand years of Arctic vegetation and megafaunal diet. Nature 506: 47–51.

Williams, J. W., J. L. Blois, and B. N. Shuman. 2011. Extrinsic and intrinsic forcing of abrupt ecological change: case studies from the late Quaternary. Journal of Ecology 99:664–677.

Wilson, J. B. 2012. Species presence/absence sometimes represents a plant community as well as species abundances do, or better. Journal of Vegetation Science 23:1013–1023.

Woodward, F. I. 1987. Climate and plant distribution. Cambridge University Press, Cambridge, UK.

Woodward, F. I., T. M. Smith, and H. H. Shugart. 2011. Defining plant functional types: the end view.

Pages 355–359 *in* T. M. Smith, H. H. Shugart, and F. I. Woodward, editors. Plant functional types. Cambridge University Press, New York, New York, USA.

Young, J. A., and R. A. Evans. 1973. Downy brome: intruder in the plant succession of big sagebrush communities in the Great Basin. Journal of Range Management 26:410–415.

11

Large river floodplain as a natural laboratory: non-native macroinvertebrates benefit from elevated temperatures

Amael Paillex,[1,2,†] Emmanuel Castella,[3] Philine zu Ermgassen,[4] Belinda Gallardo,[5] and David C. Aldridge[2]

[1]Eawag, Swiss Federal Institute of Aquatic Science and Technology, 8600 Dübendorf, Switzerland
[2]Aquatic Ecology Group, Department of Zoology, University of Cambridge, The David Attenborough Building, Pembroke Street, Cambridge CB2 3QZ UK
[3]Department F.-A. Forel for Environmental and Aquatic Sciences & Institute for Environmental Sciences, University of Geneva, Boulevard Carl-Vogt 66, 1211 Geneva 4, Switzerland
[4]School of Geosciences, University of Edinburgh, The King's Buildings Alexander Crum Brown Road, Edinburgh EH9 3FF UK
[5]Applied and Restoration Ecology Group, Pyrenean Institute of Ecology (IPE-CSIC), Avda. Monañana 1005, 50059 Zaragoza, Spain

Abstract. Water temperature is known to influence individual animal metabolism, development, and reproduction. However, in situ studies aiming to demonstrate the link between water temperature and community structure in complex ecosystems such as large river floodplains are still rare. In particular, we have little indication about how an increase in temperature affects the density of native and invasive species within a community. Large river floodplains cover a varied range of environmental conditions, are rich in species, and therefore potentially useful ecosystems to study the effect of water temperature at the community level. Moreover, as freshwater communities are increasingly impacted by global warming and biological invasions, an improved understanding of the possible interaction between these drivers would be beneficial. First, we studied during two years the thermal heterogeneity of 36 sites in a large river (Rhone) floodplain. Second, we compared the thermal regimes of sites having different levels of hydrological connectivity with the main river channel. Third, we studied the combined and separated effects of the thermal regime and the hydrological connectivity on the presence and densities of native and non-native species of macroinvertebrates. The studied large river floodplain covered a wide range of thermal regimes, with some sites displaying a yearlong constant temperature of about 10°C, whereas others experienced thermal amplitude of over 25°C. The thermal regime was independent of the level of hydrological connectivity of the sites. The increase in hydrological connectivity had a significant and positive effect on the richness of non-native species within sites. The thermal regime had a positive influence on the density of non-native species but no effect on the total density of native taxa within communities. This study showed that large river floodplains possess a wide range of thermal conditions and that the increase in water temperature can have a positive influence on the presence of non-native populations of macroinvertebrates. This study provides a first set of empirical results to establish models predicting the effect of increasing temperatures on the establishment of non-native and native species in a complex ecosystem and underline the problem of biological invasions under climate change.

Key words: air–water temperature; alien species; biological invasions; climatic change; ecosystem management; generalized additive models; lateral connectivity; Rhone River; species distribution model; thermal pollution.

† **E-mail:** amael.paillex@swissonline.ch

INTRODUCTION

Temperature can have a profound effect on the growth, development, and metabolism of organisms (Brown et al. 2004). In a world experiencing global change, both the composition and abundance of communities can be expected to change in freshwater environments that are especially vulnerable to compounded effects of climate warming and biological invasions (Sala et al. 2000, Rahel and Olden 2008). For instance, Milner et al. (2001) showed in glacier-fed streams that the composition of macroinvertebrates changed when water temperature increased by two degrees. Likewise, Castella et al. (2001) showed that the maximum water temperature in a glacier-fed stream was an explanatory variable for changes in macroinvertebrate composition. Fish abundance has also been shown to be related to flow and temperature conditions in the lower part of the Rhone River, with higher abundance in warm habitats (Daufresne et al. 2004).

Increasing temperatures have been shown to especially favor the spread and establishment of non-native species (e.g., Domisch et al. 2013, Gallardo and Aldridge 2013a), which may be better suited to thermal changes than are native taxa. Non-native species can be a major threat to biodiversity and continue to proliferate at an increasing rate (Chapin et al. 2000, Sala et al. 2000). In freshwater systems, warmer temperatures have been shown to facilitate the spread of invasive species at both the continental and within-catchment scales. For instance, Gallardo and Aldridge (2013a) showed that the spread of Ponto-Caspian invaders into western Europe was highly correlated with elevated temperatures, while Verbrugge et al. (2012) demonstrated that non-native mollusks in the Rhine River tend to tolerate warmer habitats than native species. Native aquatic macroinvertebrates are highly vulnerable to the negative effects of invasive species, which often display strong competitive abilities (Statzner et al. 2008, DAISIE 2009).

To date, studies have mostly focussed on the effects of climate change and invasive species on the communities within the main channel of rivers (e.g., Daufresne et al. 2004). However, large river floodplains provide an ideal and important system for exploring the effects of invasive species as they cover a wide range of thermal conditions (Tonolla et al. 2010), and support rich aquatic macroinvertebrate communities which themselves display multiple strategies to adapt to extreme conditions (Statzner et al. 2001, Verberk et al. 2008).

In large river floodplains, the lateral, vertical, longitudinal, and temporal dimensions provide a wide diversity of habitat conditions (Gray and Harding 2009). Water temperature varies temporally due to seasonal changes in air temperature and melting snow in spring (Tonolla et al. 2010) and longitudinally in the main river channel with colder water in the highlands compared to the lowlands. Groundwater-fed channels with a high vertical connectivity have a more constant surface temperature than channels with low vertical connectivity (Caissie 2006). In the lateral dimension of floodplains, channels can be permanently connected to the main river channel, partially or totally disconnected (Amoros and Bornette 2002). This gradient of lateral hydrological connectivity can generate differences in water temperature of 16°C (Ward et al. 2001, Webb et al. 2008) and has been shown to influence the distribution of animal communities (Reckendorfer et al. 2006, Lasne et al. 2007). The link between water temperature and lateral hydrological connectivity is, however, poorly studied in large river floodplains. This is due primarily to the challenges in undertaking the long-term temperature monitoring necessary to develop a clear understanding of spatial and temporal patterns.

While temperature is believed to influence non-native species distribution and is commonly used to model their potential range of establishment (Domisch et al. 2013, Gallardo and Aldridge 2013a), very few studies have examined the impact of temperature on the distribution of non-natives in situ. An existing study by Verbrugge et al. (2012), however, showed that non-native mollusks in the Rhine River tend to tolerate warmer habitats than native species. As such, it is possible that water temperature is an important factor influencing the success of non-native populations in floodplain channels, and this topic is worthy of further investigation. A better understanding of the influence of water temperature on the establishment of non-native species in river floodplains can have implication for ecosystem management, biological conservation and to forecast the effect of global warming on the spread of aquatic non-native species.

The aims of our current research were (1) to measure the thermal regime of the Rhone River and its floodplain, (2) to differentiate the effects of water temperature and hydrological connectivity on macroinvertebrates, and (3) to model the effect of water temperature on native and non-native macroinvertebrate abundance and richness. We hypothesized that floodplain channels would encompass a large diversity of thermal regimes independently from their level of lateral hydrological connectivity with the main river channel. We also hypothesized that the hydrological connectivity of a site would influence macroinvertebrate composition and richness such as described in previous studies (Reckendorfer et al. 2006, Paillex et al. 2015), while water temperature would influence the abundance of the species (e.g., Verbrugge et al. 2012). Finally, we hypothesized that native and non-native taxa would respond differently to water temperature and in particular that higher water temperature would encourage non-native species.

MATERIALS AND METHODS

Measurements of temperature

The study was carried out in two sectors of the French Rhone River (i.e., Belley and Brégnier Cordon; Fig. 1). Thirty-seven data loggers (Onset HOBO data loggers) were installed in the Rhone River and its floodplain to record hourly water temperature. Temperature loggers were installed in different habitats (i.e., lotic or lentic sites) to cover the thermal diversity that exists in the floodplain. Four loggers were installed in the main river channel (Fig. 1). In addition to these four loggers, the data of two permanent stations in Belley and Brégnier-Cordon measured by Compagnie Nationale du Rhône (CNR; authority managing the Rhone River) were used to represent the conditions in the main river channel (Fig. 1). Five loggers were installed in secondary floodplain channels permanently connected with the main river channel (i.e., eupotamal channels; Fig. 1; Appendix S1: Fig. S1). Eleven were installed in secondary channels with a permanent connection downstream with the main river channel (i.e., parapotamal channels; Fig. 1; Appendix S1: Fig. S1). Fifteen were installed in disconnected channels

that are in direct connection with the main river only during floods (i.e., plesiopotamal channels; Fig. 1; Appendix S1: Fig. S1). Two loggers were installed, respectively, in Belley and Brégnier-Cordon sectors to measure air temperature (Fig. 1). All loggers recorded the water and air temperature hourly in °C with an error of ± 0.2°C, from the 1 January 2009 to 31 December 2010.

Lateral connectivity gradient

As described in Paillex et al. (2007), we summarized with a principal component analysis (PCA) five environmental variables known to represent the level of lateral hydrological connectivity in a given channel. The five variables (submerged vegetation cover, organic matter in the upper part of sediments, water conductivity, diversity of the mineral grain size measured by a Simpson index, and water NH_3-N) were measured during the same years as the monitoring of temperature and the macroinvertebrate samplings. The organic content of the sediment and the water concentration of NH_3-N were measured during the winter period. The cover of submerged vegetation, the diversity of mineral grain size, and conductivity were measured during macroinvertebrate samplings. The first axis of the PCA summarized the level of lateral connectivity for each site (Paillex et al. 2007). This ordination of the sites has the advantage of providing a value of lateral connectivity for each site, avoiding classification of the sites into discrete categories. This surrogate of lateral connectivity was shown to be well correlated with common hydrological indices (i.e., upstream overflow frequency and magnitude, and shear stress during floods, see Riquier et al. 2015).

Thermal variables and gradient

In order to describe the heterogeneity of thermal regimes among channels, we studied the relationship between water and air temperature in the main river and floodplain channels. The relationships were analyzed with logistic regressions. The best model was selected as having the smallest Akaike's information criterion (AIC; Mohseni et al. 1998, Ritz and Streibig 2005). The relationship between water and air temperature has already been modeled for rivers (Koch and Gruenewald 2010) and has been proved useful in characterizing their thermal regimes (Morrill

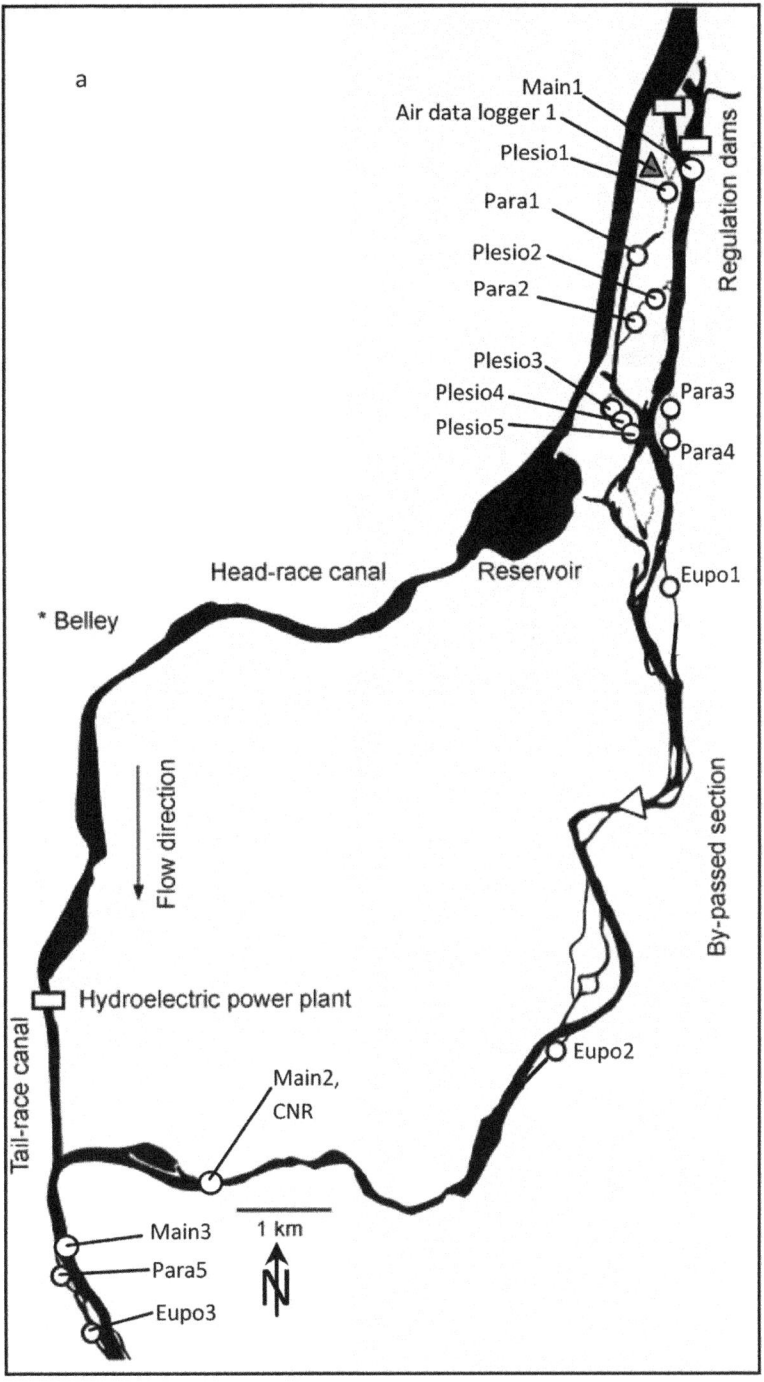

Fig. 1. Locations of sampling sites in Belley (a) and Brégnier-Cordon (b) sectors. The sites are coded according to their type of connection with the main river channel: main river channel (Main), permanently connected channels called eupotamal channels (Eupo), upstream disconnected channels called parapotamal channels (Para), upstream and downstream disconnected channels called plesiopotamal channels (Plesio), and numbered for each category along the flow direction in the two studied sectors (Main: 1–6, Eupo: 1–5, Para: 1–11, Plesio: 1–15). Triangles represent data loggers measuring air temperature.

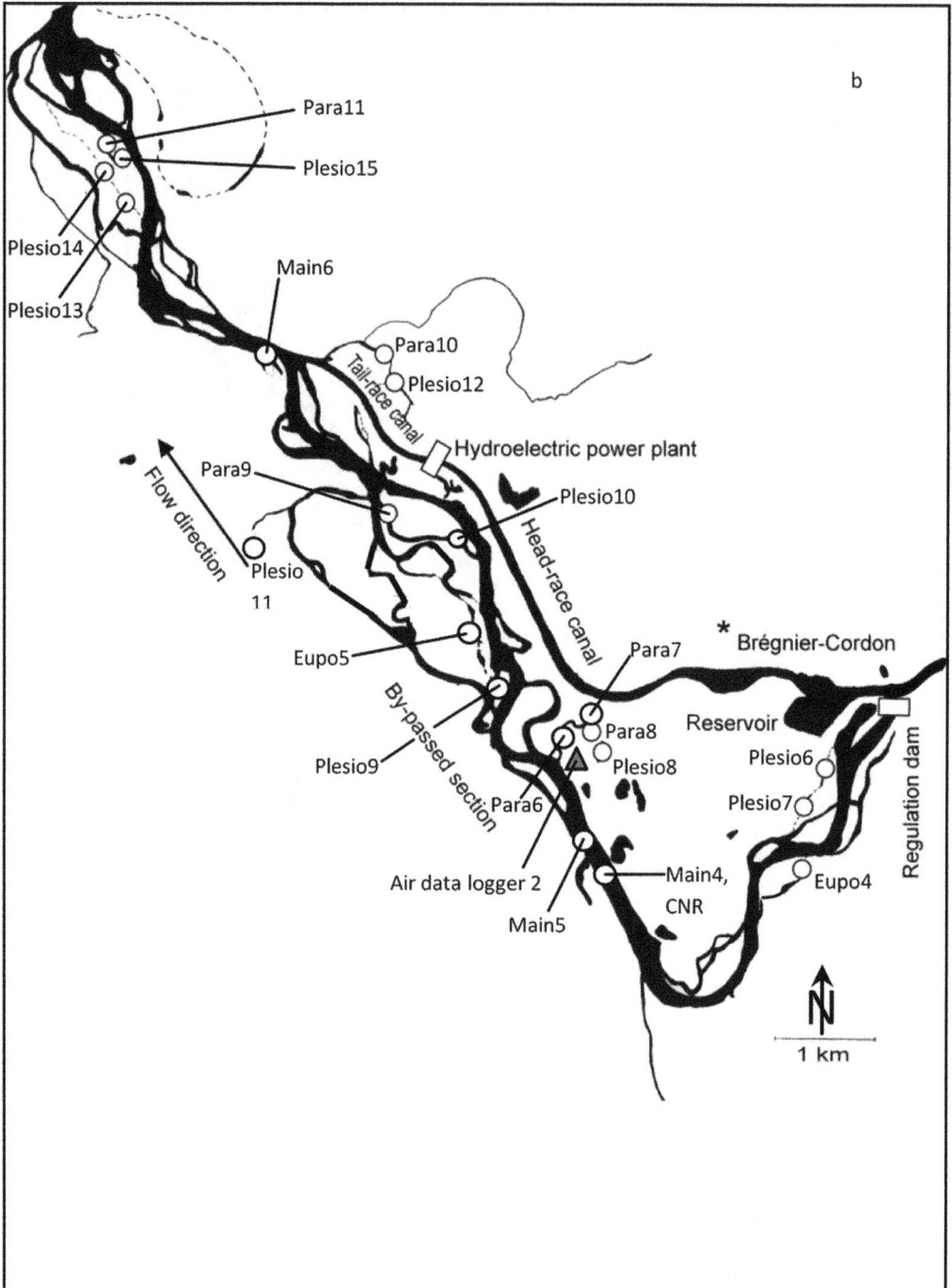

(Fig. 1. *Continued*)

et al. 2005). Logistic regressions were used to derive parameters that can be used to characterize the thermal regimes (see Appendix S1: Table S1). The following three parameters derived from the relationship between air and water temperatures were found to best describe the thermal regime: (1) the asymptotic maximum (as_max), (2) the inflexion point (infl), and (3) the slope of logistic regressions (slope; see Appendix S1: Table S1). The asymptotic maximum reflects the maximum water temperature even if the air temperature continues to increase. The inflexion point reflects the air temperature at which water temperature changes from an increasing to a decreasing trend or vice versa. The slope reflects the speed of temperature increase. The efficiency of the logistic regressions was tested by Nash–Sutcliffe efficiency tests (NSE). Nash–Sutcliffe efficiency ranks between 0 and 1, where 1 is the optimal value (Moriasi et al. 2007). A threshold of 0.70 was selected to discriminate the models correctly fitting with the observations (Moriasi et al. 2007). The root mean square error (RMSE) was also calculated to test the models. It indicates the standard deviation of the prediction error and indicates a better model performance when values are small. The package "drc" (Ritz and Streibig 2005) in R 3.1.0 (R Core Team 2014) was used to calculate and select models. The package "hydroGOF" was used to test the models with NSE and calculate RMSE. In addition, we measured a series of indicators reflecting the thermal regimes of the sites. We measured the maximum and minimum water temperature observed during a year (T_{max} and T_{min}), and the thermal amplitude ($T_{max} - T_{min}$; Appendix S1: Table S1), which reflects the thermal change during a year within a site. The Kamler coefficient was also calculated, which is the ratio of the maximum water temperature divided by the minimum value (T_{max}/T_{min}; Arscott et al. 2001). We calculated the mean water temperature of sites during the studied year (mean_year; Appendix S1: Table S1) and the mean water temperature for each of three seasons: winter (1 January–20 March); spring (21 March–20 June); and summer (21 June–22 September). These eleven variables were selected as they are known to influence macroinvertebrate species occurrence and abundance (Arscott et al. 2001, Daufresne et al. 2004) and included in multivariate analyses to describe the thermal regime of the sites.

Following the same approach of the hydrological gradient, a PCA was performed with the eleven thermal variables to identify composite variables expressing thermal regimes within and among the studied sites (e.g., Vaughan and Ormerod 2005). Site scores along the first PCA axis were used to express a gradient of thermal regime. This approach summarizes the complexity of the data into one unique variable (Vaughan and Ormerod 2005) and permits to test its independence from the gradient of hydrological connectivity. However, the variables can also be analyzed individually to test the specific response of biota, thereby providing a more mechanistic explanation for the distribution of the species. Based on the results of the PCA, Euclidean distances between sites were calculated and sites were classified with a Ward clustering method, which allowed identification of groups of sites with similar thermal regimes. The package "ade4" in R 3.1.0 (R Core Team 2014) was used to perform the PCA, to calculate the Euclidean distances and perform the classification (Chessel et al. 2004). Correlation between the thermal gradient and hydrological gradient was tested with a Spearman rank test, using the "Hmisc" package in R 3.1.0 (R Core Team 2014).

Macroinvertebrate samplings and metrics

Macroinvertebrate communities were sampled in 19 floodplain sites, where water temperature was recorded. Nine sampling sites were sampled in 2009 and 10 in 2010. The macroinvertebrates were sampled in spring and summer to account for seasonal differences and pooled for statistical analyses. In each site, macroinvertebrates were sampled within a quadrat (quadrat size: 0.25 m^2) with a hand net (mesh size: 500 μm) at three random positions along a 15-m stretch. Macroinvertebrates were identified to genus or species level, except Diptera that were identified to family level (see Data S1).

We divided the macroinvertebrates into two groups: the native and the non-native taxa, as they potentially respond differently to the thermal and hydrological gradients (e.g., Townsend et al. 1997, Besacier-Monbertrand et al. 2010, Verbrugge et al. 2012). We tested the effect of hydrological connectivity and temperature gradients on (1) the partition of the variance in taxa abundance, (2) the abundance of communities within sites, and (3) the taxonomic richness of macroinvertebrate communities. Partition of variance for native and non-native assemblages was tested to assess the independent and joint effect of hydrological connectivity and the thermal regime. Individual variables included in the PCA and significantly correlated with the first axis of the PCA were used to assess the partition of the variance. In addition, the partition of the variance permits the assessment of the residual variability not accounted for by the variables used to explain the faunal data (Peres-Neto et al. 2006). The function varpart from the "vegan" package was used in R (Oksanen et al. 2011, R Core Team 2014). In addition, we tested the response of the macroinvertebrate abundance and richness along the thermal and hydrological gradients for a subgroup of native taxa comprising the Ephemeroptera, Plecoptera, and Trichoptera (EPT), because EPT are known to be favored by a high connectivity of the secondary channels with the main river channel (Usseglio-Polatera and Tachet 1994, Gallardo et al. 2014).

We tested the influence of the environmental variables on the richness and the abundance of macroinvertebrates using generalized additive models (GAMs) with a Poisson link function, a thin plate regression spline as smooth term and a chi-square significance test (Wood 2006). Generalized additive model was chosen instead of other regression procedures because of its ability to deal with nonlinear relationships between the response and the set of explanatory variables (Guisan et al. 2002). When necessary, we used AIC to identify the best model. We used the fitted values of the GAM to plot the relationships between the explanatory and the response variable of the models. Generalized additive model was calculated with the gam function from "mgcv" package (Wood 2006) in R (R Core Team 2014).

RESULTS

Temperatures in the river and floodplain channels

In the floodplains of Belley and Brégnier-Cordon, air temperature oscillated between 0 in winter and 30°C in summer (Fig. 2a for Belley sector). Water temperature recorded in the river and floodplain channels showed contrasting patterns with either a unimodal response similar to the air temperature or a flatter response with little seasonal change (see a selection of examples in Fig. 2 for the year 2009). Water temperatures in the main river and connected channels increased from a minimum around 2.9°C in winter to maximum of 23.4°C in summer (Fig. 2b, c). In contrast, disconnected channels showed a larger variability, with either a more constant

yearly temperature and a maximum of 15.5°C in summer (Fig. 2d, e) or larger changes among seasons and a maximum reaching 26.4°C (Fig. 2f).

The water temperature in the main river channel showed an S-shaped relationship with air temperature (Appendix S1: Fig. S2), with an asymptotic minimum and maximum at 3.3°C and 23.4°C, respectively. The minima and maxima were similar among most of the sites in the main river channel (Appendix S1: Fig. S2). However, some disconnected sites experienced a less pronounced S-shaped relationship between the water and air temperatures, with a smaller amplitude (see for example PARA1 in Appendix S1: Fig. S3 and PLES1 in Appendix S1: Fig. S4). Overall, models relating air temperature with water temperature showed good results with NSE > 0.70, with similar patterns between years, and a higher NSE and a lower RMSE when calculated on daily average (Appendix S1: Table S2).

Thermal heterogeneity among sites

The three-first PCA axes obtained with the thermal data from 2009 explained 81.3% of the between-site variability. The first and second axes explained, respectively, 47.5% and 23.4% of the variability. The first axis of the PCA was highly correlated with the yearly maximum water temperature and with the mean summer water temperature, reflecting a gradient of thermal regime from warmer to colder sites (Fig. 3a). A classification of the channels according to their Euclidean distances based on their thermal regimes showed four groups: (1) sites with buffered conditions, (2) sites similar to the main river, (3) sites warmer than the main river, and (4) a group of contrasted sites that did not fit in the previous groups (Fig. 3b; Appendix S1: Fig. S5). The same patterns were observed in 2010 within and among the same floodplain channels. The stability in the ordination between the years was measured as the correlation between site scores along the first (Spearman, $r = 0.83$, $P < 0.001$) and second axes (Spearman, $r = -0.89$, $P < 0.001$).

Hydrological connectivity and thermal regime

The PCA of the lateral connectivity variables ordinated sites from totally disconnected sites (Appendix S1: Fig. S6a, b) to totally connected sites (Appendix S1: Fig. S6a, b). The correlation between the thermal gradient (first PCA axis;

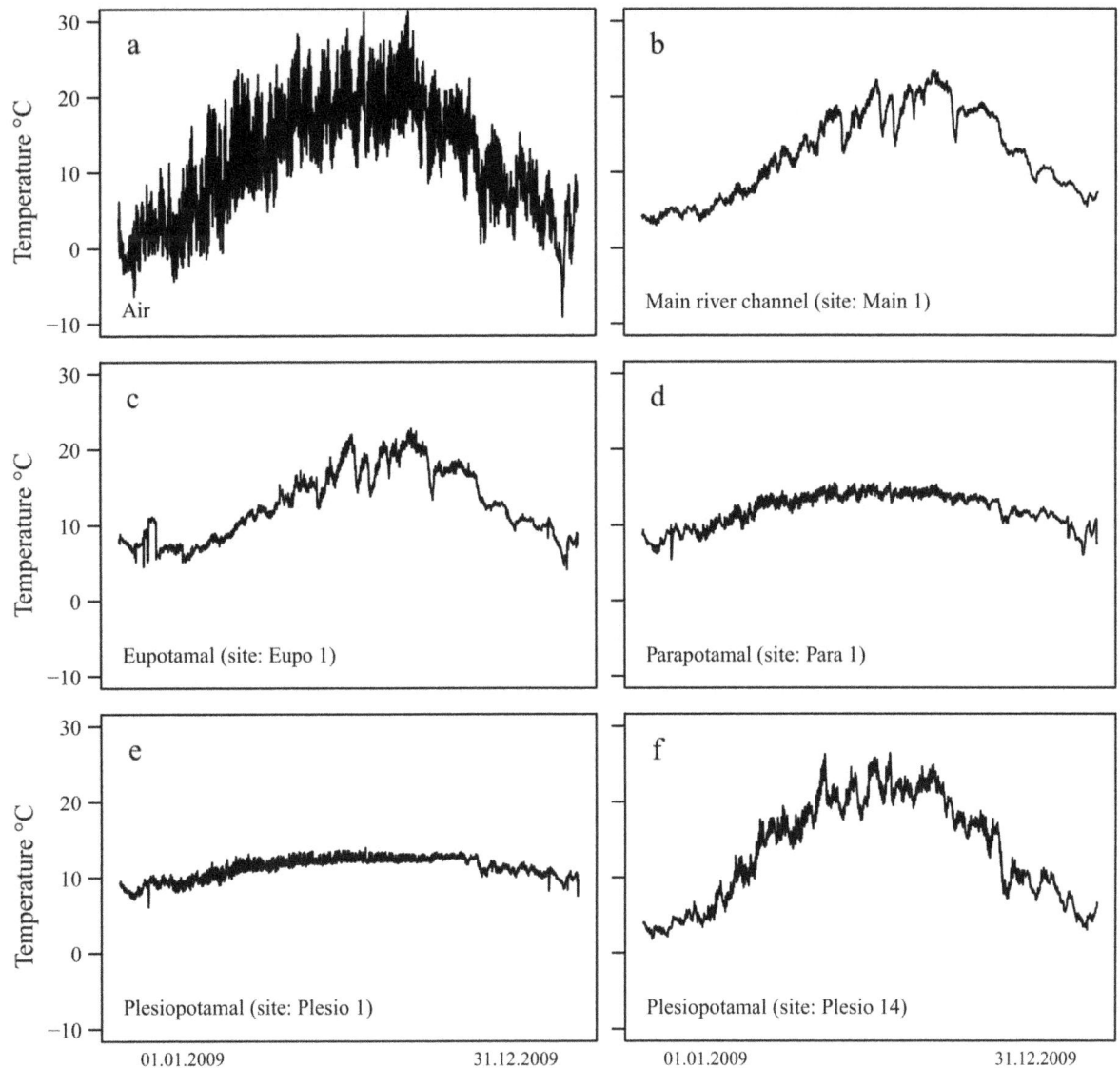

Fig. 2. Hourly air temperature in Belley sector (a) and hourly water temperature in five distinct channels (b–f). Hourly air and water temperatures are represented from the 1st of January 2009 until the end of December 2009. In panel (a), we can observe that the daily fluctuations of air temperature are much larger than for the water temperature in the main river channel (b) and the water temperature in secondary floodplain channels (c–f).

Appendix S1: Fig. S6c) and the gradient of hydrological connectivity (first PCA axis; Appendix S1: Fig. S6a) was not significant (Spearman, $r = -0.35$, $P = 0.11$). The absence of correlation between the two gradients is partially due to the thermal heterogeneity among disconnected channels. Some disconnected sites are warmer in summer whereas others are colder. Moreover, while some disconnected channels had a similar temperature to the main river channel and connected channels in winter (Fig. 2b, f), others had a temperature nearly 5°C higher (e.g., Fig. 2e). Nevertheless, the thermal gradient (first PCA axis; Appendix S1: Fig. S6c) was correlated with the second axis of the gradient of hydrological connectivity driven by water conductivity (Spearman, $r = 0.67$, $P < 0.001$).

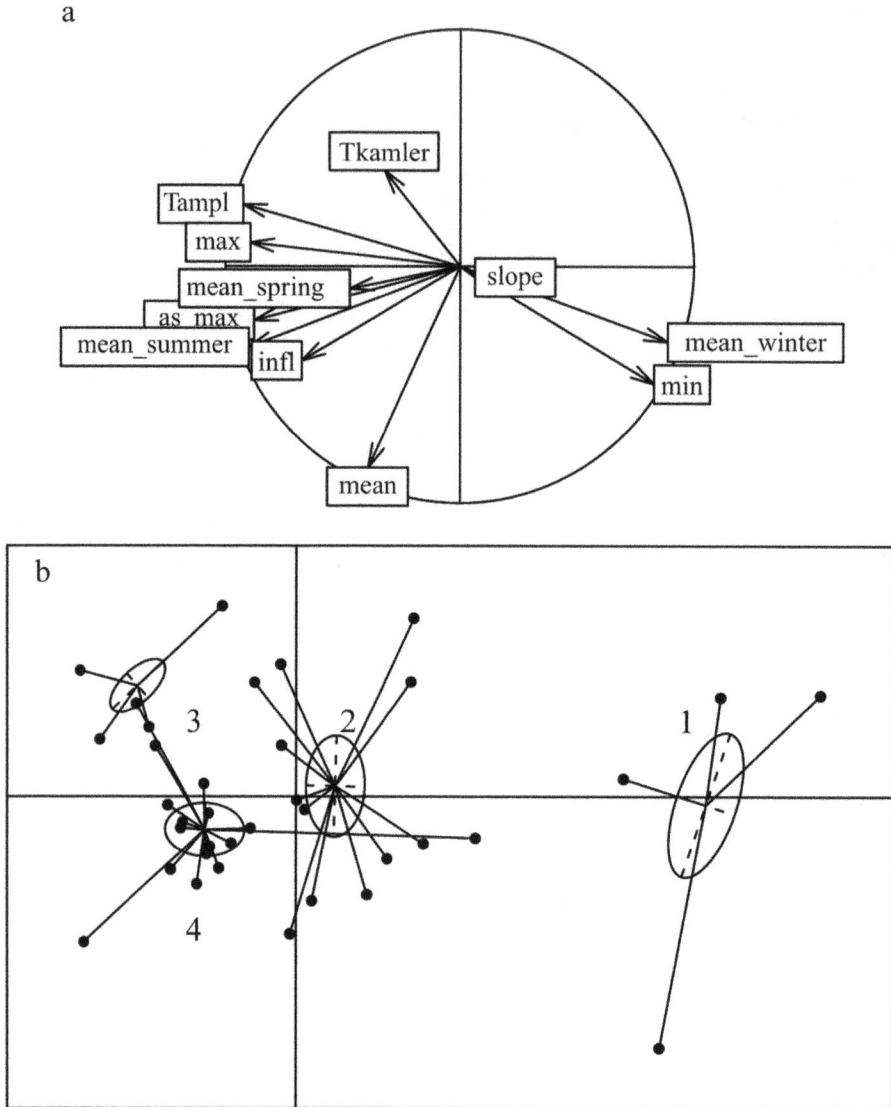

Fig. 3. Principal component analysis based on thermal variables recorded within sites in 2009. (a) Correlation circle of the 11 selected variables. as_max: asymptotic maximum for air–water relationship; infl: inflexion value for air–water relationship; slope: maximum slope for air–water relationship; Tmax: maximum water temperature; Tmin: minimum water temperature; Kamler: TKamler coefficient; Tampl: amplitude; mean: year average temperature; mean_summer: average temperature in summer; mean_spring: average temperature in spring; mean_winter: average temperature in winter. (b) Ordination of the sites according to their thermal regimes and sites grouped according to cluster analysis in Appendix S1: Fig. S5. Group 4 contains sites from the main river channel and eupotamal channels, as well as a few (n = 3) plesiopotamal channels. Group 3 contains sites with a similar thermal regime to sites in group 4. Group 2 contains para and plesiopotamal channels, and group 1 contains sites a different thermal regime from the rest of the sites.

Macroinvertebrate response

One hundred eighty-five native taxa and 12 non-native species were identified in the study area (Table 1). Fifty percent of non-native species were mollusks that colonized up to 65.5% of the sites (see *Physella acuta*; Table 1). Other non-native species present in the study area were arthropods, platyhelminthes, and annelids, and

Table 1. Non-native species observed in 2009 and 2010 in the studied sites.

Species	Phylum	Class	Abundance	No. of invaded sites	Percentage of sites invaded
Corbicula fluminea	Mollusk	Bivalvia	531	11	37.9
Crangonyx pseudogracilis	Arthropoda	Malacostraca	84	10	34.5
Dikerogammarus villosus	Arthropoda	Malacostraca	85	3	10.3
Dreissena polymorpha	Mollusk	Bivalvia	17	7	24.1
Dugesia tigrina	Platyhelminthes	Turbellaria	86	10	34.5
Ferrissia clessiniana	Mollusk	Gasteropoda	40	3	10.3
Gyraulus parvus	Mollusk	Gasteropoda	964	18	62.1
Hemimysis anomala	Arthropoda	Malacostraca	1	1	3.5
Hypania invalida	Annelida	Polychaeta	646	5	17.2
Orconectes limosus	Arthropoda	Malacostraca	3	2	6.9
Physella acuta	Mollusk	Gasteropoda	1748	19	65.5
Potamopyrgus antipodarum	Mollusk	Gasteropoda	3185	17	58.6

Note: Species in boldface are non-native species among the 100 most invasive species in Europe (DAISIE 2009).

found to colonize up to 34.5% of the sites (see *Dugesia tigrina*; Table 1). Three of the observed non-native species are invasive species in Europe, *Corbicula fluminea*, *Dikerogammarus villosus*, and *Dreissena polymorpha* (Table 1). The partition of the variance in the abundance of native taxa was well explained by the thermal regime and the hydrological connectivity (60% of variance explained by both variables combined). The variance was partially explained by the interaction between the thermal regimes and the hydrological connectivity (13%), while 39% was explained by the thermal regime alone (Fig. 4a). Fifty-seven percent of the variance in non-native species abundance was explained by the thermal regime and 29% by the hydrological connectivity (Fig. 4b). For both partitions, the residuals unexplained by the variables represented 40% for native taxa and 31% for non-native species.

In contrast, the richness of macroinvertebrates was generally explained by the hydrological connectivity (Table 2; Appendix S1: Fig. S7). For example, the richness of non-native species was positively related to the gradient of hydrological connectivity (deviance explained 0.51, Table 2; Appendix S1: Fig. S7), and similarly for the richness of EPT taxa (deviance explained 0.79, Table 2; Appendix S1: Fig. S7). While the thermal regime did not explain the richness of non-native macroinvertebrates (Table 2), it contributed to the explanation of their abundance (deviance explained 0.74; Table 2), reaching a high abundance in warmer habitats and a lower abundance in sites where the temperature was buffered

(Fig. 5). The deviance explained by the thermal regime for native taxa was lower (i.e., 0.50%), and their abundance followed a humped shaped response to the temperature with a minimum in warmer habitats (Fig. 5).

Discussion

Thermal heterogeneity in a large river floodplain

We showed that the studied floodplain covers a wide range of thermal conditions, which provides a natural system for exploring the effects of temperature on invasive species. The main river channel experiences a small range of variation between the studied sites, while the more disconnected channels covered a wider range of conditions. Permanently connected channels (i.e., eupotamal) experienced the same thermal conditions as the main river channel, due to a direct and permanent inflow from the main river channel. However, upstream disconnected channels (i.e., parapotamal) were in an intermediate situation. They receive a direct inflow of surface water from the main river channel at their downstream part, while the upstream part is disconnected and can be fed by groundwater. The more disconnected channels (plesiopotamal) exhibited the widest range of thermal conditions compared to the other channel types. These findings are similar to those that were highlighted for the braided, near-pristine Tagliamento River (Arscott et al. 2001). In the context of the Tagliamento River, authors showed that thermal regime of secondary floodplain channels can be very different from

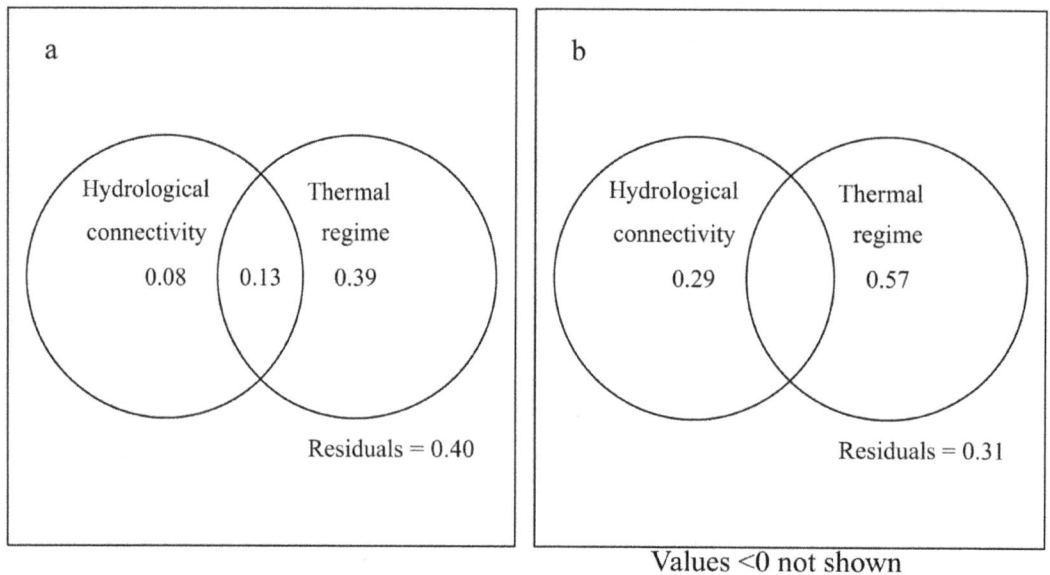

Fig. 4. Proportion of macroinvertebrate variance in the abundance matrix explained by the hydrological connectivity and thermal regime, (a) partition of variance on native taxa, (b) partition of variance on non-native species. The number between the circles indicates the interaction explained conjointly by the two sets of variables. The residuals are the information unexplained by the two sets of variables and their interaction.

each other. Similarly, Uehlinger et al. (2003) showed that the lateral heterogeneity of thermal regimes was high in a glacial floodplain. They showed that disconnected floodplain sites can have a higher average temperature throughout the year than the glacier-fed main river channel (Uehlinger et al. 2003).

The thermal regime of the floodplain channels of the Rhone River appears to be, at least in part, independent from the lateral hydrological connectivity, a major factor influencing the distribution of species at the floodplain scale (Tockner et al. 1999, Reckendorfer et al. 2006). Some isolated

floodplain sites are thermally different from the main river channel and from the most connected channels, maintaining a temperature between 10°C and 12°C throughout the year (e.g., Fig. 2d). This apparent cooling during the summer and warming during winter is well documented for sites having a good vertical connectivity with the groundwater (Amoros and Bornette 2002). A major surrogate of groundwater exfiltration in the upper-Rhone floodplain channels is the water electrical conductivity (Cellot et al. 1994). The significant correlation in our results between the thermal regimes of sites with the second axis of

Table 2. Generalized additive models explaining macroinvertebrate richness and abundance by hydrological connectivity and thermal regime.

Types	Groups	Hydrological connectivity			Thermal regime			UBRE	Dev
		edf	χ^2	P	edf	χ^2	P		
Richness	Non-native	1.0	10.5	0.001**	1.0	1.1	ns	−0.05	0.51
Richness	Native	1.0	3.9	0.047*	1.8	2.3	ns	0.70	0.31
Richness	EPT	1.8	35.9	10^{-8} ***	1.3	1.6	ns	−0.03	0.79
Abundance	Non-native	2.99	904.2	10^{-16} ***	2.96	2171.1	10^{-16} ***	76.92	0.74
Abundance	Native	2.99	2775	10^{-16} ***	2.98	3227	10^{-16} ***	614.48	0.50
Abundance	EPT	2.97	1133	10^{-16} ***	2.99	1306	10^{-16} ***	363.4	0.43

Notes: edf, estimate degree of freedom; Dev, deviance explained by models; ns, non-significant; EPT, Ephemeroptera, Plecoptera, and Trichoptera.
*$P < 0.05$, **$P < 0.01$, ***$P < 0.001$.

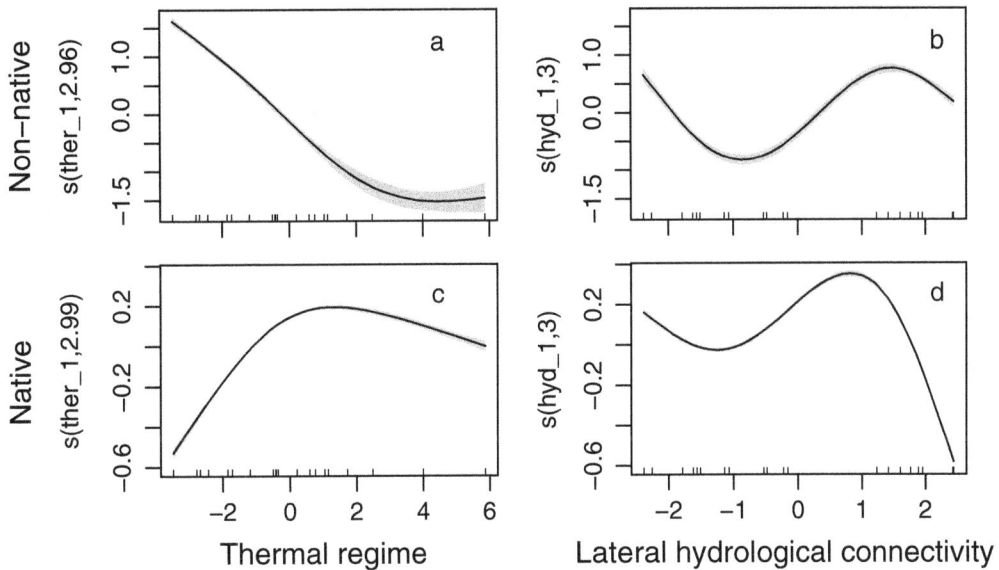

Fig. 5. Generalized additive models of non-native (a, b) and native species (c, d) density along a gradient of thermal regime from warm sites in summer (*x* axis: −2) to cold sites (*x* axis: 6), and a gradient of lateral hydrological connectivity from connected (*x* axis: −2) to disconnected sites (*x* axis: 2). The *x* values represent the scores along the first axis of principal component analysis combining variables expressing the level of lateral hydrological connectivity of the sites (Appendix S1: Fig. S6a–c) or the thermal regime of the sites (Appendix S1: Fig. S6b–d). The *y* values represent the response of non-native and native species density to the smooth of the gradient of thermal regime (ther_1) and hydrological gradient (hyd_1), with an estimated 2.99 df. Gray areas represent 95% confidence interval.

the PCA, mostly driven by conductivity, highlighted the dependence of the thermal regime with a surrogate of groundwater exfiltration. It suggests that water temperature across the Rhone floodplain is strongly influenced by exchanges with the groundwater. Other factors influencing water temperature could be the percentage of riparian shade covering the channels reducing the direct influence of solar radiation (Caissie 2006) or the more or less pronounced development of hydrophytes. The water depth of disconnected channels may also influence water temperature where stratification occurs with higher temperature at the surface and lower at the bottom. In our study, however, most of the sites had a water depth less than 2 m; therefore, vertical stratification would be unlikely.

Macroinvertebrate response to water temperature

In our study, the independence of the gradients in lateral hydrological connectivity and thermal regime allowed us to test their effects on the richness and abundance of aquatic macroinvertebrates. The gradient of hydrological connectivity seems to have a greater effect on macroinvertebrate richness than the thermal gradient. This is likely because species need to be adapted to flood disturbances in the lateral dimension of the floodplain and are therefore filtered along this gradient of hydrological connectivity (Poff 1997, Paillex et al. 2009, Gallardo et al. 2014). The thermal gradient nevertheless also showed a significant effect on the abundance of macroinvertebrates, especially non-native species density, which increased toward a maximum in the warmer sites. This observation can be related to the higher tolerance for warm habitats observed for non-native species. Indeed, Verbrugge et al. (2012) showed that the maximum temperature tolerance of native mollusks (24.0–32.0°C) was 5° lower than the tolerance of non-native species in the Rhine River. Similarly, studies using species distribution models found temperature to be a key factor in explaining the large-scale distribution of aquatic invaders (Domisch et al. 2011, Gallardo and Aldridge 2013a, b).

For native species, Milner et al. (2001) showed that the temperature in a glacier-fed stream is an important factor that drives the richness of macroinvertebrates within sites. Changes in water temperature (~2°C in the case of glacier-fed stream) can have an important effect on composition and the number of taxa per site (Milner et al. 2001). However, in an alpine glacial floodplain, hydrological connectivity was a primary factor explaining macroinvertebrate distribution while temperature had a reduced role (Burgherr et al. 2002). Studies from dry and wet Australian floodplains found temperature was not underlined as a significant factor explaining the spatio-temporal distribution of macroinvertebrates, while hydrological connectivity was (Sheldon et al. 2002, Leigh and Sheldon 2009). Similarly, macroinvertebrates in Mediterranean floodplains that can experience periodic droughts were shown to respond to changes in hydrological connectivity among floodplain channels and not necessarily to changes in temperature (Gallardo et al. 2008). Therefore, macroinvertebrates appear to be more influenced by the hydrological connectivity of a site than its thermal regime. Our results further corroborate the importance of the hydrological connectivity in explaining the distribution of taxon richness. We also find, however, that the gradient of thermal regimes is important in explaining the abundance of macroinvertebrates, especially for non-native species. While the mechanism for this apparent facilitation is unknown, we can assume that the growth rates of taxa increase with temperature and lead to denser communities in warmer sites (e.g., Brown et al. 2004). A second potential explanation could be that elevated temperatures increase productivity thus facilitating higher densities of some taxa (Brown et al. 2004). Non-native species may capitalize quicker than their native counterparts on the increase in primary productivity associated with warming temperatures, or have a superior ability to exploit local resources compared to native species, which lead to denser communities (e.g., Statzner et al. 2008). In both cases, non-native species may have considerable deleterious impacts upon native freshwater diversity through competition for habitat and resources, aggressive behavior toward native species, or transmission of pathogens affecting native populations (Gurevitch and Padilla 2004, DAISIE 2009, Gallardo et al. 2016).

Some limitations of our models should be discussed. First, results from this study refer to the Rhone river floodplain, and further studies covering a wider range of climatic and hydrological conditions are needed to generalize our results. Second, the models could be limited by the number of sites studied. However, the quality of the results and the logical shape of the responses overcome this concern. One important aspect of this success was the selection of the sites in order to cover a large range of conditions from connected to totally disconnected channels. Choosing a large river floodplain permitted to cover a large diversity of conditions and the monitoring aimed to have a series of sites well distributed along the gradient of hydrological connectivity. A third limitation concerns the unequal number of taxa included in the models. The model on native species included 185 taxa, while the model on non-native species included 12 species. The high number of native taxa included in the model could have a negative effect on its performance. Possibly, different groups of species having different levels of sensitivity to physical constraints could be mixed and reduce the level of significance of the model. One possibility would be to divide the native taxa into groups of species according to their sensitivity to physical conditions (e.g., rheophilous or lenitophilous species). Oppositely, the finer level of identification of the non-native species with relatively similar environmental preferences (mostly rheophilous species) permits to develop more robust models with a clearer response to temperature and connectivity gradients.

Future increase in water temperature in the Rhone River due to climatic change, heated effluent discharge, or greater water abstraction leading to reduced groundwater may put new pressure on native macroinvertebrates. Our findings show that sites with higher water temperatures often have higher densities of non-native species, which may in turn increase their impact on native species through increased competition for food resources and habitats, as well as alteration of food-web structure (Gallardo et al. 2016). Such modifications could impact negatively upon threatened and rare native species, which typically have a narrow range of environmental tolerance and weak competitive abilities compared to non-native species. For example, the spread of generalist top predators like

D. villosus is a threat for a wide range of native species sharing the same habitats and being target as a prey (Chapin et al. 2000, Dodd et al. 2014). This latter change is problematic as it may modify the food-web structure and the functions performed by communities at a reach scale (Chapin et al. 2000). The link between water temperature and abundance of non-native species is therefore problematic for the protection of threatened species and the functionality of aquatic communities. Within the frame of restoration measures, we recommend to diversify the level of hydrological connectivity between channels in river floodplains, in order to diversify the habitat and physical conditions. In turn, this measure will permit to maintain a high biodiversity (e.g., Reckendorfer et al. 2006, Lasne et al. 2007, Paillex et al. 2009). High level of heterogeneity could have a positive effect on maintaining sites not suitable to invasive species (Paillex et al. 2015). In turn, this could have a positive effect on the resilience of the ecosystem. Alternatively, shading river sections or deepening channels to increase vertical connectivity with groundwater could in turn buffer water temperature. Therefore, preventing waterbodies to warm-up could limit the development of dense populations of non-native species. These solutions should be tested in order to guarantee that shading or deepening a river channel has no negative side effect and ultimately limit or reduce the population of non-native species.

Conclusion

First, we showed that large river floodplains experience a wide range of thermal regimes, thereby emphasizing the importance of such environment for studies concerning the potential effect of water temperature on freshwater animal communities. Second, we showed that the thermal regime is independent from levels of hydrological connectivity between the main river channel and secondary channels, a key parameter known to influence animal community composition. Third, we showed the positive link between the water temperatures and the density of non-native species. Warm sites were found to be inhabited by abundant populations of non-native species. However, mechanisms favoring the establishment of non-native species in warmer habitats are still not elucidated. This set of empirical results based on water temperature and hydrological connectivity should aid the development of predictive models which combine key parameters for explaining the distribution of animal communities in freshwater environments. Such models would support the forecasting of the potential effects of thermal change and habitat modification on the biota of large river floodplains.

Acknowledgments

We thank the Swiss National Science Foundation (project PBGEP3 – 136309) for funding this project with a fellowship to AP, the "Rhône-Méditerrannée-Corse" Water Agency, the "Compagnie Nationale du Rhône," the "Région-Rhône-Alpes," the Rhône local collectivities, and the EU FEDER program for financial support; the University of Geneva for funding acquisition of temperature loggers, J.M. Olivier, and N. Lamouroux for the coordination of the scientific monitoring of the Rhône, A. Ramseier, S. Mermod, and T. Richter for collecting water temperature data, D. McCrae, N. Peru, A.L. Besacier-Monbertrand, O. Béguin, H. Mayor, and all those involved in collecting macroinvertebrate data, and two anonymous reviewers for constructive comments. BG is supported by a Juan de la Cierva postdoctoral fellowship (JCI-2012-11908).

Literature Cited

Amoros, C., and G. Bornette. 2002. Connectivity and biocomplexity in waterbodies of riverine floodplains. Freshwater Biology 47:761–776.

Arscott, D. B., K. Tockner, and J. V. Ward. 2001. Thermal heterogeneity along a braided floodplain river (Tagliamento River, northeastern Italy). Canadian Journal of Fisheries and Aquatic Sciences 58:2359–2373.

Besacier-Monbertrand, A. L., A. Paillex, and E. Castella. 2010. Alien aquatic macroinvertebrates along the lateral dimension of a large floodplain. Biological Invasions 12:2219–2231.

Brown, J. H., J. F. Gillooly, A. P. Allen, V. M. Savage, and G. B. West. 2004. Toward a metabolic theory of ecology. Ecology 85:1771–1789.

Burgherr, P., J. V. Ward, and C. T. Robinson. 2002. Seasonal variation in zoobenthos across habitat gradients in an alpine glacial floodplain (Val Roseg, Swiss Alps). Journal of the North American Benthological Society 21:561–575.

Caissie, D. 2006. The thermal regime of rivers: a review. Freshwater Biology 51:1389–1406.

Castella, E., et al. 2001. Macrobenthic invertebrate richness and composition along a latitudinal gradient of European glacier-fed streams. Freshwater Biology 46:1811–1831.

Cellot, B., M. J. Dole-Olivier, G. Bornette, and G. Pautou. 1994. Temporal and spatial environmental variability in the Upper Rhône River and its floodplain. Freshwater Biology 31:311–325.

Chapin, F. S., et al. 2000. Consequences of changing biodiversity. Nature 405:234–242.

Chessel, D., A. B. Dufour, and J. Thioulouse. 2004. The ade4 package—I: one-table methods. R News 4: 5–10.

DAISIE, editor. 2009. Handbook of alien species in Europe. Springer, Dordrecht, The Netherlands.

Daufresne, M., M. C. Roger, H. Capra, and N. Lamouroux. 2004. Long-term changes within the invertebrate and fish communities of the Upper Rhone River: effects of climatic factors. Global Change Biology 10:124–140.

Dodd, J. A., J. T. A. Dick, M. E. Alexander, C. MacNeil, A. M. Dunn, and D. C. Aldridge. 2014. Predicting the ecological impacts of a new freshwater invader: functional responses and prey selectivity of the 'killer shrimp', *Dikerogammarus villosus*, compared to the native *Gammarus pulex*. Freshwater Biology 59:337–352.

Domisch, S., M. B. Araújo, N. Bonada, S. U. Pauls, S. C. Jähnig, and P. Haase. 2013. Modelling distribution in European stream macroinvertebrates under future climates. Global Change Biology 19:752–762.

Domisch, S., S. C. Jähnig, and P. Haase. 2011. Climate-change winners and losers: stream macroinvertebrates of a submontane region in Central Europe. Freshwater Biology 56:2009–2020.

Gallardo, B., and D. Aldridge. 2013a. The 'dirty dozen': Socio-economic factors amplify the invasion potential of 12 high-risk aquatic invasive species in Great Britain and Ireland. Journal of Applied Ecology 50:757–766.

Gallardo, B., and D. C. Aldridge. 2013b. Priority setting for invasive species management: risk assessment of Ponto-Caspian invasive species into Great Britain. Ecological Applications 23:352–364.

Gallardo, B., M. Clavero, M. I. Sanchez, and M. Vila. 2016. Global ecological impacts of invasive species in aquatic ecosystems. Global Change Biology 22:151–163.

Gallardo, B., S. Doledec, A. Paillex, D. B. Arscott, F. Sheldon, F. Zilli, S. Merigoux, E. Castella, and F. A. Comin. 2014. Response of benthic macroinvertebrates to gradients in hydrological connectivity: a comparison of temperate, subtropical, Mediterranean and semiarid river floodplains. Freshwater Biology 59:630–648.

Gallardo, B., M. Garcia, A. Cabezas, E. Gonzalez, M. Gonzalez, C. Ciancarelli, and F. A. Comin. 2008. Macroinvertebrate patterns along environmental gradients and hydrological connectivity within a regulated river-floodplain. Aquatic Sciences 70:248–258.

Gray, D., and J. S. Harding. 2009. Braided river benthic diversity at multiple spatial scales: a hierarchical analysis of beta diversity in complex floodplain systems. Journal of the North American Benthological Society 28:537–551.

Guisan, A., T. C. Edwards, and T. Hastie. 2002. Generalized linear and generalized additive models in studies of species distributions: setting the scene. Ecological Modelling 157:89–100.

Gurevitch, J., and D. K. Padilla. 2004. Are invasive species a major cause of extinctions? Trends in Ecology and Evolution 19:470–474.

Koch, H., and U. Gruenewald. 2010. Regression models for daily stream temperature simulation: case studies for the river Elbe, Germany. Hydrological Processes 24:3826–3836.

Lasne, E., S. Lek, and P. Laffaille. 2007. Patterns in fish assemblages in the Loire floodplain: the role of hydrological connectivity and implications for conservation. Biological Conservation 139:258–268.

Leigh, C., and F. Sheldon. 2009. Hydrological connectivity drives patterns of macroinvertebrate biodiversity in floodplain rivers of the Australian wet dry tropics. Freshwater Biology 54:549–571.

Milner, A. M., J. E. Brittain, E. Castella, and G. E. Petts. 2001. Trends of macroinvertebrate community structure in glacier-fed rivers in relation to environmental conditions: a synthesis. Freshwater Biology 46:1833–1847.

Mohseni, O., H. G. Stefan, and T. R. Erickson. 1998. A nonlinear regression model for weekly stream temperatures. Water Resources Research 34:2685–2692.

Moriasi, D. N., J. G. Arnold, M. W. Van Liew, R. L. Bingner, R. D. Harmel, and T. L. Veith. 2007. Model evaluation guidelines for systematic quantification of accuracy in watershed simulations. Transactions of the ASABE 50:885–900.

Morrill, J. C., R. C. Bales, and M. H. Conklin. 2005. Estimating stream temperature from air temperature: implications for future water quality. Journal of Environmental Engineering-ASCE 131:139–146.

Oksanen, J., et al. 2011. vegan: community ecology. R package version 2.0-0. https://cran.r-project.org/web/packages/vegan/index.html

Paillex, A., E. Castella, and G. Carron. 2007. Aquatic macroinvertebrate response along a gradient of lateral connectivity in river floodplain channels. Journal of the North American Benthological Society 26:779–796.

Paillex, A., E. Castella, P. Zu Ermgassen, and D. Aldridge. 2015. Testing predictions of changes in alien and native macroinvertebrate communities and their interaction after the restoration of a large river floodplain (French Rhône). Freshwater Biology 60:1162–1175.

Paillex, A., S. Dolédec, E. Castella, and S. Merigoux. 2009. Large river floodplain restoration: predicting species richness and trait responses to the restoration of hydrological connectivity. Journal of Applied Ecology 46:250–258.

Peres-Neto, P. R., P. Legendre, S. Dray, and D. Borcard. 2006. Variation partitioning of species data matrices: estimation and comparison of fractions. Ecology 87:2614–2625.

Poff, N. L. 1997. Landscape filters and species traits: towards mechanistic understanding and prediction in stream ecology. Journal of the North American Benthological Society 16:391–409.

R Core Team. 2014. R: a language and environment for statistical computing. R Foundation for Statistical Computing, Vienna, Austria.

Rahel, F. J., and J. D. Olden. 2008. Assessing the effects of climate change on aquatic invasive species. Conservation Biology 22:521–533.

Reckendorfer, W., C. Baranyi, A. Funk, and F. Schiemer. 2006. Floodplain restoration by reinforcing hydrological connectivity: expected effects on aquatic mollusc communities. Journal of Applied Ecology 43:474–484.

Riquier, J., H. Piégay, and M. Michalkova. 2015. Hydromorphological characterization of restored floodplain channels along a large river (French Rhône). Freshwater Biology 60:1085–1103.

Ritz, C., and J. C. Streibig. 2005. Bioassay analysing using R. Journal of Statistical Software 12:1–22.

Sala, O. E., et al. 2000. Global biodiversity scenarios for the year 2100. Science 287:1770–1774.

Sheldon, F., A. J. Boulton, and J. T. Puckridge. 2002. Conservation value of variable connectivity: aquatic invertebrate assemblages of channel and floodplain habitats of a central Australian arid-zone river, Cooper Creek. Biological Conservation 103:13–31.

Statzner, B., N. Bonada, and S. Doledec. 2008. Biological attributes discriminating invasive from native European stream macroinvertebrates. Biological Invasions 10:517–530.

Statzner, B., A. G. Hildrew, and V. H. Resh. 2001. Species traits and environmental, constraints:

entomological research and the history of ecological theory. Annual Review of Entomology 46:291–316.

Tockner, K., F. Schiemer, C. Baumgartner, G. Kum, E. Weigand, I. Zweimuller, and J. V. Ward. 1999. The Danube restoration project: species diversity patterns across connectivity gradients in the floodplain system. Regulated Rivers Research and Management 15:245–258.

Tonolla, D., V. Acuna, U. Uehlinger, T. Frank, and K. Tockner. 2010. Thermal heterogeneity in river floodplains. Ecosystems 13:727–740.

Townsend, C. R., M. R. Scarsbrook, and S. Dolédec. 1997. The intermediate disturbance hypothesis, refugia, and biodiversity in streams. Limnology and Oceanography 42:938–949.

Uehlinger, U., F. Malard, and J. V. Ward. 2003. Thermal patterns in the surface waters of a glacial river corridor (Val Roseg, Switzerland). Freshwater Biology 48:284–300.

Usseglio-Polatera, P., and H. Tachet. 1994. Theoretical habitat templets, species traits, and species richness: Plecoptera and Ephemeroptera in the Upper Rhône River and its floodplain. Freshwater Biology 31:357–375.

Vaughan, I. P., and S. J. Ormerod. 2005. Increasing the value of principal components analysis for simplifying ecological data: a case study with rivers and river birds. Journal of Applied Ecology 42:487–497.

Verberk, W., H. Siepel, and H. Esselink. 2008. Life-history strategies in freshwater macroinvertebrates. Freshwater Biology 53:1722–1738.

Verbrugge, L. N. H., A. M. Schipper, M. A. J. Huijbregts, G. Van der Velde, and R. S. E. W. Leuven. 2012. Sensitivity of native and non-native mollusc species to changing river water temperature and salinity. Biological Invasions 14:1187–1199.

Ward, J. V., K. Tockner, U. Uehlinger, and F. Malard. 2001. Understanding natural patterns and processes in river corridors as the basis for effective river restoration. Regulated Rivers Research and Management 17:311–323.

Webb, B. W., D. M. Hannah, R. D. Moore, L. E. Brown, and F. Nobilis. 2008. Recent advances in stream and river temperature research. Hydrological Processes 22:902–918.

Wood, S. 2006. Generalized additive models: an introduction with R. Chapman & Hall/CRC, Boca Raton, Florida, USA.

12

Host plant cyanotype determines degree of rhizobial symbiosis

Adrienne L. Godschalx,† Vy Tran, and Daniel J. Ballhorn

Department of Biology, Portland State University, Portland, Oregon 97201 USA

Abstract. Plants with nitrogen-fixing bacteria, such as legumes with rhizobia, can tap the atmospheric nitrogen pool to obtain resources for defense compounds. Cyanogenesis, a nitrogen-based plant defense against herbivores, increases in response to rhizobial colonization, but depends on plant genotype. Here, we tested whether genotypic differences in host plant cyanogenesis influence symbiotic reliance on nitrogen-fixing rhizobia. Using thin, clear soil containers, we counted nodules on live root systems of distinct high (HC) and low (LC) lima bean (*Phaseolus lunatus*) cyanotypes across the duration of an eight-week study. We measured changes in cyanogenic potential (HCNp) and protein content to reveal quantitative interactions between nodule number and both leaf traits. High cyanogenic plants maintained consistently twice as many nodules as LC plants. Including both cyanotypes, nodule number correlated positively with HCNp, but negatively with foliar protein content. However, within-cyanotype interactions between nodule number and plant traits were not significant except for foliar protein in HC plants, which decreased with increasing nodule number. Our results imply that while genotypes with higher levels of nitrogen-based defense invest more in the rhizobial partner, the costs involved in maintaining the symbiosis may cause resource allocation constraints in the plants' primary nitrogen metabolism.

Key words: cyanogenesis; genotype; lima bean; mutualism; nitrogen fixation; *Phaseolus lunatus*; plant defense; rhizobia; symbiosis.

† **E-mail:** adrg@pdx.edu

Introduction

Plant defenses emerged from a long coevolutionary history with herbivores and pathogens (Ehrlich and Raven 1964), but also by coevolving alongside beneficial symbiotic partners that influence plant–herbivore interactions (Weber and Agrawal 2014). Mutualists that aid plants in defense include predators, but also nutrient-provisioning microbes that enhance the availability of certain resources for plant growth and defense (Herms and Mattson 1992, Stamp 2003, Thamer et al. 2011). In particular, under widespread nitrogen-limited terrestrial conditions (Vitousek et al. 2002), microbes facilitating resource acquisition can mediate plant–insect interactions (Pineda et al. 2010, Thamer et al. 2011). Symbioses between plants and nitrogen-fixing microbes range from loose associations with rhizobacteria in the soil (Dean et al. 2009, 2014, Pineda et al. 2010, Algar et al. 2014, Pangesti et al. 2015) to highly regulated interactions within root nodules. Nodule-based endosymbiosis involves substantial resource exchange and has evolved in multiple taxa of plants and microbes (Vessey et al. 2004, Kempel et al. 2009). Receiving symbiotic nitrogen in exchange for photoassimilates can enable plants to increase growth and defense traits simultaneously, even for relatively costly nitrogen-based defenses (Thamer et al. 2011, Ballhorn et al. 2017). Access to nitrogen is a significant advantage as producing such costly defense compounds can reduce plant–plant competitive ability and fitness (Herms and Mattson 1992, Marak et al. 2003).

Cyanogenesis, the release of toxic hydrogen cyanide from wounded cells, is an example of a costly nitrogen-based defense. In several species, including lima bean (*Phaseolus lunatus*), the total amount of cyanide-containing precursors in a given tissue (cyanogenic potential [HCNp]) is mostly constitutive and varies by genotype. Cyanogenic genotypes (cyanotypes) produce consistent levels of HCNp that range from high to low HCNp (Ballhorn et al. 2013, Kautz et al. 2014). High expression can incur both biochemical and ecological costs (Ballhorn et al. 2008, 2010, Kautz et al. 2017). Biochemical costs include producing cyanide-containing compounds (glucosides) from proteinogenic amino acids, as well as two enzymes, β-glucosidases and hydroxynitrile lyases. In various plant species, these enzymes work sequentially to efficiently release cyanide from these precursors (Frehner and Conn 1987, Kakes 1990, Poulton 1990, Vetter 2000, Gleadow and Møller 2014). Further resources are required to transport and store vacuolar cyanogenic glucosides, spatially separating them from apoplastic β-glucosidases to prevent autotoxicity (Frehner and Conn 1987). Cyanogenesis is ecologically costly because free cyanide interferes with the function of metal-containing enzymes, including enzymes critically involved in resistance to pathogens (Ballhorn et al. 2010). Consequently, highly cyanogenic plants are generally well defended against herbivores (Ballhorn et al. 2005), but weakly defended against pathogens (Lieberei et al. 1989, Ballhorn et al. 2008, 2010). Taken together, the associated ecological consequences and biochemical mechanism involved in releasing cyanide make cyanogenesis a relatively costly plant defense. Differences in defense costs imply potential for high cyanogenic (HC) and low cyanogenic (LC) cyanotypes to have different nitrogen requirements. However, cyanotype-driven differences in symbiotic investment to obtain fixed nitrogen have not been previously explored.

While rhizobia may alleviate nitrogen-related costs of defense, maintaining this beneficial relationship introduces another set of physiological and ecological costs. Rhizobia can consume 16–30% of a plant's total photosynthate pool (Peoples et al. 1986, Kaschuk et al. 2009) and require plant-synthesized essential amino acids for their own metabolism and nitrogen fixation (Lodwig et al. 2003). The cost for plants to increase nodule

numbers includes nodule biomass and metabolic demand to maintain nitrogenase activity, both of which demand carbohydrates. Carbohydrate demands serve as carbon sinks that stimulate photosynthetic rates (Kaschuk et al. 2009). Yet despite increased photosynthetic rates, carbon-rich traits such as indirect defense via extrafloral production and predator recruitment can be influenced by rhizobial carbon demands (Godschalx et al. 2015). Mutualistic carbon sinks play an important role in terrestrial carbon cycles (Pringle 2015), implying that increasing degree of rhizobia colonization to obtain nitrogen is not trivial.

Despite greater carbon sink and ecological costs, plant cyanotypes with constitutively high levels of cyanogenesis that require large inputs of nitrogen for defense could exert a pressure for plants to form more nodules. Here, we tested for quantitative differences in rhizobia colonization between HC and LC lima bean cyanotypes and differential responses in leaf chemical phenotype, including HCNp and soluble protein content to assess overall leaf quality. We collected all three metrics weekly—nodule number, HCNp, and protein content—over the course of a two-month study using a nondestructive method for nodule counting through clear, thin soil containers. If costs of direct chemical defense impose demand for stronger rhizobial association (measured as nodule number), we expected more nodules would form on roots of HC compared with LC plants. If leaf phenotype is directly influenced by nitrogen available from symbiosis, we would expect a positive correlation between nodule number and both nitrogen-containing traits. While nitrogen-fixing symbioses benefit plant productivity and provide a competitive edge in costly defense investment, understanding the factors facilitating or limiting the degree to which plants engage in this symbiotic exchange remains limited. Here, we used two different cyanotypes of the same plant species to test the impact of investment into nitrogen-based defense on the legume–rhizobia relationship.

MATERIALS AND METHODS

Experimental setup

To determine rhizobia colonization differences between plant cyanotypes, we created two treatments by inoculating low and high cyanotypes with rhizobia. We used lima bean (Fabaceae: *Phaseolus lunatus* L.) genotypes previously established as HC or LC cyanotypes based on consistent HCNp (Ballhorn et al. 2008). These accessions, HC_8078 and LC_8071, were provided by the Institute of Plant Genetics and Crop Plant Research in Gatersleben, Germany. Six seeds per cyanotype were germinated on moist paper towels. Once germinated, lima bean plants were individually planted 0.5 cm below the substrate surface level (greenhouse mix #3; SunGro Horticulture, Bellevue, Washington, USA) in soil containers that were custom-designed to facilitate rhizobia nodule counts on the intact root system. Soil containers (15 × 20 × 1.25 cm) were clear plastic containers wrapped in sheets of aluminum foil to block light, thus simulating belowground conditions. Plants were arranged randomly, and positions were rotated bi-weekly to account for potential position effects. Plants were watered daily with no additional nutrient solutions and cultivated under greenhouse conditions according to Ballhorn et al. (2014) at Portland State University (Portland, Oregon, USA) from March to April 2015.

Rhizobia inoculation

To identify rhizobia, several nodules were surface-sterilized, lysed, and plated to isolate colonies before the 16S gene was sequenced using 27f/1492r(I) primers. Using Geneious software, colonies were identified as *Bradyrhizobium elkanii* (Accession DJB1033-Ballhorn Lab; Portland State University). Inoculum was prepared by grinding 10 nodules, 0.5–5 mm in diameter, with a micropestle in a 1.5-mL centrifuge tube, and suspending the slurry in 600 mL H_2O. Both cyanotypes were inoculated with rhizobia once seedlings developed at least two true leaves by pouring 50 mL of rhizobia inoculum on the soil at the base of each seedling. Two weeks after inoculation, all plants showed root nodules.

Plant trait analysis

Nodulation was quantified as total nodule number per root system display. All five surfaces of the clear, thin soil containers were included in the root system display, including both sides, both narrow edges, and the narrow base, enabling most of the plant's root system to be included in the survey. Nodule assessments took place weekly

for 8 weeks. The same collection schedule was followed for leaf trait determination in order to relate rhizobia nodule counts to the quantitative expression of chemical leaf traits. Cyanogenic potential was quantified using the Spectroquant cyanide test (Ballhorn et al. 2005). Briefly, leaves were removed, and three leaf punches from each individual leaf were weighed to the nearest 0.001 g, ground with a mortar and pestle at 4°C in 2 mL ice-cold Na_2HPO_4 buffer, and centrifuged. Samples were analyzed for HCNp through enzymatically hydrolyzing cyanogenic precursors in gas-tight glass Thunberg vessels and spectrophotometrically assaying released cyanide at 585 nm (Ballhorn et al. 2005, 2013). Foliar soluble protein was quantified from the same leaf extracts; using Bradford's reagent and a calibration curve from 50 μg/mg to 1000 μg/mg bovine serum albumin (Amresco, Solon, Ohio, USA), soluble protein was measured spectrophotometrically at 595 nm (Bradford 1976).

Statistical analysis

Weekly quantified nodulation, HCNp, and protein content were all analyzed with repeated-measures ANOVAs with cyanotype and time as factors. All within-cyanotype data met assumptions of ANOVA and were not transformed. Relationships among trait means in response to nodule number means were analyzed with a linear model to determine significant relationships and Pearson's coefficients. All analyses were conducted using the software R (version 3.0.2; R Core Team, 2016).

Results

Rhizobia colonization varied greatly by cyanotype in repeated nodule counts throughout an experimental period of 2 months. On average, HC plants formed 60% more nodules than LC plants ($F_{1,10} = 21.27$, $P < 0.001$; Fig. 1A). For any given sampling date, HC plants maintained consistently higher numbers of nodules ($F_{6,60} = 20.72$, $P < 0.001$), with this lead ranging from 56 to 143 more mean nodules than LC plants. Nodule numbers varied by a significant interaction between cyanotype and sampling date ($F_{6,60} = 2.64$, $P < 0.05$).

To test the effects of rhizobia colonization on aboveground plant traits, we measured HCNp

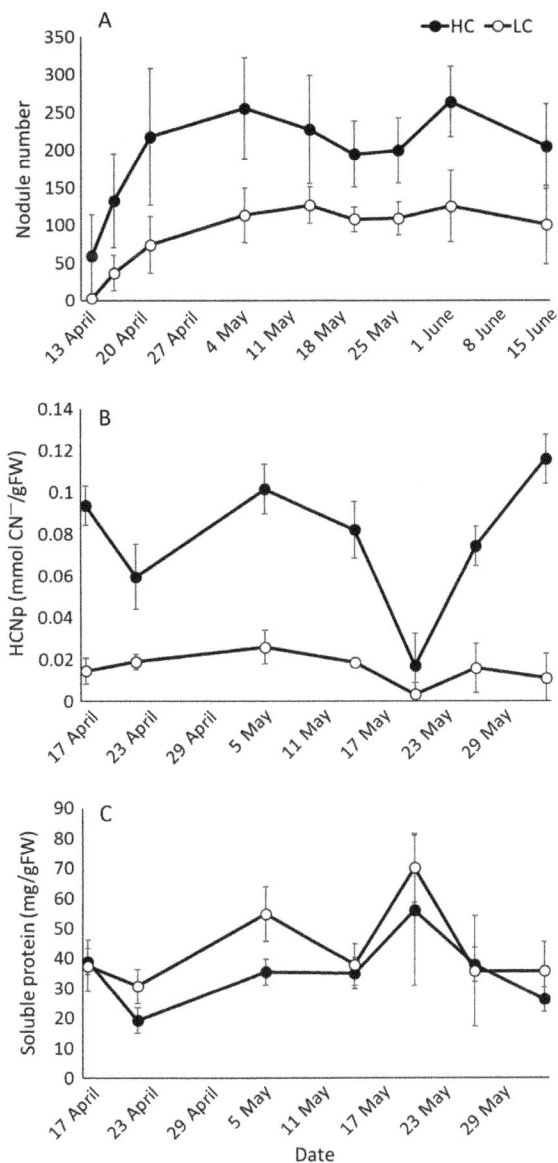

Fig. 1. Nodulation and leaf traits differences between cyanotypes across time. Low cyanogenic plants (LC; white circles) and high cyanogenic plants (HC; black circles) measured repeatedly across a two-month span to determine (A) extent of nodule colonization, (B) cyanogenic potential (HCNp), and (C) soluble protein content as a nutritive trait. Points show mean values, and bars represent standard deviation of the mean.

Fig. 1B); however, HC leaves contained 23% less average protein than LC leaves ($F_{1,10}$ = 13.44, $P < 0.05$; Fig. 1C). Sampling date significantly affected both HCNp ($F_{6,60}$ = 46.09, $P < 0.001$) and soluble protein ($F_{6,60}$ = 15.461, $P < 0.001$). Further, significant interaction effects between sampling date and cyanotype affected HCNp ($F_{6,60}$ = 29.06, $P < 0.001$), but not protein content ($F_{6,60}$ = 1.739, $P = 0.128$). Both cyanotypes were flowering on 17th May, eight weeks after being planted, which corresponded with varying expression of plant traits (Fig. 1).

To test for quantitative relationships among leaf chemistry in relation to nodule number across and within cyanotype, we regressed chemical trait averages for each plant replicate against nodule number averages for that same plant individual to test for significant correlations. We found a significant positive relationship between nodule number and HCNp when we included both cyanotypes ($F_{1,10}$ = 15.2, $P = 0.003$, adjusted R^2 = 0.564; Fig. 2A). This positive correlation between nodule number and HCNp did not hold true for within-cyanotype correlations. Within HC plants only, HCNp did not form a significant correlation with nodule number, and the slope of the trendline was slightly negative ($F_{1,4}$ = 0.834, $P = 0.413$, adjusted R^2 = −0.034). Within LC plants, the positive trendline was not significant ($F_{1,4}$ = 3.817, $P = 0.122$, adjusted R^2 = 0.337).

Foliar protein content also responded to increasing nodulation. Including both HC and LC plants, as plants formed greater numbers of nodules, protein content significantly decreased ($F_{1,10}$ = 21.67, $P < 0.001$, adjusted R^2 = 0.653; Fig. 2B). Unlike HCNp, which did not form a significant correlation within either cyanotype, a significant negative correlation between nodule number and leaf protein was present within HC plants ($F_{1,4}$ = 20.25, $P = 0.01$). By contrast, such correlation was not significant within LC plants, which showed a positive trendline between nodules and protein content ($F_{1,4}$ = 3.542, $P = 0.133$, adjusted R^2 = 0.337).

Discussion

Incorporating symbiotic interactions into plant secondary metabolism patterns has been an important challenge because advantages afforded by symbioses can drastically influence plant

and soluble protein content in leaves of a defined developmental stage from HC and LC plants. HC leaves produced an average of 77% higher HCNp than LC leaves ($F_{1,10}$ = 329.4, $P < 0.001$;

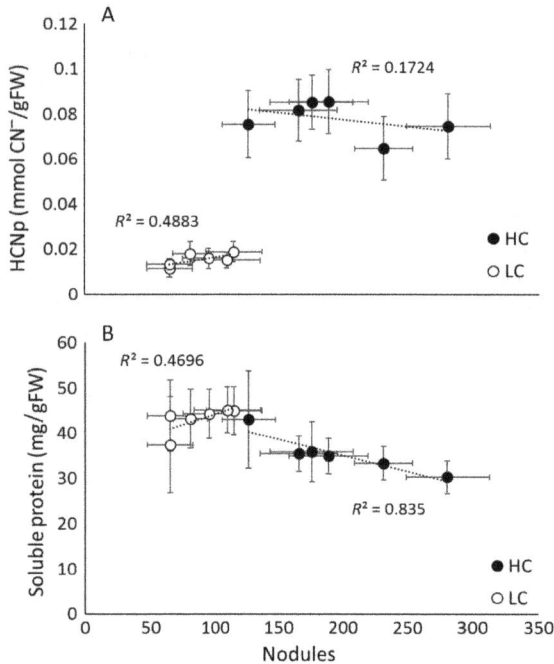

Fig. 2. Quantitative relationships between nodule number and variation in plant traits. Plant trait values including (A) cyanogenic potential (HCNp) and (B) soluble protein content were averaged across the duration of the time series experiment and regressed against nodulation means to assess putative correlations. Low cyanogenic plants (LC) and high cyanogenic plants (HC) are represented by white and black circles, respectively. Points show mean values from repeated assays and nodule counts for each plant across the time series; bars represent standard error of the mean.

resource allocation patterns (Kempel et al. 2009, Heath et al. 2014). Here, we show how the legume–rhizobia symbiosis interacts with leaf trait expression quantitatively in HC and LC cyanotypes of lima bean. While HC plants formed more root nodules and produced constitutively higher HCNp than LC plants, this positive relationship between nodulation and cyanogenesis did not result in quantitatively higher HCNp within either cyanotype. These data support our hypothesis that cyanotype may have played a role in selecting for the degree of rhizobial colonization based on constitutive nitrogen demands inherent in cyanogenesis. Surprisingly, our data do not support degree of nodulation quantitatively benefitting defense phenotype. While

nodule number did not translate into an increase in HCNp, HC plants expressed quantitatively less soluble protein in plants with greater nodule numbers. By contrast, LC plants formed fewer nodules overall and foliar protein was not constrained by a negative correlation with nodule number. Our findings suggest that symbiotic investment plays a role in plant defense and nutritive phenotype, but also that genotypic defense levels may simultaneously shape the plant's obligatory investment in maintaining the symbiosis.

Genotypic nitrogen requirements and nodule formation

Our hypothesis that HC plants require more fixed nitrogen for cyanogenesis and would therefore form more nodules than LC plants was supported by our nodulation data. In another system with polymorphic cyanogenesis, *Trifolium repens*, acyanogenic strains did not form more nodules than cyanogenic strains (Kempel et al. 2009). However, the nature of cyanogenesis in *T. repens* is qualitative with presence or absence of either cyanogenic glucosides or β-glucosidases, which may impose different resource demands compared to the quantitative variation in the lima bean system, with LC plants that are cyanogenic but at lower levels than HC plants (Ballhorn et al. 2005). If degree of colonization depends on plant nitrogen availability and demand, one potential mechanism for differential nodulation could involve the autoregulated negative feedback loop inhibiting further nodulation. Autoregulation of nodulation involves an interplay of root- and shoot-derived signals in the presence of excess soil nitrate (Oka-Kira and Kawaguchi 2006). While the chemical nature of these signals is still largely unknown (Kouchi et al. 2010), our data may present evidence for shoot-derived signals to be differentially regulated in high and low cyanotypes based on nitrogen requirements.

Putative mechanisms for nodule differences between cyanotypes

Nodule formation and regulation is a highly controlled process, involving crosstalk of several plant hormones, plant signaling molecules, and bacterial Nod factors (Sun et al. 2006). Therefore, cyanotypic differences in nodule numbers may be connected to cyanotype-specific biology. In addition to a myriad of traits regulated differentially between cyanotypes (Ballhorn et al. 2013), the most obvious difference between cyanotypes would be the presence of high concentrations of cyanogenic glucosides and potentially free cyanide surrounding damaged leaf cells. However, regulation of nodulation is not likely a product of direct cyanide exposure because free cyanide, which is released in the ethylene biosynthesis pathway, acts as a positive feedback loop for further ethylene synthesis (Smith et al. 2000), and ethylene is well known to inhibit nodulation (Penmetsa and Cook 1997). Inhibiting or reducing ethylene synthesis could be one way HC plants enable more nodules to form if cyanotypes differentially regulate this pathway, although this remains to be tested.

Is there a nodule number optimum? Nitrogen benefit vs. cost of maintaining nodules

If HC plants form more nodules to attain sufficient nitrogen for cyanogenesis, we would expect HCNp to correlate positively with nodule number. We see this effect across both genotypes, as HCNp increased with increasing nodule number. However, once these data are examined within cyanotype, plants do not produce more cyanide with higher nodule numbers. Interestingly, foliar protein concentration also correlated with nodule number overall, but in the opposite direction, decreasing as plants formed more nodules, which could point to the cost of maintaining symbiosis. In a study comparing symbiotic plants against nitrogen-fertilized plants, plants with rhizobia had lower protein levels, along with tannins and overall biomass (Briggs 1990). Our within-cyanotype data support this notion as HC plants, which likely have tighter allocation budgets, had quantitatively reduced protein levels as colonization intensity grew—as opposed to LC plants, in which nodule number and protein levels show a positive trend. If neither cyanotype's HCNp responded to rhizobia colonization, but protein trends correlated with nodules in opposite directions, HC plants may allocate more of the total symbiotic nitrogen pool to HCNp, reducing overall soluble protein levels. Alternatively, the number of nodules HC plants formed may have passed a threshold from which plants quantitatively benefit from greater colonization, contributing to why plants regulate nodulation (Oka-Kira and Kawaguchi 2006). Although it remains to be tested, HC plants may have selected for relaxed autoregulation in order to attain nitrogen for constitutive cyanogenic levels, which resulted in resource allocation constraints as the cost of maintaining higher numbers of nodules may limit this additional colonization from directly benefitting leaves.

In conclusion, cyanotype influences HCNp more strongly than input of nitrogen from increased nodulation. This finding is consistent with previous work in which nitrogen treatments increased both foliar nitrogen and cyanogenesis in *Eucalyptus cladocalyx*, but cyanogenic levels were restricted within genetically determined constraints (Simon et al. 2010). Despite nitrogen benefits, providing carbohydrates (Kaschuk et al. 2009) and specific amino acids (Prell et al. 2009) to nodules, our plant trait data demonstrate how symbiotic maintenance contributes to plant resource allocation challenges (Herms and Mattson 1992). Additionally, because high and low cyanotypes differentially engage in symbiosis, our data show the potential for plant defense schemes to influence the degree of symbiotic resource exchange.

ACKNOWLEDGMENTS

We thank Marc Nisenfeld from the Science Support Center at Portland State University for designing and building custom soil containers. We greatly appreciate molecular work by Nathan Stewart to identify rhizobia. We appreciate Maureen Daschel for facilitating a mentorship program through her Science Research Methods course at St. Mary's High School in Portland, OR. We thank Whitney Korenek, Vy Nguyen, Barbara Olvera, JG Bradner, Tanya Smagula, Omar Cazares-Alvarez, and Fidel Gonzales-Ferrer for participation in data collection and helpful discussion. Funding by the National Science Foundation (NSF) to DJB (Grants IOS 1457369 and 1656057 as well as DEB 1501420) and ALG (Fellow ID: 2014159631) is gratefully acknowledged.

LITERATURE CITED

Algar, E., F. J. Gutierrez-Mañero, A. Garcia-Villaraco, D. García-Seco, J. A. Lucas, and B. Ramos-Solano. 2014. The role of isoflavone metabolism in plant protection depends on the rhizobacterial MAMP that triggers systemic resistance against *Xanthomonas axonopodis* pv. glycines in *Glycine max* (L.) Merr. cv. Osumi. Plant Physiology and Biochemistry 82:9–16.

Ballhorn, D. J., J. D. Elias, M. A. Balkan, R. F. Fordyce, and P. G. Kennedy. 2017. Colonization by nitrogen-fixing Frankia bacteria causes short-term increases in herbivore susceptibility in red alder (*Alnus rubra*) seedlings. Oecologia 184:497–506.

Ballhorn, D. J., A. L. Godschalx, and S. Kautz. 2013. Co-variation of chemical and mechanical defenses in lima bean (*Phaseolus lunatus* L.). Journal of Chemical Ecology 39:413–417.

Ballhorn, D. J., A. L. Godschalx, S. M. Smart, S. Kautz, and M. Schädler. 2014. Chemical defense lowers plant competitiveness. Oecologia 176:811–824.

Ballhorn, D. J., S. Kautz, U. Lion, and M. Heil. 2008. Trade-offs between direct and indirect defences of lima bean (Phaseolus lunatus). Journal of Ecology 96:971–980.

Ballhorn, D. J., R. Lieberei, and J. U. Ganzhorn. 2005. Plant cyanogenesis of *Phaseolus lunatus* and its relevance for herbivore–plant interaction: the importance of quantitative data. Journal of Chemical Ecology 31:1445–1473.

Ballhorn, D. J., A. Pietrowski, and R. Lieberei. 2010. Direct trade-off between cyanogenesis and resistance to a fungal pathogen in lima bean (*Phaseolus lunatus* L.). Journal of Ecology 98:226–236.

Bradford, M. 1976. A rapid and sensitive method for the quantitation of microgram quantities of protein utilizing the principle of protein-dye binding. Analytical Biochemistry 72:248–254.

Briggs, M. A. 1990. Chemical defense production in *Lotus corniculatus* L. I. The effects of nitrogen source on growth, reproduction and defense. Oecologia 83:27–31.

Dean, J. M., M. C. Mescher, and C. M. De Moraes. 2009. Plant–rhizobia mutualism influences aphid abundance on soybean. Plant and Soil 323:187–196.

Dean, J. M., M. C. Mescher, and C. M. De Moraes. 2014. Plant dependence on rhizobia for nitrogen influences induced plant defenses and herbivore performance. International Journal of Molecular Sciences 15:1466–1480.

Ehrlich, P. R., and P. H. Raven. 1964. Butterflies and plants: a study in coevolution. Evolution 18:586–608.

Frehner, M., and E. E. Conn. 1987. The linamarin beta-glucosidase in Costa Rican wild lima beans (*Phaseolus lunatus* L.) is apoplastic. Plant Physiology 84:1296–1300.

Gleadow, R. M., and B. L. Møller. 2014. Cyanogenic glycosides: synthesis, physiology, and phenotypic plasticity. Annual Review of Plant Biology 65:155–185.

Godschalx, A. L., M. Schädler, J. A. Trisel, M. A. Balkan, and D. J. Ballhorn. 2015. Ants are less attracted to the extrafloral nectar of plants with symbiotic, nitrogen-fixing rhizobia. Ecology 96:348–354.

Heath, J. J., A. Kessler, E. Woebbe, D. Cipollini, and J. O. Stireman. 2014. Exploring plant defense theory in tall goldenrod, *Solidago altissima*. New Phytologist 202:1357–1370.

Herms, D. A., and W. J. Mattson. 1992. The dilemma of plants: to grow or defend. The Quarterly Review of Biology 67:283–335.

Kakes, P. 1990. Properties and functions of the cyanogenic system in higher plants. Euphytica 48:25–43.

Kaschuk, G., T. W. Kuyper, P. A. Leffelaar, M. Hungria, and K. E. Giller. 2009. Are the rates of photosynthesis stimulated by the carbon sink strength of rhizobial and arbuscular mycorrhizal symbioses? Soil Biology and Biochemistry 41:1233–1244.

Kautz, S., J. A. Trisel, and D. J. Ballhorn. 2014. Jasmonic acid enhances plant cyanogenesis and resistance to herbivory in lima bean. Journal of Chemical Ecology 40:1186–1196.

Kautz, S., T. Williams, and D. J. Ballhorn. 2017. Ecological importance of cyanogenesis and extrafloral nectar in invasive English Laurel, *Prunus laurocerasus*. Northwest Science 91:214–221.

Kempel, A., R. Brandl, and M. Schädler. 2009. Symbiotic soil microorganisms as players in aboveground plant-herbivore interactions – the role of rhizobia. Oikos 118:634–640.

Kouchi, H., H. Imaizumi-Anraku, M. Hayashi, T. Hakoyama, T. Nakagawa, Y. Umehara, N. Suganuma, and M. Kawaguchi. 2010. How many peas in a pod? Legume genes responsible for mutualistic symbioses underground. Plant and Cell Physiology 51:1381–1397.

Lieberei, R., B. Biehl, A. Giesemann, and N. T. Junqueira. 1989. Cyanogenesis inhibits active defense reactions in plants. Plant Physiology 90:33–36.

Lodwig, E., A. Hosie, A. Bourdes, K. Findlay, D. Allaway, R. Karunakaran, J. Downie, and P. Poole. 2003. Amino-acid cycling drives nitrogen fixation in the legume-*Rhizobium* symbiosis. Nature 422:722–726.

Marak, H. B., A. Biere, and J. M. M. Van Damme. 2003. Fitness costs of chemical defense in *Plantago lanceolata* L.: effects of nutrient and competition stress. Evolution; International Journal of Organic Evolution 57:2519–2530.

Oka-Kira, E., and M. Kawaguchi. 2006. Long-distance signaling to control root nodule number. Current Opinion in Plant Biology 9:496–502.

Pangesti, N., B. T. Weldegergis, B. Langendorf, J. J. A. van Loon, M. Dicke, and A. Pineda. 2015. Rhizobacterial colonization of roots modulates plant volatile emission and enhances the attraction of a parasitoid wasp to host-infested plants. Oecologia 178:1169–1180.

Penmetsa, R. V., and D. Cook. 1997. A legume ethylene-insensitive mutant hyperinfected by its rhizobial symbiont. Science 275:527–530.

Peoples, M. B., J. S. Pate, C. A. Atkins, and F. J. Bergersen. 1986. Nitrogen nutrition and xylem sap composition of peanut (*Arachis hypogaea* L. cv Virginia Bunch). Plant Physiology 82:946–951.

Pineda, A., S.-J. Zheng, J. J. A. van Loon, C. M. J. Pieterse, and M. Dicke. 2010. Helping plants to deal with insects: the role of beneficial soil-borne microbes. Trends in Plant Science 15:507–514.

Poulton, J. E. 1990. Cyanogenesis in plants. Plant Physiology 94:401–405.

Prell, J., J. P. White, A. Bourdes, S. Bunnewell, R. J. Bongaerts, and P. S. Poole. 2009. Legumes regulate *Rhizobium* bacteroid development and persistence by the supply of branched-chain amino acids. Proceedings of the National Academy of Sciences of the United States of America 106:12477–12482.

Pringle, E. G. 2015. Integrating plant carbon dynamics with mutualism ecology. New Phytologist 210:71–75.

R Core Team. 2016. RStudio: Integrated development environment for R (Version 0.99.489). Boston, Massachusetts, USA. http://www.rstudio.org

Simon, J., R. M. Gleadow, and I. E. Woodrow. 2010. Allocation of nitrogen to chemical defence and plant functional traits is constrained by soil N. Tree Physiology 30:1111–1117.

Smith, J. M., R. N. Arteca, J. McMahon Smith, and R. N. Arteca. 2000. Molecular control of ethylene production by cyanide in *Arabidopsis thaliana*. Physiologia Plantarum 109:180–187.

Stamp, N. 2003. Out of the quagmire of plant defense hypotheses. The Quarterly Review of Biology 78: 23–55.

Sun, J., V. Cardoza, D. M. Mitchell, L. Bright, G. Oldroyd, and J. M. Harris. 2006. Crosstalk between jasmonic acid, ethylene and Nod factor signaling allows integration of diverse inputs for regulation of nodulation. The Plant Journal 46: 961–970.

Thamer, S., M. Schädler, D. Bonte, and D. J. Ballhorn. 2011. Dual benefit from a belowground symbiosis: Nitrogen fixing rhizobia promote growth and defense against a specialist herbivore in a cyanogenic plant. Plant and Soil 341:209–219.

Vessey, J. K., K. Pawlowski, and B. Bergman. 2004. Root-based N2—fixing symbioses: legumes, actinorhizal plants, *Parasponia* sp. and cycads. Plant and Soil 266:205–230.

Vetter, J. 2000. Plant cyanogenic glycosides. Toxicon: Official Journal of the International Society on Toxinology 38:11–36.

Vitousek, P. M., S. Hättenschwiler, L. Olander, and S. Allison. 2002. Nitrogen and nature. Ambio 31: 97–101.

Weber, M. G., and A. A. Agrawal. 2014. Defense mutualisms enhance plant diversification. Proceedings of the National Academy of Sciences of the United States of America 111:16442–16447.

PERMISSIONS

All chapters in this book were first published in ECOSPHERE, by John Wiley & Sons Ltd.; hereby published with permission under the Creative Commons Attribution License or equivalent. Every chapter published in this book has been scrutinized by our experts. Their significance has been extensively debated. The topics covered herein carry significant findings which will fuel the growth of the discipline. They may even be implemented as practical applications or may be referred to as a beginning point for another development.

The contributors of this book come from diverse backgrounds, making this book a truly international effort. This book will bring forth new frontiers with its revolutionizing research information and detailed analysis of the nascent developments around the world.

We would like to thank all the contributing authors for lending their expertise to make the book truly unique. They have played a crucial role in the development of this book. Without their invaluable contributions this book wouldn't have been possible. They have made vital efforts to compile up to date information on the varied aspects of this subject to make this book a valuable addition to the collection of many professionals and students.

This book was conceptualized with the vision of imparting up-to-date information and advanced data in this field. To ensure the same, a matchless editorial board was set up. Every individual on the board went through rigorous rounds of assessment to prove their worth. After which they invested a large part of their time researching and compiling the most relevant data for our readers.

The editorial board has been involved in producing this book since its inception. They have spent rigorous hours researching and exploring the diverse topics which have resulted in the successful publishing of this book. They have passed on their knowledge of decades through this book. To expedite this challenging task, the publisher supported the team at every step. A small team of assistant editors was also appointed to further simplify the editing procedure and attain best results for the readers.

Apart from the editorial board, the designing team has also invested a significant amount of their time in understanding the subject and creating the most relevant covers. They scrutinized every image to scout for the most suitable representation of the subject and create an appropriate cover for the book.

The publishing team has been an ardent support to the editorial, designing and production team. Their endless efforts to recruit the best for this project, has resulted in the accomplishment of this book. They are a veteran in the field of academics and their pool of knowledge is as vast as their experience in printing. Their expertise and guidance has proved useful at every step. Their uncompromising quality standards have made this book an exceptional effort. Their encouragement from time to time has been an inspiration for everyone.

The publisher and the editorial board hope that this book will prove to be a valuable piece of knowledge for researchers, students, practitioners and scholars across the globe.

LIST OF CONTRIBUTORS

Thomas W. Bonnot
School of Natural Resources, University of Missouri, 302 Natural Resources Building, Columbia, Missouri 65211 USA

Frank R. Thompson Iii
United States Forest Service, Northern Research Station, University of Missouri-Columbia, 202 Natural Resources Building, Columbia, Missouri 65211 USA

Joshua J. Millspaugh
Wildlife Biology Program, Department of Ecosystem and Conservation Sciences, W. A. Franke College of Forestry and Conservation, University of Montana, Missoula, Montana 59812 USA

Nigel V. Gale, Mark Horsburgh and Sean C. Thomas
Faculty of Forestry, University of Toronto, 33 Willcocks Street, Toronto, Ontario, Canada

Md Abdul Halim
Faculty of Forestry, University of Toronto, 33 Willcocks Street, Toronto, Ontario, Canada
Department of Forestry and Environmental Science, School of Agriculture and Mineral Sciences, Shahjalal University of Science and Technology, Sylhet 3114 Bangladesh

Nathan S. Gill
Graduate School of Geography, Clark University, 950 Main Street, Worcester, Massachusetts 01610 USA
Pacific Island Ecosystems Research Center, 344 Crater Rim Drive, Volcano, Hawaii 96718 USA

Daniel Jarvis
Graduate School of Geography, Clark University, 950 Main Street, Worcester, Massachusetts 01610 USA
Vermont Technical College, 124 Admin Drive, Randolph Center, Vermont 05061 USA

Thomas T. Veblen
Geography Department, University of Colorado-Boulder, Guggenheim 110, 260 UCB, Boulder, Colorado 80309 USA

Steward T. A. Pickett
Cary Institute of Ecosystem Studies, Box AB, 2801 Sharon Turnpike, Millbrook, New York 12545 USA

Dominik Kulakowski
Graduate School of Geography, Clark University, 950 Main Street, Worcester, Massachusetts 01610 USA

M. Van Regteren
Wageningen Marine Research, Wageningen University & Research, Ankerpark 27, 1781 AG Den Helder, The Netherlands
Environmental Sciences Group, Wageningen University & Research, 6700 AA Wageningen, The Netherlands

R. Ten Boer, E. H. Meesters and A. V. De Groot
Wageningen Marine Research, Wageningen University & Research, Ankerpark 27, 1781 AG Den Helder, The Netherlands

James Weiwang
National Parks Board, 1 Cluny Road, Singapore S259569
Department of Geography, National University of Singapore, 1 Arts Link, Kent Ridge, Singapore S117570

Choon Hock Poh
National Parks Board, 1 Cluny Road, Singapore S259569

Chloe Yi Ting Tan, Vivien Naomi Lee, Anuj Jain and Edward L. Webb
Department of Biological Sciences, National University of Singapore, Science Drive 4, Singapore S117543

Chao Wu
Ministry of Education Key Laboratory for Earth System Modeling, Department of Earth System Science, Tsinghua University, Beijing 100084 China
College of Life and Environmental Sciences, University of Exeter, Exeter EX4 4QF UK

Sergey Venevsky, Yang Yang, Menghui Wang, Lei Wang and Yu Gao
Ministry of Education Key Laboratory for Earth System Modeling, Department of Earth System Science, Tsinghua University, Beijing 100084 China

Stephen Sitch
College of Life and Environmental Sciences, University of Exeter, Exeter EX4 4QF UK

Guoyong Yan and Qinggui Wang
College of Agricultural Resource and Environment, Heilongjiang University, 74 Xuefu Road, Harbin 150080 China
School of Forestry, Northeast Forestry University, 26 Hexing Road, Harbin 150040 China

Yajuan Xing
College of Agricultural Resource and Environment, Heilongjiang University, 74 Xuefu Road, Harbin 150080 China
Institute of Forestry Science of Heilongjiang Province, 134 Haping Road, Harbin 150081 China

Lijian Xu, Xiongde Dong, Wenjun Shan and Liang Guo
College of Agricultural Resource and Environment, Heilongjiang University, 74 Xuefu Road, Harbin 150080 China

Jianyu Wang
School of Forestry, Northeast Forestry University, 26 Hexing Road, Harbin 150040 China

Lionel R. Hertzog, Wolfgang W. Weisser and Sebastian T. Meyer
Terrestrial Ecology Research Group, Department of Ecology and Ecosystem Management, Center for Food and Life Sciences Weihenstephan, Technische Universit€at M€unchen, Hans-Carl-von-Carlowitz-Platz 2, DE-85354 Freising Germany

Anne Ebeling
Institute for Ecology, Friedrich-Schiller University Jena, Dornburger Strasse 159, DE-07743 Jena Germany

Laura A. McMahon and Janet L. Rachlow
Department of Fish and Wildlife Sciences, University of Idaho, Moscow, Idaho, USA

Lisa A. Shipley
School of the Environment, Washington State University, Pullman, Washington, USA

Jennifer S. Forbey
Department of Biological Sciences, Boise State University, Boise, Idaho, USA

Timothy R. Johnson
Department of Statistical Science, University of Idaho, Moscow, Idaho, USA

Robert S. Nowak
Department of Natural Resources & Environmental Sciences, University of Nevada Reno, MS 186, 1664 North Virginia Street, Reno, Nevada 89557 USA

Cheryl L. Nowak and Robin J. Tausch
U.S. Forest Service Great Basin Research Laboratory, 920 Valley Road, Reno, Nevada 89521 USA

Amael Paillex
Eawag, Swiss Federal Institute of Aquatic Science and Technology, 8600 D€ubendorf, Switzerland

Aquatic Ecology Group, Department of Zoology, University of Cambridge, The David Attenborough Building, Pembroke Street, Cambridge CB2 3QZ UK

Emmanuel Castella
Department F.-A. Forel for Environmental and Aquatic Sciences & Institute for Environmental Sciences, University of Geneva, Boulevard Carl-Vogt 66, 1211 Geneva 4, Switzerland

Philine Zu Ermgassen
School of Geosciences, University of Edinburgh, The King's Buildings Alexander Crum Brown Road, Edinburgh EH9 3FF UK

Belinda Gallardo
Applied and Restoration Ecology Group, Pyrenean Institute of Ecology (IPE-CSIC), Avda. Monañana 1005, 50059 Zaragoza, Spain

David C. Aldridge
Aquatic Ecology Group, Department of Zoology, University of Cambridge, The David Attenborough Building, Pembroke Street, Cambridge CB2 3QZ UK

Adrienne L. Godschalx, Vy Tran and Daniel J. Ballhorn
Department of Biology, Portland State University, Portland, Oregon 97201 USA

Index